# Computers In Mathematics: A Sourcebook of Ideas

Copyright 1979 by Creative Computing

All rights reserved. No portion of this book may be reproduced— mechanically, electronically or by any other means, including photocopying—without written permission of the publisher.
Library of Congress No. 79-57487
ISBN 0-916688-16-X

Printed in the United States of America
Second printing September 1981

10 9 8 7 6 5 4 3

Creative Computing Press
P.O. Box 789-M
Morristown, NJ 07960

# Computers In Mathematics: A Sourcebook of Ideas

Edited by David H. Ahl

Creative Computing Press
Morristown, New Jersey

**ABOUT THE EDITOR**

David Ahl has a BEE from Cornell University, MBA from Carnegie-Mellon University and has done further work in educational psychology at the University of Pittsburgh.

Two years in the Army Security Agency were followed by four years with Management Science Associates working on computer models and analysis of new consumer products. He continued work in computer analysis (of vocational education graduates) with Educational Systems Research Institute.

He joined Digital Equipment Corporation in early 1970. As Education Product Line Manager he formulated the concept of an educational computer system consisting of hardware, software, and courseware (Edu-System) and helped guide DEC into a leading position in the education market.

Mr. Ahl joined AT&T in 1974 as Education Marketing Manager and was later promoted to Manager of Marketing Communications where he was responsible for the development of sales promotional strategies and materials for the Bell System. Concurrent with this move, he started Creative Computing as a hobby in late 1974.

As Creative Computing grew, Mr. Ahl left AT&T in 1978 to devote full time to it. Creative Computing magazine today is number 1 in software and applications for small computers and a leader in publishing books, cassette and disk software, and related materials.

Mr. Ahl is the author of 6 books and over 70 articles in the use of computers. He is a frequent lecturer and workshop leader at colleges and professional conferences. He is a member of ACM, AEDS, AERA, COSMEP and NCTM.

To Christian E. Mills, the toughest and best mathematics teacher I ever had.

# Contents

## Introducing the Computer and How to Buy a Microcomputer System

**2** What Is Computer Literacy ............... Moursund
Discussion and 30 question quiz

**4** Computers in Society ........................ Ahl
An interview activity with a questionnaire

**6** The 10¢ Computer and Other Games ........... Lyon
You don't need a computer for these

**7** Considerations in Buying a Personal Computer ... Zinn
What, which, when, how much?

**8** Personal Computer Comparison Chart ......... North
Features of six popular models

## Thinking Strategies and How To Solve Problems

**12** Turning a Puzzle Into a Lesson .............. Homer
Not every problem is suitable for computer solution

**14** Tac Tix and the Complications of Fallibility ................................ Koetke
Piet Hein's Tac Tix game has interesting aspects

**16** Palindromes ............................. Koetke
For those who like to end at the beginning

**19** Computing Factorials—Accurately .......... Koetke
50-digit precision from any computer

**22** Aedi, Mutab, Neda and Sogal ................ Koetke
Which civilization will rule the galaxy?

**24** What the Computer Taught Me About My Students ............. Pasquino
Or is binary search natural?

**26** Thinking Strategies With the Computer ................... Piele & Wood
Inference

**30** Working Backward .................... Piele & Wood
Start at the goal and return to the start

**33** Subgoals ............................ Piele & Wood
Break the problem into parts and solve each one

**36** Trial and Error ....................... Piele & Wood
Taxes problem, turkey puzzle, Newton's method

**41** Contradiction ........................ Piele & Wood
Indirect reasoning, whodunit?, cryptarithmetic

**45** On Solving Alphametrics ................. Beidler
SEND + MORE = MONEY in Fortran

**48** Brain Teaser ........................ Knippenberg
Deceptively simple but devilishly tricky

## Computer Simulations

**52** How To Write a Computer Simulation .......... Ahl
Collect data, analyze, write a model and program

**58** Greed ................................ Ragsdale
A game playing program with adjustable skill level

**60** Seeing is Believing ....................... Koetke
But simulating is convincing

## Probability

**62** Computer Simulations and Problem-Solving in Probability ............ Camp
Population planning, marketing, and system reliability

**65** Problems in Probability and Compounding ............... Johnson & Reeves
11 problems to solve

**66** What Will Happen If? ............... Lappan & Winter
Coins in a pocket, baseball cards, gambler's choice

**70** Two Million Frantic Frenchmen ............ Winkless
A study in probability

**72** Simulations in the Game of Craps .............. Tanis
A fair game—almost

**74** How Late Can You Sleep in the Morning? ........ Ahl
Two routes to work. Which is best?

## Mathematical Miscellany

**76** The World of Series ....................... Reagan
The playoffs that is

**77** How Many Ways Can You Make Change ......... Hess
For a dollar?

**78** Keeping the Loan Arranger Honest .......... Warden
Interest calculations

**80** An Analysis of Change .................... Barnett
Differential equations made simple

**83** Progression Problems ..................... Reeves
Five fascinating problems to solve

**84** Multiple Regression Analysis—Simplified ..... Chereb
Analyzing statistical data

**88** A Roman's Assignment Problem ........... Anderson
Minimizing the boos can be a challenge

**91** Days and Dates ......................... Reagan
Four problems for computer solution

**93** Non-usual Mathematics .................... Reagan
Infinitely many primes

**94** Computerized Sports Predictions............. Smith
Predict the NFL games

**98** Sports Judging Made Faster and Easier..... Winkless
Calculate the scores from multiple judges

**100** Distance and Error-Correcting Codes..... Project Solo
My code is 1101. What is yours?

**103** Accuracy Plus............................ Barnett
Multiprecision multiplication

**106** Double Precision........................ Hinrichs
Use it for sine, cosine, tangent, roots, logs and powers

**110** Doubling Up............................. Tapson
How thick is a sheet of newspaper folded 100 times?

**111** Magic Squares on the Computer............. Piele
Franklin magic squares are more magic than you think

**117** Pocket Calculator Tricks.................... Ahl

**118** Circular Functions.................. Project Solo
Sine, cosine and tangent by computer

**121** Trigonometric Functions.............. Project Solo
Tchebychev approximations
and transcendental functions

**124** Phantom Vortac..................... Project Solo
Program an on-board navigational computer

**127** Pascal's Triangle....................... Mechner
APL and Basic programs

## Art, Graphics and Mathematics

**130** Art and Mathematical
Structures................. Chandhok & Critchfield
Polygons, symmetry and the computer

**134** Patterns................................Games
Overlaying exponential functions

**138** Inkblot.................................. Costello
Overlapping ellipses

**140** A Picture in 20 Lines..................... Young
The variety of programming methods is surprising

**142** Superose............................... Zorn
A computer spiralograph

**144** Computer Planned Snowmen............... McLean
Model of a snowman

## Computer Assisted Instruction

**147** Mathematics Drill and Practice................. Ahl
Horizontal and vertical addition and subtraction

**151** Structuring the Lesson to the Student........... Ahl
The object is to always keep the student challenged

**154** Multiple Problem Types.................McLaughlin
Incorporating a sliding grade level

**158** Interaction Between
Student and Computer................McLaughlin
Alignment, personalization, variety

**165** Theorem-Proving With Euclid............. Kelanic
Geometry proof by computer

## Programming Style

**170** How To Hide Your Basic Program........... Nevison
Four rules of good programming style

**172** How to Hide...Round 2................... Jaquiss
Five more rules

## Short Programs

**176** Convergence.........................Ahl & Ball

**178** Systematic Savings,
Compound Interest.................... Denenberg

**179** Prime Factoring, Powers of 2......... Jeffery & Ahl

**180** Double Subscripts,
Common Birthdays................. Garon & Smith

**181** Rorschach, Date Conversion........... Ligori, et al

**182** Indefinite Articles....................Hammerstein

## Puzzles, Problems and Programming Ideas

**184** Reading, Writing and Computing............ Koetke
Four fascinating problems

**186** Programming Contest..................... Piele
Five problems with easy and hard versions

**188** 40 Programming Ideas....................Cletheroe
Some easy, some difficult

**189** Computer Conversations............ The Math Group
Seven problems from a series of cards

**190** The Mechanical Mouse......... Maniotes & Quasney
Can you flowchart his path through four mazes?

**191** Pinball Maze............................. Quinn
Bumper pinball with a goal of 100

**192** Learning by Doing..................... Gruenberger
Ten challenging problems

**195** Puzzles and Problems........................ et al
A baker's dozen of new ones

**197** Grid Addressing....................... Akkerhuis
Use this scheme to control appliances

**198** Puzzles and Problems, Computer Recreations
Thinkers' Corners, and more.................. et al
A whopping 72 problems to sink your teeth in

VII

# Preface

In working with government bodies, computer manufacturers, colleges, schools and, most important, kids over the past twenty years, I have several observations on education and learning. Some of these are not mine exclusively, but have been put forth also by groups and individuals such as the Cambridge Conference on School Mathematics (CCSM), the National Council of Teachers of Mathematics (NCTM), Project SOLO, Art Luehrmann, Bob Taylor and others. Some years ago I published a list of 11 of these observations which some educators regarded as rather cynical. Since then I have "cleaned up my act" by trying to present everything in the most optimistic light possible and eliminating the negative items. So here is my little list.

1. Kids' minds are unencumbered by constraints of what can and can't be done. Kids will try just about anything.
2. Learning by manipulating, doing or discovering sticks with one far better than learning by reading about a subject.
3. "Learning to learn" is infinitely more important than learning facts and data.
4. Motivation of the learner is much more important than teaching method or style of delivery.
5. Computers are the most powerful tool man has ever invented and the most awesome responsibility he has ever faced.
6. Thought-provoking activities should not be done as a special treat or reward, but sufficiently often that scientific thought will be habitual.
7. Thinking, *per se*, is worthwhile, particularly when it is in imaginative and creative directions. Rote attention to details is rarely justified.
8. Every child must be convinced that *his* thinking is worthwhile.
9. Education has become relatively less efficient than practically any other facet of our social or economic system.

A brief example: some high school students in the Philadelphia area heard about Weizenbaum's "Eliza" program in which the computer carries on a conversation with the user in the style of a non-directive therapist. They thought the program would be neat to have on their PDP-8 minicomputer so they set out to write it. They didn't realize that the program was written in Lisp on a huge mainframe at MIT and could not possibly be converted to Basic for an 8K PDP-8, even with disk storage for the files. Any adult or teacher would have told them not to waste time on a futile project. Needless to say, these kids successfully wrote a version of "Eliza" for the '8. In 1976, I gave this Basic version of "Eliza" by Jeff Schrager to Steve North, then a college freshman, with the challenge to convert it to run on a microcomputer. Three short weeks later he was back waving the finished program.

This one example, in some sense, illustrates the first eight points on my list. Combining the motivation of doing something "neat" with an unencumbered mind and access to a computer will lead to analytical thinking, learning by discovery, and learning to learn. The accomplishment that will inevitably come out of this process will lead to a feeling of fulfillment, satisfaction and a positive self image.

Forward-looking curriculum researchers, educators and government administrators have known the benefits of using computers in the instructional process since the late 60's. Why then, you might ask, doesn't every child in every school in the world have a computer available for learning? Teachers will say it's lack of money, administrators will say the teachers aren't interested, the teacher unions will say there's no time for additional responsibilities, researchers will say it's lack of curriculum materials and software, government program administrators will say that proper objectives have not been established, and on and on. As in some multiple choice questions, the real answer is "all of the above."

Meaningful objectives probably are the best place to start. I don't propose to establish new objectives for the entire educational process in a piece such as this. Suffice it to state that objectives must be broadly stated and must not only be relevant to the world of today but to the world of the future. Knowledge is changing and advancing so rapidly that we must expect objectives to change or, conversely, be formulated in a broad enough fashion to keep up with technological advances. The Mager-style behavioral objectives don't begin to meet the need. Education today needs more than small behavioral steps. It needs objectives that are dynamic and can be expected to change over time; objectives that are stated in an entirely new way.

Objectives must be devised to lead young minds through an imaginative exploration of the jungle of political, social, psychological, and ethical issues that will confront them as adults. What is the objective that will foster decision-making under uncertainty? What are the objectives which elucidate the ways in which technology and values will interact in the world of tomorrow? How does one measure whether one has learned to learn? I'm not saying these objectives are impossible to devise. Indeed, these are the kinds of things upon which education must focus and the kind of objectives that must be devised.

Not only must objectives evolve; so must the teaching/learning process. It should no longer be a question of whether or not computers should be in the curriculum, but rather how many. The worth of the computer in the educational process has been proved a hundred times over in practically every discipline at every grade level. Prices of microcomputers are within department budget levels, indeed many PTA's and student groups have purchased computers for their schools. Consequently, today many former excuses for not using computers are no longer valid.

Why should the computer be singled out as the one device that **must** be incorporated into the curriculum? Why not the videodisc, tachistoscope, or portable conference telephone? Because computers are profoundly different from any previous or current device or innovation. As Margot Critchfield noted, "A computer can best be thought of as a machine that can transform itself into any other machine—that is, it can carry out any procedure that can be fully described to it. Therefore, it can be a medium of instruction, a lunar lander, a game player, a mathematical formula cruncher, an ecological system, and so on, ad infinitum."

Art Luehrmann likens computers to reading and writing. His fundamental philosophical premise is that, like reading and writing, "computing constututes a new and fundamental educational resource. To use that resource as a mere delivery system for instruction, but not to give the student instruction in how he might use the resource itself, has been the failure of the CAI effort. What a loss of opportunity if the skill of computing were to be harnessed for the purpose of turning out masses of students who were unable to use computing! We ask, how much longer will a computer illiterate be considered educated?"

**Purpose of This Book**

One of the reasons given for not using computers is lack of curriculum material and software. Admittedly the publishers in the field are not the giants, at least not so far, but I see signs of them entering. Also, much of the best material to date has been published in periodicals and conference proceedings rather than in traditional textbooks. The purpose of this collection of reprints is partially to bridge the gap. This is a collection of 77 of the best articles and activities that have appeared in **Creative Computing** magazine from 1974 through 1979.

It should be emphasized that every single article in this collection contains material for immediate use in the classroom or for self teaching at home or in school. The book has problems for assignment—over 200 of them altogether. It has nearly 100 programs. It has descriptions of the underlying techniques, algorithms and principles. Everything necessary for effectively using the computer in mathematics instruction at all grade levels is found between the covers.

The reprints are divided into nine subject areas starting with the very fundamentals of computer literacy, computers in society, and selecting a system for educational use. None of the activities in this first section requires a computer.

The second section is titled "Thinking Strategies and How to Solve Problems." The 13 pieces in this section focus mainly on procedures for attacking problems. The five articles by Piele and Wood focus on five specific analytical techniques such as inference, working backwards, breaking a problem into parts, trial and error, and contradiction. The four articles by Walter Koetke from the "Problems for Creative Computing" series present challenging problems in creative settings.

The section on "Computer Simulations" discusses how to write a simulation and provides several examples. These examples continue into the next section on "Probability" with six more articles. Camp's article, "Computer Simulations and Problem-Solving in Probability" and the piece "What Will Happen If?" by Lappan and Winter are excellent presentations of concepts and problems that are familiar to every student yet are almost impossible to solve without the use of a computer.

"Mathematical Miscellany" contains 21 articles on circular functions, regression analysis, flowcharts, magic squares, Pascal's triangle, differential equations, series, double precision and much more. Truly a miscellany of motivation and fascination.

Often the interest of the "non-mathematically inclined" student can be captured by art and music. Unfortunately, there is no music in this book, but seven articles on art and graphics and their relationship to mathematics are included.

Although I do not believe in using the computer soley or even mainly for CAI (Computer Assisted Instruction), one cannot argue with the research showing the effectiveness of CAI. The five articles on CAI show several ways of humanizing drill and practice as well as keeping it challenging and interesting.

The section on "Programming Style" presents nine rules for writing good programs while the section on "Short Programs" contains ten shorties for various tasks.

The last section,"Puzzles, Problems and Programming Ideas," contains a whopping 159 problems that can be used for assignments, homework or simply posted as a challenge for interested students. It is surprising how many students will set a goal for themselves to do every one.

I could go on describing the content of the book, but you'll be better off plunging in without further ado. Remember, computers are the most powerful tool ever invented by man and every person should, indeed must, learn to use them effectively and without fear. It is my hope that this book brings us one step closer to that goal.

Morristown, New Jersey
January, 1980                      David H. Ahl

# Introducing the Computer and How to Buy a Microcomputer System

# WHAT IS COMPUTER LITERACY?

by David Moursund
University of Oregon

The concept of "computer literacy" is receiving much mention today. Over a period of time, we have developed a definition.

Computer literacy refers to a knowledge of the non-technical and low-technical aspects of the capabilities and limitations of computers, and of the social, vocational, and educational implications of computers. While such a definition can provide a focus for thought and discussion, it still does not pinpoint what is meant by computer literacy. Among other things it does not provide a measure of computer literacy nor a method for improving a person's level of computer literacy.

Most of you are familiar with the question "What is IQ?" and the answer "IQ is what is measured by an IQ test." It seems to me that we are at a similar stage of development for CL (computer literacy). Lately, many course outlines for computer concepts or computer literacy courses at the college level have been developed at Oregon and elsewhere. These courses are designed to raise a person's level of CL, and a knowledge of the content of such courses constitutes a certain level of computer literacy.

The University of Oregon's computer concepts course is a no-prerequisite, low level, introductory computer science course. Its major goal is to raise a student's level of computer literacy. Over a period of six years the course has evolved to the current point, where its content is approximately 1/3 computer programming and 2/3 non-programming materials. A Venn diagram of the course content is given below.

In the diagram the computer programming, computer usage, and hands-on experience provides a foundation upon which the non-programming aspects of the course are built. Each of these four areas strongly overlaps the other three, and each supports the other three. A well balanced course needs aspects of each of these four areas.

It seems difficult to develop a course that is coherent and well integrated, and still preserves a reasonable balance among the four major areas. Probably the computer programming and related computer usage and hands-on experience is the major source of trouble. Most computer programming texts are designed to teach computer programming. That is, their major goal is to move a student rapidly along the computer programmer path. Most such books contain little information on the capabilities, limitations, or implications of computers. The material is not organized in a manner to make it fit in well with non-programming, computer literacy materials.

To overcome this difficulty in the UO's course, I have written a 150 page book, *BASIC Programming for Computer Literacy* (McGraw-Hill). This book is currently being used in the course, and seems to be a satisfactory text.

The non-programming content of a CL course can range over a wide variety of topics, and will depend to a certain extent upon the interests and knowledge of the instructor. One cannot tell if a person is computer literate on the basis of a single true-false or multiple choice question. That is, CL refers to a broad, integrated knowledge of low level computer science. Such knowledge must include many facts and how these facts interrelate. But it is difficult to isolate a single fact that is indispensable, or fundamental.

On the non-programming content of the course, I use an objective-type final exam. In fall 1974 this exam consisted of 150 questions. An item analysis was run on these questions to determine which were the more difficult and which best differentiated the students who scored high on the test from those who scored lower on the test. Thirty of the better questions (harder, and good differentiators) have been selected and appear at the end of this article. A student making an A or high B on the exam probably answered at least 3/4 of these questions correctly.

The answers in most cases are not obvious. The 30 question test was administered to students on the first day of the winter term 1975 course. The class average was 14.75. Random guessing by all students would have produced a class average of about 12.

Taken individually, the merits of any single question are certainly subject to debate. One can easily argue that the question is not relevant to his concept of what constitutes computer literacy. Taken as a whole, however, such a group of questions provides a reasonably broad measure of many parts of the non-programming content of a computer literacy course. Try the test yourself. Try it on your students. Individual questions can provide a good basis for class discussion or individual student reading/study projects.

# COMPUTER LITERACY QUIZ

1. All computers understand the language BASIC because, as its name implies, it is the most fundamental of computer languages.
2. For any problem within its capability, a computer can always solve it more quickly and cheaper than can be done manually.
3. Example of random access storage devices include:
   1. core and disk
   2. magnetic tape and punch cards
   3. disk and magnetic tape
   4. paper tape and punch cards
   5. all of the above
4. M.I.C.R. stands for magnetic ink character recognition, and is used on bank checks in the United States.
5. Which of the following does not manufacture and sell computers?
   1. Control Data Corporation
   2. IBM
   3. Digital Equipment Corp.
   4. Honeywell
   5. American Telephone and Telegraph Company
6. A typical CAI drill and practice program:
   1. works only when one is teaching elementary arithmetic
   2. aks the student questions and checks his answers
   3. forces all the students to work the same set of problems
   4. allows three incorrect responses before going on to the next problem
   5. none of the above
7. Although learning a machine language is difficult, once one has mastered it, he can write programs that will be understood by any machine.
8. The best computer programs for playing chess and checkers are based upon having the computer memorize tens of thousands of board positions (i.e. rote memory).
9. It is now possible to manufacture a single large-scale integrated circuit, called a chip, which contains all of the circuitry for a CPU.
10. The concept and use of punched cards was developed:
    1. before 1900
    2. about 1920
    3. about 1940
    4. about 1960
11. PLATO is an educational computer system which uses a gas plasma display terminal.
12. In the early days of computers, all programming was done:
    1. in FORTRAN
    2. in BASIC
    3. in machine language
    4. in UNIVAC
13. The Turing "Imitation Game":
    1. has a computer imitate a business environment to train executives in decision-making.
    2. has a person imitate a computer to find program errors.
    3. has a computer simulate a complex situation providing a detailed study of alternative effects.
    4. has a computer pretend to be human, demonstrating artificial intelligence.
14. The science of control and feedback theory is called cybernetics, and Norbert Weiner contributed a lot to this area.
15. One threat to privacy comes from the willingness of most people to provide information about themselves voluntarily.
16. Which of the following is a characteristic of a problem which is well-suited to solution by the computer?
    1. Problem solution involves value judgments
    2. All necessary decisions are quantifiable
    3. The problem is ill-defined
    4. The solution to the problem is needed only one time
17. The largest user of computers in the U.S. Government is:
    1. The Internal Revenue Service
    2. The Census Bureau
    3. The military
    4. Congress
    5. None of these.
18. When one is buying a computer system, he might purchase hardware and software from two different companies.
19. Magnetic tape is an effective medium in operations requiring frequent access to data on a random basis.
20. Very large computer programs are apt to contain undetected errors even after the programs have been used for several years.
21. NCIC is a method whereby checks printed with a special ink can be machine read.
22. By 1950 about 1000 electronic digital computers had been manufactured and placed into service.
23. The fastest core memories have retrieval times of about one millisecond.
24. Using an 8 bit code (such as on magnetic tape), how many different characters can be represented?
    1. 8
    2. 16
    3. 32
    4. 256
    5. 512
25. A computer's memory can think about and solve a problem much in the same way as a person's brain works on a problem.
26. A major problem with computerized data banks is guarding against erroneous data getting into the system.
27. A disadvantage of punched card machines is that the speed of processing is limited by the movement of mechanical parts and devices.
28. Which of the following is not an example of the administrative application of computers in education?
    1. Payroll
    2. Scheduling
    3. Student records
    4. Computer-assisted instruction
29. Why do computer scientists write computer programs to play games?
    1. Computer scientists have lots of fun doing this.
    2. To communicate the ability of the computer.
    3. To study the nature of problem solving.
    4. All of the above.
    5. None of the above.
30. Computer costs (measured in terms of computations per dollar) have leveled off in the last five years.

**Answers on page 92**

# INTERVIEW

Here's an informative activity to do in a "Computers in Society" or "Computer Appreciation" course, or for that matter, in a social studies or sociology course.

## EXERCISE 1

Make up copies of the interview form on the next page and give each student two copies. (You may want to use a subset of the questions instead of the entire list.) Each student should fill out one copy of the questionnaire himself. Then each student should interview an adult on these issues. Try to obtain interviews with a diverse cross-section of people. Students may feel more comfortable working in groups to get interviews; if so, let them pair off. But no more than two students to a group; more than that tends to overwhelm interviewees.

## EXERCISE 2

Tabulate the results, compare the various answers obtained, and discuss in class. Can you draw any conclusions about the attitude of the general public toward computers? Do students' attitudes generally agree or disagree with those of the interviewees? Are there any obvious relationships between the attitudes expressed and the demographic characteristics (age, sex, etc.) of the respondents?

## OPTIONAL EXERCISE 1

Write a computer program to tabulate the results and compute average scores for each question as well as percentage distributions.

## OPTIONAL EXERCISE 2

If your computer system has file capabilities, write a program to administer the questionairre via a terminal, store the results and merge them with all the previous results and then print out the scores to date. For real pizzazz, print the results graphically like this:

COMPUTERS WILL IMPROVE HEALTH CARE

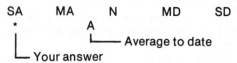

Or, show a bar chart of answers like this:

COMPUTERS DEHUMANIZE SOCIETY BY TREATING EVERYONE AS A NUMBER

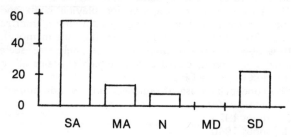

# Computers and Society Questionaire

Statement

|   | Strongly Agree (1) | Mostly Agree (2) | Neutral or No Opinion (3) | Mostly Disagree (4) | Strongly Disagree (5) |
|---|---|---|---|---|---|

1. Computers will improve health care.

2. Computers will improve education.

3. Computers will improve law enforcement.

4. Computers slow down and complicate simple business operations.

5. Computers are best suited for doing repetitive, monotonous tasks.

6. Computers make mistakes at least 10 percent of the time.

7. Programmers and operators make mistakes, but computers are, for the most part, error free.

8. Computers dehumanize society be treating everyone as a number.

9. It is possible to design computer systems which potect the privacy of data.

10. Credit rating data stored on computers have prevented billions of dollars of fraud. This is a worthwhile use of computers.

11. In the U.S. today, a person cannot escape the influence of computers.

12. Computers will create as many jobs as they eliminate.

13. Computers will replace low skill jobs and create jobs needing specialized training.

14. Computers are a tool just like a hammer or a lathe.

15. Computers are beyond the understanding of the typical person.

16. Computer polls and prediction influence the outcome of elections.

17. Computers isolate people by preventing normal social inteactions among people who use them.

Age_____ Sex_____ Education_____
Occupation_____ Location_____
Name (optional)_____

**creative computing**

# The 10¢ Computer and Other Games

by Gwyn Lyon
Gates Elementary School
Acton, Massachusetts

The teaching of math to kids is traditionally divided into lecture and drill. In the re-thinking of a second grade math program, I decided to incorporate some different approaches to aspects of math computation. A game such as Input-Output can be noisily exciting, and effectively teach the relationships between sets of numbers.

## INPUT-OUTPUT

A game for the Blackboard or Overhead Projector. Divide the acetate into two columns or make two columns on the blackboard. Place a numeral in the left or Input column. Through the function of the magic-black box-computing-teaching machine, the number is transformed into a new number.

|  | teaching machine = ? |
|---|---|
| Input | Output |
| 4 | 7 |

Place a new number in the input column, and using the same teaching machine function, a new number appears in Output.

|  | teaching machine = ? |
|---|---|
| Input | Output |
| 4 | 7 |
| 0 | 3 |
|  | answer: t.m. = +3 |

Continue to add numbers until a student guesses what function the teaching machine is set for.

Second graders can do problems of this level of difficulty:

|  | teaching machine = ? |
|---|---|
| Input | Output |
| 4 | 11 |
| 7 | 17 |
| 10 | 23 |
|  | answer: t.m. = 2x + 3 |

The game can be adapted for many types of algebraic equations:

|  | teaching machine = ? |
|---|---|
| Input | Output |
| 9 | 4 |
| 25 | 6 |
| 4 | 3 |
|  | answer: t.m. = $\sqrt{x}$ +1 |

As an added bonus, a child with such learning problems as would exclude him from the successful completion of a traditional worksheet can often excel in "head" games that require no written response.

## THE 10¢ COMPUTER

The teacher must also begin to rely more on manipulative materials to move the child securely from the concrete to the abstract. Hence, the ten cent computer:

*Materials*: a large grocery box. 1 sheet of acetate, 8" x 10".

*Method*: Cut the box up so that you have one large sheet of cardboard, without seams, that will completely cover the stage of the overhead projector. Mine is 12" x 11½", but measure yours to be sure.

Cut four holes in the top third of your cardboard. Save the cut-outs! Tape the sheet of acetate so that it covers the holes. Then hinge the covers back over the holes with tape.

*A Simple Binary Game*: Beginning at the left side, mark the numerals 1, 2, 4, 8 on each acetate-covered hole.* Close all the covers. Place the computer on the stage of the projector, and turn on the projector. Lift the first cover on the left. Computer now shows "1". Write on the blackboard, 1000.

Close all the covers. Lift cover marked "2". Write on blackboard, 0100, and explain that because the light is on in the "2" position, the binary notation is 0100.

Close all covers. Explain that all numbers up to 15 can be shown with only the four lighted positions. How can you make 3? 7? 14?

### Binary Numers for the 10¢ computer

| | | | |
|---|---|---|---|
| ●●●● | 0 | ●●●○ | 8 |
| ○●●● | 1 | ○●●○ | 9 |
| ●○●● | 2 | ●○●○ | 10 |
| ○○●● | 3 | ○○●○ | 11 |
| ●●○● | 4 | ●●○○ | 12 |
| ○●○● | 5 | ○●○○ | 13 |
| ●○○● | 6 | ●○○○ | 14 |
| ○○○● | 7 | ○○○○ | 15 |

By this time, the children will have grasped the idea, and can work quite competently up to 32, using only the blackboard notations. If you want to carry on, you can make an "advanced" computer with 6 holes that can record numbers all the way up to 63!

*Normal binary number notation goes right-to-left, i.e. 1 = 0001, 6 = 0110, 8 = 1000, etc. If you think your kids can grasp this, give it a try.

---

Suggested Bibliography

Ahl, David. *Getting Started in Classroom Computing.* Digital Equipment Corporation, Maynard, Massachusetts.

Adler, Irving and Ruth. *Numerals, New Dresses for Old Numbers.* John Day, Company. New York, N.Y.

Kenyon, Raymond. *I Can Learn about Calculators and Computers.* Harper and Row. New York, N.Y.

# Considerations in Buying A Personal Computer

## What, Which, When, How Much?
### by Karl L. Zinn

Many people ask me for advice on whether to buy a personal computer now or wait for the price to go down. Some are asking which one to buy. A few recognize that the inexpensive machines available today won't do everything one might be led to believe from the advertising. But then, these portable, low-cost, convenient, and personal computers will do a lot of things for which we presently turn to very expensive machines. This outline is a first attempt to assemble a checklist of considerations. Of course, the entries need elaboration, and they tend to interact.

(I would appreciate questions, comments and suggestions which will help elaborate this checklist for use by teachers. Call (313) 763-4410 or write me at CRLT, Univ. of Michigan, 109 East Madison Street, Ann Arbor, MI 48104.)

---

### A Note to Educators

With the prices of microcomputer systems coming down so rapidly, and with the availability of integrated, high-quality system software, it is now possible for many educators to consider the use of micros to supplement or to replace the use of a larger timesharing system. Nevertheless many educators who would like to begin using microcomputers find themselves in a quandary, not having the technical expertise to select a complete microcomputer system. The following list, compiled by Karl Zinn, should at least help these educators get an idea of what to look for. My own experience in talking to educators who want to use microcomputers in high school indicates that three other things are of special importance:

1. Compatibility between BASIC used on the microcomputer and the timesharing system. Obviously if the BASICs are similar, it won't be necessary to develop new course materials, programs, etc.
2. The microcomputer should have an integrated package of system software which can be used by novices, not just computer freaks. In other words, it should be very easy to bring up BASIC (possibly by typing "BASIC," or just turning the power on). If the system has several different peripherals (such as a CRT, printer, etc.) it should be possible to direct output to a particular device without going into the assembly language I/O routines. One should also be aware that programming special-purpose I/O devices such as music boards, speech synthesis and recognition units, and some graphics interfaces often requires linking BASIC to assembly language subroutines.
3. Quality of the construction of the microcomputer is very important. Fewer components are better. — Steve North

---

1) For what purposes? Priorities? Distribution of uses?
    entertainment (Who are the users?)
        ages
        interests
    education (What kind of learning?)
        general literacy about computers
        computer programming
        computer applications
        incidental aid in study of other subjects
        other?
    scientific and creative work (How serious?)
        modeling a process
        simulation
        literary
        graphics
        music
        other?
    personal or professional or small business
        information handling (records, correspondence)
        finances (checking, budgeting, . . .)
        other
    other uses?

2) With what capabilities? Options? Peripherals?
    processor speed
    storage size and access
    programming languages
    application packages
    keyboards
    printer quality and speed
    printer quality and speed
    graphics display
    graphics input
    audio input
    speech output
    music output
    communication
      other small machines
      large machines (systems)

3) Which kind or type or style?
    kits
    components
    peripherals
    portability
    expandability
    other

4) When, and for what period of use?
    watch for predicted and dramatic changes in:
        price
        capability
        other?
    begin now to gain experience
        trade up later
        early experience worthwhile
    other considerations
    . . .

5) How much?
    total outlay
        initial
        later additions
    amortization
        actual life
        period of preferred use
    on service contract
    carry in service as needed
        by individual
    maintenance
    tax deduction
        professional
        educational
    gift
    other
    other considerations?

# Personal Computer Comparison Chart

**Steve North**

Although the accompanying comparison is self-explanatory, a few additional remarks are appropriate. To begin, the descriptions of the Atari and TI computers are somewhat sketchy since we only had manufacturer's literature from which to work. At the time of this writing (late August) neither company had actually delivered production models.

The comparison is also a bit subjective in a few categories, such as hardware quality and portability, but these ratings were arrived at after serious consideration. The availability of software from second sources is also not included but this is generally an important consideration unless you plan to write every program you want to use.

Finally, be aware that each computer is best for certain applications and it's impossible to select a computer that's the "absolute best" for everyone. The TRS-80 is one of the least expensive computers, has an excellent BASIC, and might be thought of as "every man's BASIC-speaking computer;" however, it seems less attractive when expanded beyond the basic 16K cassette-based system. The Apple is more expensive to begin with but is more sophisticated and is more competitive when more memory and disks are added. The PET made quite a splash when it was introduced and is still a good value but has a number of drawbacks. The Compucolor is an excellent color graphics machine and has found its own niche. The newer Atari and TI computers, with ROM cartridges and mass marketing, seem to be the wave of the future.

| Computer | Radio Shack TRS-80 Level II | Apple II |
|---|---|---|
| Case | grey and black plastic | beige plastic |
| Keyboard | typewriter, 2-key rollover, upper-lower case, number pad | typewriter, n-key rollover upper case only |
| Memory | 12K ROM (BASIC) 16K RAM (32-48K with expansion interface) | 12K ROM (BASIC, monitor) 16-48K RAM |
| Language | Level II BASIC: 14 commands, 33 statements, 36 functions, 23 error messages | Integer BASIC: 20 commands, 30 statements, 8 functions, 15 error messages Applesoft BASIC: 12 commands, 48 statements, 27 functions, 19 error messages |
| Resident Machine Language Monitor | no | yes |
| Graphics | 64 x 128 pixels B & W | 40 x 40, 16 colors 280 x 160, 6 colors |
| Text | 16 lines x 64 characters wide upper case only | 24 x 40, upper case only video reverse and blink |
| Expansion | expansion interface, 1-4 disk drives, modem, voice synthesizer, printers | 8 general-purpose I/O slots for disks, printers, speech boards, clocks, etc. |
| Realtime or Hardware Clock | In expansion interface | optional plug-in board |
| I/O Ports built-in | none | game paddles |
| Built-in Audio | no | 1 voice |
| Audio Cassette | 500 baud, motor control | 1200 baud |
| Disk Capacity | drive 1 - 56K drive 2, 3, 4 - 82K | 108K, 4 drives |
| Video Display | B & W TV monitor included | requires color TV |
| Factory Support and Repair | excellent | good |
| Portability | fair | good |
| Hardware Quality | fair | excellent |
| Price (16K RAM) | $849 | $1195 |

| Commodore PET | Compucolor II | Atari 800 | TI 99/4 |
|---|---|---|---|
| beige metal case | simulated wood grain and beige plastic | beige plastic | grey and black plastic |
| typewriter, n-key rollover, upper-lower case | typewriter, upper case and graphics | typewriters, upper-lower case | typewriter layout with space between keys |
| 14K ROM (BASIC and OS) 8-32K RAM | 16K ROM (BASIC, DOS, CRT mode) 8-32K RAM | 8K ROM internal 8-32K cartridge 8-48K RAM | 26K ROM internal 30K ROM cartridge 16K RAM |
| Commodore BASIC: 7 commands, 26 statements, 17 functions, 23 error messages | Disk BASIC 8001: 3 commands, 29 statements, 27 functions | Atari BASIC | TI BASIC: ANSI BASIC with sound and graphics 14 commands, 33 statements, 19 functions |
| no | no | no | no |
| 64 graphics characters B & W | 128 x 128, 8 colors 64 graphics characters | 380 x 192, 16 colors | 192 x 256, 16 colors |
| 25 x 40, upper/lower or upper/graphics video reverse | 32 x 64, in color | 24 x 40, upper-lower case | 24 x 32 |
| IEEE-488 Bus, printer | 2nd disk drive expanded keyboards | printer, disks, modem | speech synthesizer, modem, printer, cassette recorder, disk drives |
| yes | yes | interval timer | interval timer |
| parallel port | RS-232 | game paddles, light pen | general-purpose I/O port, RS-232 option |
| no | no | 4 voices | 3 voices, noise generator |
| 500 baud (?) | no (floppy disk built-in) | 600 baud | optional, 1300 baud |
| ? | 51K, 2 drives | 80K, 4 drives | 80K, 4 drives |
| built-in B&W monitor | color TV monitor included | requires color TV | color TV monitor included |
| fair | good | ? | ? |
| good | fair | good | fair |
| fair to good | good | ? | ? |
| $795 | $1695 | $1000 | $1150 |

# Thinking Strategies and How to Solve Problems

'Not every problem is one to be solved by computer programming.'

# TURNING A PUZZLE INTO A LESSON

Eugene D. Homer
C. W. Post College, Greenvale, NY

The second problem in the feature column "Puzzles and Problems For Fun", *Creative Computing*, 1, 4, (May-June, 1975) proved to be an ideal nucleus for class discussion in a course in advanced programming, although the results were not what, I presume, the author intended.

The problem was stated thusly:

"Mr. Karbunkle went to the bank to cash his weekly paycheck. In handing over the money, the cashier, by mistake, gave him dollars for cents and cents for dollars.

"He pocketed the money without examining it and spent a nickel on candy for his little boy. He then discovered the error and found he possessed exactly twice the amount of the check.

"If he had no money in his pocket before cashing the check, what was the exact amount of the check? One clue: Mr. Karbunkle earns less than $50 a week."

I assume the intent of the author was to have readers write a computer program to solve the problem by trial and error. My intent was to show the class how analysis of the problem before coding could simplify the program. I would like to share this lesson and the resulting conclusion with the readers of *Creative Computing*.

Our first step was to state whatever relationships we could from the problem in mathematical form.

Let D be the integer number of dollars and C the integer number of cents on Mr. Karbunkle's paycheck. The total amount printed on his check, expressed in cents, is:

$$A = 100D + C \qquad (1)$$

Since the teller reversed D and C, the amount of cash Mr. Karbunkle received, again expressed in cents, is:

$$R = 100C + D \qquad (2)$$

We are told that

$$R - 5 = 2A \qquad (3)$$

Substituting Equations 1 and 2 in Equation 3, we obtain:

$$100C + D - 5 = 2(100D + C),$$

which can be simplified to:

$$199D = 98C - 5, \qquad (4)$$

We have one equation, in two integer unknowns, which does look like a problem for trial and error solution. If we were to code at this point, we might come up with something like this, remembering that D is less than 50:

```
      INTEGER D,C
      DO 1 D = 1, 49
      N = 199*D
      DO 1 C = 1, 100
      L = 98*(C-1)-5
      IF (N-L) 1,2,1
    2 A = (100*D+C)/100
      WRITE (5,3) A
      CALL EXIT
    1 CONTINUE
      WRITE (5,4)
      CALL EXIT
    3 FORMAT (F7.2)
    4 FORMAT (1X,'NO SOLUTION')
      END
```

Although this looks like a fairly simple program, I pointed out that it would require a maximum of 4,900 repetitions of the main loop. (In the following, it becomes convenient to measure interatives by the number of times the IF statement is executed.) In view of this large amount of computation we agreed that the analyst should attempt two things:

a) Reduce the amount of computation in the loop, and
b) Reduce the number of times the program must loop.

Tackling the first idea, the class came up with such suggestions as replacing lines 2 and 3 of the above program with the line:

DO 1 N = 199, 9751, 199

This led to a similar discussion about simplifying lines 4 and 5 of the above program. In this round, it became apparent that C must be greater than 1, since for C=1, L = −5, and L would never be equal to N. It also became apparent that our inner loop could be terminated as soon as L became greater than N. We now had our program down to something like this:

```
      INTEGER D,C
      DO 1 N=199, 9751, 199
      DO 5 L=93, 9697, 98
      IF (N-L) 5, 2, 1
    2 A=(100*D+C)/100.
      WRITE (5, 3) A
      CALL EXIT
    5 CONTINUE
    1 CONTINUE
      etc.
```

We had reduced our maximum number of executions of the IF statement by almost one-half (to 2,499 executions) and had removed all arithmetic calculation from the loops.

We were still not happy with the program, since the inner loop, on L, was too repetitive. We saw that if a particular value of L was less than a particular value of N, there was no need to try that value of L again for the next value of N. This led us to the removal of the inner loop altogether:

```
    INTEGER D,C
    L = 93
    DO 1 N = 199, 9751, 199
    IF (N−L) 1, 2, 5
2   A = (100*D+C)/100.
    WRITE (5, 3) A
    CALL EXIT
5   L=L+98
1   CONTINUE
    etc.
```

A few "runs" by hand of this program indicated that we would try only two values of L for each value of N, reducing the maximum number of IF statement executions to 98.

Before we left this approach, we took another look at Equation 4. For any non-negative integer value of C, the right side of the equation will be odd. Therefore the term 199 D must be odd, and therefore D must be odd. Thus, D may assume only the values 1,3,5,...,49, and N = 199 D will increase in increments of 2(199)=398. Our last program, then, can have its DO statement changed to

$$\text{DO 1 N = 199, 9751, 398,}$$

resulting in another halving of the iterations.

It was now time to take another tack. I reminded the class of last week's work with modular numbers, and showed them that Equation 4 satisfied the first definition of a modular number,

$$N = q \cdot m + r,$$

but failed the second definition,

$$0 < r < (m-1)$$

where   N = 199D
        q = C
        m = 98
        r = −5

However, we could rewrite Equation 3 as:

$$199D = 98C - 98 - 5 + 98$$

or

$$199D = 98(C-1) + 93$$

which satisfied both definitions of modular numbers with

        N = 199D
        q = C−1
        m = 98
        r = 93

We could then write, from the familiar expression

$$r \equiv N \bmod m,$$
$$93 \equiv 199D \bmod 98 \qquad (5)$$

Our strategy then could be to use the MOD function as we increment N. If we find a value of N satisfying Equation 5, we can solve for

$$D = \frac{N}{199}$$

and, by rewriting equation 4

$$C = \frac{N+5}{98}$$

Our new program follows. Note that we have also dropped the integer declaration as being unnecessary.

```
    DO 1 N = 199, 9751, 398
    IR = MOD(N,98)
    IF (IR−93) 1, 2, 1
2   D = N/199
    C = (N+5)/9800.
    A = D+C
    WRITE (5, 3) A
    CALL EXIT
1   CONTINUE
    WRITE (5,4)
    CALL EXIT
3   FORMAT (F7.2)
4   FORMAT (1X,'NO SOLUTION')
    END
```

This looked like a reasonable program, requiring only 24 repetitions of the IF statement, maximum. We set out to run it by hand, with these results:

| N    | IR |
|------|----|
| 199  | 3  |
| 597  | 9  |
| 995  | 15 |
| 1393 | 21 |

At this point the class saw that as N increased by 199, IR increased by 6, and that it might be possible to "figure out" when IR would hit 93.

It took only a few minutes to work out the fact that
for    N = 199+398i; i=0,1,2,...
       IR = 3+6i

If IR is to be equal to 93:
       93 = 3+6i
or,
       i = 15

This would occur when
       N = 199+398(15)=6169,

and
$$D = \frac{6169}{199} = 31,$$
$$C = \frac{N+5}{98} = 63,$$
$$A = \$31.63$$

To be sure, we checked with equations 2 and 3:
       R = $63.31
       $63.31 − $.05 = $63.26 = 2($31.63).

Our "program" now has reduced to:

```
    WRITE (5,5)
    CALL EXIT
5   FORMAT (1X,'$31.63')
    END
```

We spent about two hours going over this puzzle. While much of our work was useless in terms of the final solution, the class did learn some valuable lessons from the discussion. They learned that careful analysis of a problem can lead to a startling reduction in the amount of computing to be done. They learned that it pays to run through a program "by hand" a few times to discover hidden relationships. Finally, they learned that not every problem presented in a computer-oriented environment is a problem to be solved by computer programming.

# Problems for Creative Computing
by Water Koetke

The problems to be discussed in this column are those that seem particularly well suited not just for computing, but for creative computing. They will cover a wide variety of topics and subjects, and all are intended for both students and teachers — for anyone turned on by challenging problems, games or programs.

Your reactions will be very much appreciated. Suggestions for future columns, solutions to problems discussed, new problems, extensions and experiences with problems discussed are all solicited. Please address all correspondence to Walter Koetke in care of *Creative Computing.*

The challenge of creative thought is before all of us — this column is intended for those who choose to demonstrate that creative thought is also behind them. I hope you find the ideas rewarding.

# Tac Tix and the Complications of Fallibility

The game of Tac Tix was created by Piet Hein, also the inventor of Hex, in the late forties. A first impression of the game is likely to be that it is indeed simple, but first impressions themselves are over-simplifications aren't they? The rules of Tac Tix are few, a desirable characteristic of games to be used in the classroom.

Each game begins with 25 markers arranged in a 5 x 5 square formation as in the diagram.

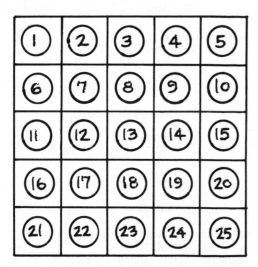

Two players then alternate turns. On each turn a player may take as many markers as he chooses from any single row or column, provided that the markers are next to each other. For example, markers 1 and 3 cannot both be removed on a single turn unless marker 2 is present and is also removed with the others. The player who removes the last marker is the winner.

**STOP READING** this article. Put it down and take a few minutes to analyze the game. The first player has an easily described winning strategy. Can you find it?

Assume that the first player can play without error. On his first turn he should remove marker 13, the center marker. On each subsequent turn he should remove the markers symmetrically opposite those removed by his opponent. By playing in this manner he is assured of winning the game. After the center marker is removed, a typical game might be:

| Second Player | First Player |
|---|---|
| 2 - 5 | 21 - 24 |
| 15, 20, 25 | 1, 6, 11 |
| 8 - 10 | 16 - 18 |
| 7 | 19 |
| 14 | 12 (wins) |

When playing this game, an equally infallible second player is likely to be bored to death.

Since Tac Tix is played on a small board, has only a few easily stated rules, and requires only a short time to play, it is a very good game to implement on a computer. Writing a program that will play Tac Tix with a user by following a well defined strategy is an excellent problem at two different levels.

First, try a program in which the computer is the first player. To do this, one must be able to create a program that: represents the 5 x 5 board using single or double subscripted variables; makes symmetrical moves; determines if the second player is making a legal move; and realizes that the game is over. When writing the program one faces many of the likely difficulties encountered in far more complex problems.

Second, try a program in which the computer is the second player. If all users were infallible, then this really isn't worth writing. However, somewhere there may be a student or teacher who occasionally makes an error. Assume that you're writing the program for him. By considering this small bit of reality, a trivial case in the world of perfect people has become a rather challenging, interesting problem.

Consider each of the following opening plays of the first player. What is the best counter play for the second player?

| Opening Play | Counter Play |
|---|---|
| 3, 8, 13 | ? |
| 16 - 18 | ? |
| 11 - 15 | ? |

In general, the second player should attempt to play so that for every missing marker the symmetrically opposite marker is also missing. The center marker must also be missing. If the second player succeeds in obtaining this board configuration at the end of any turn, he has successfully taken advantage of the first player's error and has a winning strategy. On all subsequent turns he should remove only those markers symmetrically opposite those removed by his opponent. Following this strategy, if the first player's opening play is 3, 8, 13 the second player's play should be 18, 23.

But what should the second player do if the opening play is 16 - 18? That's part of the challenge of the problem! Perhaps play 9 - 10, but that seems to increase the first player's chances of making a winning move next time. We do assume the first player is smart even if he does err on his first play. Perhaps play at random, but that seems to decrease the second player's chance of obtaining a winning board configuration.

The complexity of the problem is indeed increased by letting the first player be human. The problem is very good because it is a mini-version of what one often faces in much larger problems: the solution is not trivial; although each step of a solution can be well defined, some definitions will reflect the problem solver's best judgment rather than an absolute truth; once a solution is well defined, a program can be written that plods through many cases while another can be written that uses reflections and rotations of the board to reduce the number of cases. The challenge of writing a program that plays Tac Tix with a smart but fallible user who is given the first move properly belongs under the title "Creative Computing." And those who write such a program are likely to have done some "creative analysis" before they finish.

A modified version of Tac Tix that looks easier but is actually much more complex is played on a 4 x 4 board rather than a 5 x 5 board. The only other change is that the player who removes the last marker is the loser. Is there a winning strategy for either player? After trying to define a winning strategy for one of the players, one may well become interested in writing a program that develops its strategy by learning as it plays. By repeating successful plays and avoiding the repetition of unsuccessful plays, the computer can improve its strategy with each successive game. The writing of such cybernetic programs will be the subject of a future column.

Related References

Gardner, Martin; *Mathematical Puzzles and Diversions*; New York: Simon and Schuster; 1959; Chapter 15.

Spaulding, R. E.; "Recreation: Tac Tix"; *The Mathematics Teacher*; Reston, Virginia: National Council of Teachers of Mathematics; November 1973; pages 605-606.

# Puzzles and Problems for Fun

This puzzle is calculated to test your ability in calculus: A watchdog is tied to the outside wall of a round building 20 feet in diameter. If the dog's chain is long enough to wind halfway around the building, how large an area can the watchdog patrol?

A. G. Canne
Pittsburgh, Pa.

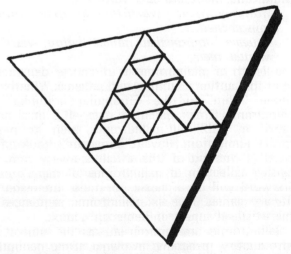

The Sheik of Abba Dabba Dhu wears this medallion, on which each equilateral triangle represents a wife in his harem. How many wives does the sheik have?

David Lydy
Cincinnati, Ohio

# Palindromes: For Those Who Like to End at the Beginning

by Water Koetke

*Egad, a base life defiles a bad age*
*Doom an evil deed, liven a mood*
*Harass sensuousness, Sarah*
*Golf; No, sir, prefer prison-flog*
*Ban campus motto, "Bottoms up, MacNab"*

A palindrome is a work, verse, number or what have you that reads the same backward as forward. The unit in a palindrome may vary. Each of the five lines in the poem at the beginning of this article is itself a palindrome using a letter as the unit. However, a palindromic poem might be written for which the entire poem is a palindrome rather than the individual lines. The unit in a palindrome might also be a word as in:

> Men wanted warning before police approached;
> squealer approached police before warning wanted men.

The length of plaindromes is, of course, dependent upon the author's cunning and patience. Whether a simple "Mom" or the "Ethopoiia Karkinikie", a palindromic Greek poem of over 400 lines published in 1802, all palindromes seem to merit special admiration. Howard Bergerson's book referenced at the end of this article is a very comprehensive collection of palindromes of many types. His work will be a classic for those interested in literary games. The six palindromic sentences in this article all appear in Bergerson's book.

Palindromes are appropriate as the subject of introductory programs involving string manipulation. For example, writing a program to recognize a palindromic sentence using a letter as the unit requires character manipulation. Writing a program to recognize a palindromic sentence or paragraph using a word as the unit requires both character and word manipulation. This is particularly interesting if the sentence or paragraph is entered one line rather than one word at a time. The creation of palindromes is likely to remain in the province of a human and not a machine endeavor. One may devise clever programs to assist that endeavor, but human creativity supported by a well thumbed dictionary shall remain the most essential resources.

Numeric palindromes are those numbers that read the same backward as forward. The examination of these numbers is a field rich with possibilities for creative computing.

Consider all palindromes that can be written in the form $N^k$ where N and k are positive integers. Would you expect that N is also a palindrome; Let's consider the case of k=2. 14641 and 484 are both palindromes that can be written in the form $N^2$:

$$14641 = 121^2$$
$$484 = 22^2$$

And both 121 and 22 are also palindromes! Is this always true; And what about other values of k; Is the cube root of a palindromic cube also a palindrome; Is the fourth root of palindromic fourth power also a palindrome?

These questions are of interest because only partial answers have been given. When k=2—if $N^2$ is a palindrome then N is often, but not always, a palindrome. When k=3—the only known palindromic cube without a palindromic cube root is 10,662,526,601, but there may be others yet undiscovered. When k=4 —all known palindromic fourth powers have fourth roots that are palindromes, but what about those that are yet unknown; When k=5 —this one's a little harder as there are no known palindromes that can be written in the form $N^5$. Clearly the answers to these questions are a proper subject for creative computing—and just as clearly, the fundamental principles required to generate a formidable attack on the answers require no more than high school algebra and the resource of computing facilities.

There is a conjecture concerning palindromes that raises another unanswered question. Begin with any positive integer. If it is not a palindrome, reverse its digits and add the two numbers. If the sum is not a palindrome, treat it as the original number and continue. The process stops when a palindrome is obtained. For example, beginning with 78:

```
     78
+    87
    165
+   561
    726
+   627
   1353
+  3531
   4884
```

The conjecture, often assumed true, is that this process will always lead to a palindrome. And indeed that is just what usually happens. Most numbers less than 10000 will produce a palindrome in less than 24 additions. But there's a real thorn in the side of this conjecture—196. No one really knows whether a palindrome will be produced if the beginning number is 196.

Writing a program that explores this conjecture can be a valuable experience on several levels. A program that examines the integers 1 through 10000 is a worthwhile student project because it requires the ability to deal with numbers of up to 14 digits. The numbers 196, 691 and the resulting sums and their reversals would, of course, have to be excluded from this program. The exploration of 196 really should be a category of its own. Pursuit of this problem will lead the student down several interesting side roads, lure him into doing some original mathematics, and certainly teach him much about computing. Solution of this problem should certainly merit an A since it will bring him recognition that extends well beyond his classroom. Personally, I'm quietly hoping that the problem of 196 is solved by a secondary school student just as the three largest known perfect numbers were discovered by a secondary student with access to computing facilities, but that's a different subject isn't it?

Related References

Bergerson, Howard W.; *Palindromes and Anagrams;* Dover Publications, New York; 1973.

Gardner, Martin; "Mathematical Games"; *Scientific American;* New York; August 1970; pages 110-114.

Kordemsky, Boris A.; *The Moskow Puzzles;* Charles Scribner's Sons, New York; 1972, page 172.

## MORE ABOUT PALINDROMES

The January and June 1974 issues of *Games & Puzzles* contained some additional discussion about palindromic numbers on Darryl Francis' Puzzle Pages and in letters from R. Hamilton and Jonathan Kessel.

R. Hamilton notes in his letter, "Of the 900 three digit numbers, 90 are themselves palindromic, 228 require just one reversal to form a palindromic number, 270 require two reversals, 143 require three reversals, 61 require four reversals, 33 require five reversals and 75 require more than five. These remaining 75 numbers could be classed into just a few groups, the members of which after one or two reversals each produce the same number and are therefore essentially the same. One of these groups consists of the numbers 187, 286, 385, 583, 682, 781, 869, 880 and 968 each of which when reversed once or twice form 1837 and eventually form the palindromic number 8813200023188 after 23 reversals (The nos. 89 and 98 also belong to this group.) The most interesting group consists of the numbers 196, 295, 394, 493, 592, 689, 691, 788, 790, 887 and 986 which form 1675 after a few reversals but after 100 reversals fail to produce a palindromic number forming the non-palindromic 44757771534490515- 6172906992715615084436277746 44."

Jonathan Kessell notes, "However, what you didn't mention, maybe because it is rather obvious, is that if 78 and 96 both yield 4884, then 87 and 69 will do so, too. Thus, not only 89 gives 8,813,200,023,188 after 24 reversals, but so does 98. You may also be interested to know that 249 integers less than 10,000 fail to produce a palindrome after 100 reversals. The smallest of these numbers is 196; indeed, even after as many as 4,147 reversals, this number still fails to generate a palindromic number. (Just how many reversals are necessary for 196 to produce a palindromic result? — DF.) The numbers 6,999 and 7,998 produce the longest palindrome: 16,668,488,486,661 — out of all the numbers from 1 to 10,000, that is. It takes twenty steps to produce this palindrome from both of the numbers.

Also, there is an infinity of palindromic squares, most of which have palindromic square roots. The smallest nonpalindromic root is 26 — the square root of 676. Similarly with cubes and cube roots. The smallest nonpalindromic cube root is 2,201, the cube being 10,662,526,601. The number 836 may be of interest, too. It is the largest three-digit integer whose square root (698,896) is palindromic. Further, 698,896 is the smallest palindromic square with an even number of digits; also, when turned upside down, the number remains palindromic. The next largest palindromic square with an even number of digits is 637,832,238,736, which is the square of 798,644."

Anyone care to take the study of palindromic numbers further still?

# Palindromes (con't)

Tom Karzes, an eighth grader at Curtis Jr. High School, Sudbury, MA wrote a program to take any number and test whether it is a palindrome; if it is not the program goes on to form the palindrome. The program fails with greater than a 7-digit number. Can you write one that doesn't?

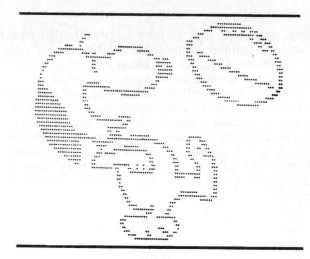

```
LISTNH
10 REM *** TOM KARZES, CURTIS JR HS, SUDBURY, MA
11 PRINT \INPUT "GIVE ME A NUMBER";A\PRINT\B=0
20 B=B+1\A=A/10\ IF INT(A)>0 THEN 20
30 FOR C=B TO 1 STEP -1\A=A*10\B(C)=INT(A-10*INT(A/10))\NEXT C
40 D=0\FOR C=1 TO INT(B/2)\ IF B(C)<>B(B+1-C) THEN D=1\NEXT C
50 FOR C=B TO 1 STEP -1\ PRINT CHR$(B(C)+48);\NEXT C
60 IF D=1 THEN 69
65 PRINT " IS A PALINDROME."\GOTO 10
69 PRINT " IS NOT A PALINDROME,"
70 IF B/2>INT(B/2) THEN B(INT(B/2)+1)=2*B(INT(B/2)+1)
72 FOR C=1 TO INT(B/2)\B(C)=B(C)+B(B+1-C)\NEXT C
75 FOR C=1 TO INT(B/2)\B(B+1-C)=B(C)\NEXT C
80 B(B+1)=0\FOR C=1 TO B\B(C+1)=B(C+1)+INT(B(C)/10)
90 B(C)=B(C)-10*INT(B(C)/10)\NEXT C
100 IF B(B+1)>0 THEN B=B+1
110 GOTO 40
120 END

READY

RUNNH
GIVE ME A NUMBER? 19

19 IS NOT A PALINDROME,
110 IS NOT A PALINDROME,
121 IS A PALINDROME.

GIVE ME A NUMBER? 38

38 IS NOT A PALINDROME,
121 IS A PALINDROME.

GIVE ME A NUMBER? 79

79 IS NOT A PALINDROME,
176 IS NOT A PALINDROME,
847 IS NOT A PALINDROME,
1595 IS NOT A PALINDROME,
7546 IS NOT A PALINDROME,
14003 IS NOT A PALINDROME,
44044 IS A PALINDROME.

GIVE ME A NUMBER? 96

96 IS NOT A PALINDROME,
165 IS NOT A PALINDROME,
726 IS NOT A PALINDROME,
1353 IS NOT A PALINDROME,
4884 IS A PALINDROME.
```

### 1675 — NON-PALINDROMIC?

Mike Lean in England took the number 1675 and reversed it 4850 times with the help of a computer, of course. Those 4850 reversals produced a 2000-digit number which still was not palindromic. Darryl Francis of *Games and Puzzles* thinks it's reasonable to assume that 1675 will never become palindromic however many times it is reversed. Do you agree?

## Spell them backward and they stay the same

Read it from right to left and it will be the same as when you read the usual left to right. What will be? A palindrome, that is what.

Here is a sample: WON'T PEWS FILL IF SWEPT NOW?

You can have a lot of fun creating your own palindromes. But before you get started on yours, read these:

TOO HOT TO HOOT.
A POTATO PA?
NO, IT IS OPEN ON ONE POSITION.
STRAP ON NO PARTS.
WAS IT A BAR OR A BAT I SAW?

And this one is more difficult to read aloud:
OH HO HAH HAAHA AHAH HAH OH HO!

You can do the same with numbers. Example: 25952.

When your friends ask you what a palindrome is, tell them, "A palindrome looks and spells exactly the same from left to right or right to left, backward or forward, or forward or backward." In a popular dictionary this example is printed: ABLE WAS I ERE I SAW ELBA.

A few commercial names are palindromes. In California a city is named Yreka; a merchant calls his bakery Yreka Bakery.

And don't overlook single words: HUH. PEP. EYE. ADA. POP. WOW.

At first, making your own will be a slow process, but as you work with them it will become easier. Many adults know at least a few such words, so ask for their help. Friends at school may have one or two. And how about your teachers?

Write all of them down before you forget some. Try to make a long list, so you can show it to friends.

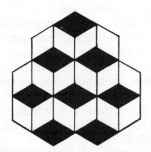

HOW MANY BLOCKS DO YOU SEE? 6 OR 7 ?

# Computing Factorials -- Accurately

by Walter Koetke

Multiple precision arithmetic is a topic that can easily capture the imagination of almost anyone interested in computing. Today's programming languages and even hand held calculators normally provide enough precision to satisfy the requirements of most users, so this topic is really most appropriate for those intrigued by the challenge of creative computing. Perhaps because multiple precision arithmetic is not studied by all students, introductory literature relating to the topic is very sparse. If you encounter a good reference, be sure to note it as the topic is rarely given more than two or three cursory pages.

Calculating factorials is a standard example in elementary programming courses. Although a good example of the technique required to compute a product, the fact that only a few factorials can be calculated exactly before being subjected to round-off error is usually ignored. Actually, not too many factorials can be computed before the arithmetic limits of BASIC are reached. A typical program that correctly calculates the factorial of an entered value is:

```
10 INPUT N
20 LET F=1
30 FOR M=1 TO N
40 LET F=M*F
50 NEXT M
60 PRINT F
70 END
```

Given a non-negative integer N, then N! (N factorial) is defined as:
  if N>0, N! = N(N-1)(N-2) . . . 1
  if N = 0, N! = 1

If only 6 significant digits are available, the results of this program are subject to round-off error for all values of N greater than 11. If $10^{38}$ is the upper limit of the available numbers, then this program can not even approximate the factorial for any N greater than 33. Even if $10^{99}$ is the upper limit available, the factorial can not be approximated for any N greater than 69. However, using multiple precision arithmetic, we can extend these limits to whatever extreme we choose.

To compute factorials more accurately, we must develop an algorithm for multiple precision multiplication. The most straightforward algorithm is that which we use when multiplying with pencil and paper. Hand calculation has many stumbling blocks, but a limited number of digits or a limited range of values are not among them.

Consider computing the product 7 x 259. You begin by multiplying 7 x 9, and although the product is 63 you only write down the 3 and "carry" the 6. After next multiplying 7 x 5, you add this "carry" to the product and obtain 41 — and again you write down the 1 and "carry" the 4. And so forth . . . After each individual multiplication, you record the units digit and "carry" those that remain.

To write a BASIC program that does multiple precision arithmetic using this same algorithm, one need only be able to separate the units digit of a product from the "carry". If P represents the product of two positive integers, then:
  carry = INT(P/10)
  and
  units digit of P = P - 10*(carry)

Let's now apply this algorithm to the larger problem of computing the factorial of any positive integer. To do this we will write a program similar to the very brief example already given. However, the product shall be represented by the subscripted variable F, *each subscripted value representing a single digit of the product.* One program that does this is:

```
10 DIM F(50)
20 LET L=50
30 INPUT N
40 FOR I=2 TO L
50 LET F(I)=0
60 NEXT I
70 LET F(1)=1
80 FOR M=1 TO N
90 LET C=0
100 FOR I=1 TO L-1
110 LET F(I)=F(I)*M+C
120 LET C=INT(F(I)/10)
130 LET F(I)=F(I)-10*C
140 NEXT I
150 NEXT M
160 FOR I=L TO 1 STEP -1
170 PRINT F(I);
180 NEXT I
190 END
```

Notice that:

1. The program will compute factorials that can be expressed using no more than 50 digits. This restriction can be decreased by using a larger value in the DIM at line 10 and making a corresponding change in the value of L at line 20.

2. The product of two integers and the addition of the previous carry is completed at line 110. The next carry is calculated in line 120 and the unit's digit of the product is obtained in line 130. If you understand these three lines, you understand the fundamental idea of multiple precision multiplication.

3. All 50 (or L) digits of the product are always printed. This isn't wrong, but leading zeroes look peculiar.

Two sample runs of the program appear as:

```
RUN
? 5
0 0 0 0 0 0 0 0 0 0 0 0 0 0 0 0 0 0 0 0 0 0 0 0
0 0 0 0 0 0 0 0 0 0 0 0 0 0 0 0 0 0 0 0 0 0 0 1
2 0
READY
RUN
? 40
0 0 8 1 5 9 1 5 2 8 3 2 4 7 8 9 7 7 3 4 3 4 5 6
0 1 1 2 6 9 5 9 6 1 1 5 8 9 4 2 7 2 0 0 0 0 0 0
0 0
READY
```

Now stop reading and try running this program. Can you improve it? Increase the number of digits in the product. Print only those digits of the product that are significant. Print the product without spaces between each digit. Try to do these things *before* you continue reading — and if you can't use a terminal you can still write the required program changes.

If you were successful in completing the suggested improvements, then read fast for awhile. Increasing the number of digits in the product from 50 to 150 can be done with:

```
10 DIM F(160)
20 LET L=160
```

The only limit to the number of digits is the upper limit of the subscripts available in the BASIC you are using.

Printing the product without spaces between digits can be done in several different ways — most of which are a function of the version of BASIC you are using. Since this has little to do with multiple precision arithmetic, removing the spaces remains your problem.

Deleting leading zeroes in the printed product doesn't have much to do with multiple precision arithmetic either, but let's delete them anyway. This is not being done arbitrarily, but because it provides a very good example of the use of a "flag" within a program. Quite simply, we will use one variable, say S, as a flag to indicate whether a non-zero digit has been printed. All zeroes can then be ignored rather than printed unless a non-zero digit has been printed. This is represented in flow chart form as:

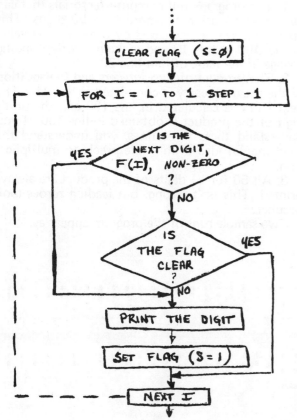

This algorithm for omitting leading zeroes is added to the program by:

```
160 LET S=0
170 FOR I=L TO 1 STEP -1
180 IF F(I)>0 THEN 200
190 IF S=0 THEN 220
200 PRINT F(I);
210 LET S=1
220 NEXT I
230 END
```

Two sample runs of the modified program appear as:

```
? 5
  1  2  0
READY
RUN
? 100
9  3  3  2  6  2  1  5  4  4  3  9  4  4  1  5  2  6  8  1  6  9  9  2
3  8  5  6  2  6  6  7  0  0  4  9  0  9  7  1  5  9  6  8  2  6  4  3
8  1  6  2  1  4  6  8  5  9  2  9  3  8  9  5  2  1  7  5  9  9  9
9  3  2  2  9  9  1  5  6  8  9  3  4  1  4  6  3  9  7  6  1  5  6  5
1  8  2  8  6  2  5  3  6  9  7  9  2  0  8  2  7  2  2  3  7  5  8  2
5  1  1  8  5  2  1  0  9  1  6  8  6  4  0  0  0  0  0  0  0  0  0  0
0  0  0  0  0  0  0  0  0  0  0  0  0  0
READY
```

If you tried running this program as suggested, you probably discovered something else that needs to be improved — the speed of computation. The present form of the program always multiplies each of the integers 1 through N by each of the variables F(1) through F(L-1). Thus when N = 100 and L = 160 as in the last sample run, the computation loop (lines 110 - 130) is repeated 15,900 (100*159) times. Even when N = 5 this loop is repeated 795 (5 x 159) times, and that's a lot of work to compute 1*2*3*4*5. This excessive computation can be eliminated by making use of *a pointer*, a very important idea in computing. Essentially, we will use another variable, say P, to point at the left most non-zero digit in the product. We can then multiply each of the integers 1 through N by each of the variables F(1) through F(P). When N = 5, this reduces the number of repetitions of the computation loop from 795 to 6, or more than 99%. When N = 100, the reduction is from 15900 to 6834, or about 57%. Clearly a pointer provides a worthwhile savings. Try to verify these reduced counts before you leave this topic.

If you are unfamiliar with the concept of a pointer, be sure you read these paragraphs very carefully. After you become familiar with this idea, teach your students about pointers if you're a teacher, or teach your teacher about pointers if you're a student. Who learns a significant idea first is not nearly so important as having everyone eventually understand the idea.

To implement the use of a pointer in our factorial program, begin with the initial value, P = 1. We then want to multiply each of the integers 1 through N by each of the variables F(1) through F(P). Thus line 100 should be changed to read FOR I = 1 TO P. The pointer does not alter the multiplication algorithm in lines 110 through 130, but after the NEXT I in line 140, we must examine the carry to see if the pointer is to be incremented. If the carry is non-zero, we increment the pointer, perform the carry, and then repeat the procedure. If the carry is zero, we can continue with NEXT M.

The instructions needed to do this are:

```
142 IF C=0 THEN 150
143 LET P=P+1
144 LET F(P)=C
145 LET C=INT(F(P)/10)
146 LET F(P)=F(P)-10*C
147 GOTO 142
```

Finally, since the product contains exactly P digits, then the print loop beginning at line 170 can be changed to read FOR I = P TO 1 STEP -1.

Our multiple precision factorial program, complete with all modifications discussed, now appears as:

```
10 DIM F(160)
20 LET L=160
30 INPUT N
40 FOR I=2 TO L
50 LET F(I)=0
60 NEXT I
65 LET P=1
70 LET F(1)=1
80 FOR M=1 TO N
90 LET C=0
100 FOR I=1 TO P
110 LET F(I)=F(I)*M+C
120 LET C=INT(F(I)/10)
130 LET F(I)=F(I)-10*C
140 NEXT I
142 IF C=0 THEN 150
143 LET P=P+1
144 LET F(P)=C
145 LET C=INT(F(P)/10)
146 LET F(P)=F(P)-10*C
147 GOTO 142
150 NEXT M
160 LET S=0
170 FOR I=P TO 1 STEP -1
180 IF F(I)>0 THEN 200
190 IF S=0 THEN 220
200 PRINT F(I);
210 LET S=1
220 NEXT I
230 END
```

Any additional improvements are left to you. Although several are possible, more efficient use of the variables in the F array is likely to be the most dramatic. The example program uses one variable to represent one digit. By allowing each variable to represent two or three digits, program speed is essentially doubled or tripled. But then there are additional problems when the product is printed, and possibly additional problems within the computation loop. Does the example have any theoretical limit on the factorials that can be computed? Does it have any realistic limits?

The topic of multiple precision arithmetic will be further explored in a subsequent column. Now that you understand multiple precision multiplication, only the operations of addition, subtraction and division separate you from successfully computing $\sqrt{2}$ and $\pi$ to at least half million digit accuracy.

(To further test your knowledge of multiple precision arithmetic, why not try Contest Problem 3. — Ed.)

## *Glum Glossary*

**Punched card:** A short piece of 80-channel paper tape.

**Program:** The footprints of hundreds of bugs. Once the bugs are eliminated, the program is all that's left.

# Puzzles and Problems for Fun

▶ Mrs. Canton wanted to buy all the grocer's apples for a church picnic. When she asked how many apples the store had, the grocer replied, "If you added 1/4, 1/5 and 1/6 of them, that would make 37." How many apples were in the store?

▶ Donna bought one pound of jellybeans and two pounds of chocolates for $2. A week later, she bought four pounds of caramels and one pound of jellybeans, paying $3. The next week, she bought three pounds of licorice, one pound of jellybeans and one pound of caramels for $1.50. How much would she have to pay on her next trip to the candy store, if she bought one pound of each of the four candies?

▶ Take a 3-digit number like 200, reverse it (002) and then multiply the two numbers. The result, 400, is a perfect square (20* 20 = 400). Find all such 3-digit numbers.

Bill Morrison<br>Sudbury, Mass.

▶ Can you put nine pigs in four pens so that there are an odd number of pigs in each of the four pens?

 If you have a favorite puzzle, perhaps we can print it here. Send it along!

*Problems for Creative Computing....*

# AEDI, MUTAB, NEDA and SOGAL

by Walter Koetke

Your non-terrestrial thoughts should not remain free of problems that require creative solutions. Toward that end, here are two situations that you might find interesting. After solving either one or both of these problems, please send your solution to Walter Koetke at the *Creative Computing* address. The best solutions received will be acknowledged in a future column.

If you think you've seen the first problem before, you may be correct. It's really an old problem in a new disguise.

The civilizations of the three planets Neda, Mutab and Sogal have agreed to begin a war in the year 2431. Although these societies have not eliminated such irrational actions as war, they have at least formalized the process. There are, for instance, no guerilla activities and wars are usually very brief and always decisive. Wars are fought with inter-planetary rockets each of which is powerful enough to completely destroy an entire planet. With such powerful weapons at their disposal, Neda, Mutab and Sogal have agreed to the following set of rules, for only in this way can they be assured of a single victor.

Rule 1: The fight will continue until only one civilization remains.

Rule 2: The rather primitive technique of drawing lots will be used to determine which planet may launch the first rocket, which the second and which the third.

Rule 3: After the launching rotation is established, rocket launching begins and continues in order until only one planet remains.

When contemplating the outcome of this war, the three civilizations have full knowledge of the background of their adversaries.

Mutab is clearly the technologically superior civilization. Once launched, their rockets always strike with perfect accuracy — thus disproving a modern theory that nothing is perfect. Before this war begins, both of the other civilizations are aware of the terrifying fact that if a Mutab rocket is fired at them, the probability of their being completely destroyed is 1.

Neda is the oldest civilization and long ago had the superior technology. However, the complacency of a self-centered, unchallenged mind has been eroding this superiority for many years. As a result, the technology of Neda has not advanced in over 40 years. If a Nedian rocket is fired at another planet, the probability of hitting that planet is 0.8, just as it was 40 years ago.

Sogal is by far the newest of the three civilizations. Being dedicated to producing its own technology on its own terms has resulted in a proud and purposeful civilization, but one that is technologically four to five hundred years behind its present adversaries. A missile launched by Sogal has only a 50-50 chance of reaching its intended target.

Your role in this future war is to determine each civilization's probability of winning.

The second problem is based upon an idea presented by C. Stanley Ogilvy in the text *Tomorrow's Math*, Oxford University Press, 1972. However, you should attempt your own solution before seeking Ogilvy's support.

Although the civilization on the planet of Aedi is generally considered rather advanced, its political system no longer attracts the imagination and support of the majority of citizens. In an effort to attract more capable leaders at the highest level, a new plan was formulated for selecting the president. The originators of the plan also hoped that their new idea would result in a younger president and a change of presidents at least every 10 years.

Essentially, the new plan is as follows. Once a president is selected, he holds office for at least five years. At that time he may or may not be replaced by a newly selected person. The selection process is new and is the key to this new plan. The selection process is a problem — one problem known by all citizens at all times. When a president has served for five years, all citizens of Aedi are invited to submit their solution to the problem. If a solution submitted is better than that previously submitted by the current president, then the submitter becomes the new president. If a solution submitted is equal to that previously submitted by the current president, then the submitter becomes the new president only if his solution is also different from that previously submitted by the current president.

The problem used by the Aedians to select their leader can be attacked on many different levels. The problem involves three sets of three instructions each and a board on which play is recorded. Three blank instruction sets and the playing tablet appear as:

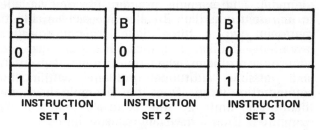

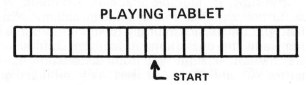

Instructions are of the form STOP (self-explanatory) or contain three elements:

1. An indication of what to record on the playing tablet. The only possibilities are 1, 0 or B (blank).

2. An indication of which direction to move on the playing tablet. One square left or right are the only choices.

3. An indication of which instruction board contains the next instruction to be followed.

Thus the instruction 1-R-3 means: record a 1 on the playing tablet, move one square right, and go to board 3 for the next instruction.

President of Aedi

The combination of the contents of your place on the playing tablet and the instruction board you are following dictate your next insturction. The left column of the instruction board indicates B (blank), 0 or 1. If your current place on the playing tablet is blank, you follow instruction B; if it is a 0, you follow instruction 0; and if it is a 1, you follow instruction 1. The play always begins with board 1.

Consider the following complete set of instructions. If you think you understand the rules, try following the instructions before reading further.

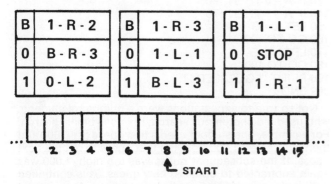

START

The infinite tablet has been partially numbered for the convenience of this discussion. Play begins on board 1 and, since square 8 is blank, our move is 1-R-2. Thus we write a 1 in square 8, move 1 square to the right (square 9) and go to board 2 for the next instruction. Since square 9 is blank, the second instruction is 1-R-3. Once again we write a 1, move to the right, and this time go to board 3 for the next instruction. Our tablet now looks like

WE'RE HERE

The next instruction is 1-L-1, which records a 1 in square 10, returns us to square 9 and indicates that the next instruction is on board 1. Because square 9 contains a 1, our instruction is 0-L-2 so we replace the 1 with a 0, move to square 8 and proceed to board 2 for the next instruction. The tablet now appears as:

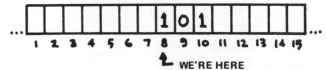

WE'RE HERE

And on we go. If you continue following these instructions until you reach STOP, the tablet will finally appear as:

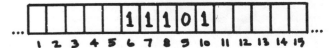

When STOP is reached, the success of the effort is measured by the longest string of consecutive ones that appear on the tablet. In the example, the longest string contained but three ones.

The Aedians problem was not to follow a particular instruction set, but to create one. Specifically, their leader would be the person who could write the series of instructions that would produce the longest *finite* sequence of consecutive ones. Since you've just seen the example used to introduce the problem to the young Aedians, you'll have to beat three consecutive ones before you're their new leader. If you generate an impressive series, be sure to send the instructions to *Creative Computing*. All worlds seem desperately in need of leaders and we'll gladly publish your name as a likely candidate.

\* \* \*

Never underestimate the importance of just fooling around.

Kenneth Boulding

\* \* \*

"The only time my education was interrupted was when I went to school."

George Bernard Shaw

\* \* \*

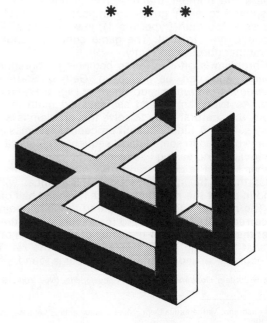

# What the Computer Taught Me About My Students...

### Anne Pasquino

The increased availability of digital computers in the classroom presents a challenge to mathematics teachers at all levels. Generally this challenge concerns how to make the most effective use of the computer in the existing syllabus. Answers to this challenge range from using the computer as a super desk-calculator to developing sophisticated materials such as those produced by the students and teachers involved in Project SOLO.

In the rush to do "impressive things" with the computer and thereby assure school administrators that their dollars have been well spent, we run the risk of overlooking an important and subtle byproduct of computers in the classroom. The student-written program provides us with the opportunity to scrutinize the thought processes of our students and gain some valuable insight into the way they attack problems. In attempting to instruct the computer to perform calculations necessary for the solution of a problem, the student's program mirrors his own problem solving technique or lack of it.

The following problem was assigned to a class of college freshman, non-mathematics majors and produced some interesting results.

> Write a computer program to play the game "I am thinking of a number." The user picks an integer between 1 and 10,000 inclusive. The computer tries to guess the number the user has in mind. The user responds to each guess by indicating whether the guess is too high (type in a 1), low (type in a -1) or correct (type in a 0). The game continues until the computer guesses correctly.

This problem is often given to beginning programming students and can be found in *Getting Started in Classroom Computing*, published by Digital Equipment Corp. As every smug and self-satisfied math teacher knows, the "natural" way to solve this problem is with a binary search. A binary-search procedure assumes a set of items ordered in some logical sequence; in this case, a numerically increasing sequence of numbers. The number sought is compared to the *midpoint* of the set; unless this is the number sought, this number will be found in either the right-hand or the left-hand half of the set. The number sought is compared to the midpoint of the correct half; if not equal to the midpoint, it is then in either the right or the left-hand half of that portion of the set, that is, in one of the two quarters of the set. This procedure is carried out until the number is found.[1]

[1] Philip B. Jordain, *Condensed Computer Encyclopedia*, (New York, 1969), p. 57.

Anne Pasquino, Mathematics Dept., State College at Westfield, Westfield, MA 01085.

True to expectations, several students did use this approach. They "taught" the computer to guess systematically, by first selecting an upper and lower bound. The upper (or lower) bound was revised each time the guess was too high (or too low). In this way the student enabled the computer to squeeze down on the correct value. Each guess was computed by averaging the upper and lower bounds. A typical student program is shown below.

```
10 PRINT "THE NAME OF THE GAME IS:"
15 PRINT "PICK A NUMBER FROM 1 TO 10000."
20 PRINT "THE RULES ARE AS FOLLOWS:"
30 PRINT "IF MY GUESS IS LOW, TYPE -1."
40 PRINT "IF MY GUESS IS HIGH. TYPE 1."
50 PRINT "IF MY GUESS IS CORRECT, TYPE 0."
60 PRINT
70 PRINT "PICK YOUR NUMBER"
80 LET U = 10000
90 LET L = 0
100 LET G = INT((U + L)/2)
110 PRINT "IS THE NUMBER"; G
120 INPUT A
130 IF A = 0 THEN 190
140 IF A = 1 THEN 170
150 LET L = G
160 GO TO 100
170 LET U = G
180 GO TO 100
190 PRINT "I GUESSED IT."
999 END
```

Not so true to expectations were a number of students who used a dichotomous search, but not a binary search. For example, one student used a first guess of 10,000 and then subtracted 1,000 from the first guess to get the next guess. If the subsequent guess was too high, 1,000 was again subtracted to obtain a new guess. This continued until a response of "too low" was obtained. At this point the last guess was increased in steps of 100 until the guess became "too high." The guess was then lowered in steps of 10 until it became "too low" and finally, increased in steps of one till it was correct. Hence, trapping the correct value was accomplished by adding or subtracting powers of ten to the upper and lower bounds rather than averaging them. The mental decision tree used by the student is pictured in Figure 1.

A similar but more elaborate decision pattern was used by another student who wrote a somewhat longer program; see Figure 2 for the pattern.

Still another student used lower and upper bounds that were adjusted by adding or subtracting an increment. The increment was calculated by a process which resembles the "limiting process" in calculus.

# ...or Is Binary Search "Natural"?

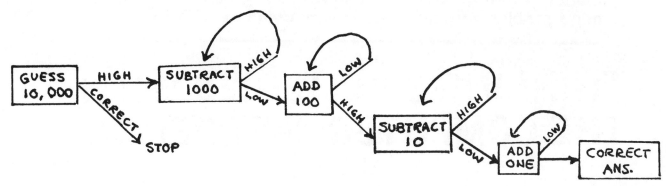

Fig. 1. One student's mental decision tree.

```
10 LET L = 0
15 LET U = 10000
20 LET G = 0
30 FOR N = 1 to 14
35 PRINT "IS THE NUMBER"; G
40 INPUT A
45 IF A = 0 THEN 98
50 LET H = INT(10000/2 ↑ N)
55 IF A = 1 THEN 75
60 LET L = G
65 LET G = L + H
70 GO TO 85
75 LET U = G
80 LET G = U - H
85 NEXT I
98 PRINT "I GUESSED IT."
99 END
```

The value of the increment H in line 50 grows successively smaller with each pass through the loop.

The techniques used by the students in solving this problem were intriguing for two reasons. First, the "natural" application of a binary-search procedure where one continually guesses the midpoint, turned out to be not so "natural." Second, although the students involved lacked formal training in calculus, they seemed to possess an intuitive understanding of the notions of upper and lower bound, convergence and limit, and were able to use complex decision trees. These observations suggest the desirability of inventing a series of "problems" such as the "number game" which might be used to introduce concepts in calculus such as limit, convergence, etc. The problems might also serve as a diagnostic tool to help assess where a student stands regarding such concepts. Further, the use of such problems may reveal that many problem-solving techniques which teachers think are "natural" to student thought processes are learned techniques which are alien to or only remotely related to the student's manner of thinking. At any rate, student-written programs are indicative of a great deal more than just the student's ability to program. ■

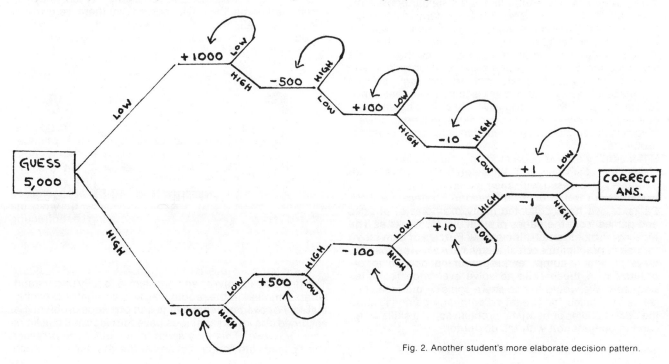

Fig. 2. Another student's more elaborate decision pattern.

This is the first in a series of articles about strategies or approaches for solving practical problems on the computer. Readers will find the heuristics and rules of thumb discussed in these articles are independent of subject matter and of great value in solving all types of programming problems from simple to very complex. —DHA.

# THINKING STRATEGIES WITH THE COMPUTER: INFERENCE

### Donald T. Piele and Larry E. Wood[*]

*Experience in solving problems and experience in watching other people solve problems must be the basis on which heuristic is built.*
                                           G. Polya

Some 32 years ago, in 1945, George Polya published a little book called *How To Solve It*. Judging from the title, one might expect to find inside special techniques and sure-fire algorithms that guarantee solutions to specific problems. But this is not what Polya's book is about. Instead, it is packed full of ideas and 'rules of thumb' that are useful in attacking any type of problem but do not guarantee a solution to any specific one. Polya's methods, which he labels heuristic, are derived from the experience of good problem solvers and are characterized by their generality, their independence of subject matter, and their common sense.

Inspired by the work of Polya and recent advances in the field of artificial intelligence (e.g. Newell and Simon, 1972), Wayne Wickelgren published a similar book, *How To Solve Problems* in 1974. This book contains detailed explanations of several general problem solving strategies along with puzzles and games to illustrate each strategy. Puzzles are well suited for the task because they require the same logical thinking processes as problems in any subject area but they do not require any special knowledge. Their only drawback is that people sometimes refuse to take them seriously. They fail to see any connection between the thinking skills needed to solve a frivolous puzzle and those needed to solve more practical problems.

Recently, we have been studying the problem solving strategies of Polya and Wickelgren and have been extremely impressed with their generality and power. In preparation for a course on Thinking Strategies at UW-Parkside, we have collected many examples of puzzles and games from the pages of such classic works as The Moscow Puzzles, and Mathematical Puzzles of Sam Lloyd for use in practicing each strategy. Now we are exploring ways in which computer programming can be incorporated with these skills to solve even more complex problems. We would like to share some of our ideas in a series of articles for *Creative Computing* demonstrating the added power of heuristic problem solving skills when used in conjunction with the computer.

[*]University of Wisconsin-Parkside, Kenosha, Wisconsin 53140

**Inference**

In this first article, we will discuss the strategy of *inference*. *Webster's New Collegiate Dictionary* defines inference as "a logical conclusion from given data or premises, a judgment derived by reasoning or implication." As a heuristic problem solving tool, inference is more broadly defined to include meanings such as explicitly stating information that is implicit in the problem, making deductions and inductions, and generating and testing hypotheses. Viewed in this expanded sense, inference becomes a basic component of most problems. Indeed, it is difficult to imagine how any problem could be solved without it. As an example, consider the following problem.

## PIGS AND CHICKENS

A boy and his sister visited a farm where they saw a pen filled with pigs and chickens. When they returned home, the boy observed that there were 18 animals in all, and his sister reported that she had counted a total of 50 legs. How many pigs were there in the pen?

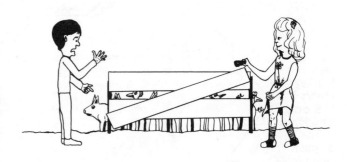

The first step in solving any problem is to fully understand what is implied in the problem as well as what is explicitly stated. For example, in the pigs and chickens problem it is assumed one knows that pigs have four legs and chickens have only two. This may seem trivial, but if the problem were posed with kangaroos and emus, the same inference might be obvious only to an Australian problem solver. The next step is to deduce relationships that may exist between elements of the problem. For example, the total of 50 legs is equal to the number of pigs times four plus the number of chickens times two. Also, the pigs plus the chickens equals 18 animals. At this stage, anyone with a knowledge of algebra would probably symbolize the above relationships in two algebraic equations (e.g., P+C=18 and 4P+2C=50) and solve for the number of pigs. This is certainly a familiar way to solve story problems, but for the purpose of emphasizing the usefulness of inference let's see how an Australian problem-solver might attack the same problem if it were posed with kangaroos and emus.

One usually visualizes a kangaroo resting on its hind legs with its smaller front legs in the air. An emu (which resembles a large turkey) has only two legs. Thus an Australian might easily infer that with 18 kangaroos and emus, a total of 36 legs are on the ground. Since there are 50 legs in all, there must be 14 legs in the air, which belong to exactly 7 kangaroos. Such a simple solution is very unlikely to occur to someone when the problem is posed with pigs and chickens, but what is to prevent pigs from standing on their hind legs—at least in our minds? It is probably true that the less likely we are to make a particular inference, the more likely we are to label it insight. However, in this problem, it may be more appropriate to call it hind-sight!

As illustrated in the pigs and chickens problem, drawing inferences depends heavily upon prior experience. Therefore, it may be difficult to make critical inferences with complex problems or with problems from an area unfamiliar to the problem solver. To overcome this difficulty, the computer can be a very effective tool. With it, one can rapidly generate important information related to a problem, which can serve as a basis for formulating and testing hypotheses about a solution. We shall illustrate this with the following example from the field of music theory.

**The Nun's Fiddle**

The Greek mathematician, Pythagoras, first discovered a basic relationship between musical harmony and number. This relationship is briefly explained by Helm (1967).

"Pluck a stretched string of any length and allow it to vibrate; it will sound a certain pitch. Allow only half of it to vibrate and the pitch will rise an octave. If two-thirds of the string vibrates, the pitch will rise a fifth above the one produced by the total length. For instance, if the total length produces C, two-thirds of the string will produce G. (The interval C-G is called a fifth because five lines and spaces on the musical staff are traversed in going from one to the other, counting C and G.) Three-fourths of the string will yield a pitch a fourth higher than the total length (F, if the total yields C) and so on. In time the fractions become more complex and the two notes represented by the resulting intervals become more dissonant if they are sounded together."

The discovery that pleasing cords correspond to exact divisions of a string by whole numbers had mystic overtones for the Pythagoreans. They inferred that if nature and number corresponded harmoniously in music it must be true that a single order, expressible in number and ratio, governed all the rhythms of nature. This led to the myth that the orbits of all heavenly bodies were related by musical intervals. "The movement of the heavens were, for them, the music of the spheres" (Bronowski, 1973). Gradually over the centuries certain ratios corresponding to musical intervals became the basis of traditional Western music. Silver (1971) explains:

The satisfying intervals were derived from natural harmonics, the frequencies of which are related to the natural number series 1:2:3.... Successive ratios 1:2, 2:3, 3:4... were favored. The lower ratios are pleasing; the higher ones tend to harshness and eventually become unacceptable. Certain ratios, although within the range of acceptable harshness, are regularly rejected, e.g., 6:7, 7:8, 10:11, 11:12.... There is no obvious reason for this empirical fact. However, an analysis of a large amount of material discloses that the ear prefers the following finite set called the *superparticular ratios:*

| | |
|---|---|
| 1:2 octave | 8:9 major tone |
| 2:3 perfect fifth | 9:10 lesser tone |
| 3:4 perfect fourth | 15:16 diatonic semitone |
| 4:5 major third | 24:25 chromatic semitone |
| 5:6 minor third | 80:81 comma of Didymus |

The relation of music to number expressed by the superparticular ratios is very beautiful, and a complete understanding of this relationship may even convince you that it is divinely inspired. The superparticular ratios are examined with the aid of the tromba marina, a late medieval bowed instrument with a single string. The instrument was frequently used by nuns and hence the German name *Nonnengeige* or nun's fiddle.

## NUN'S FIDDLE

The superparticular ratios in music are related to the prime numbers and can be defined by two simple properties. Find these properties and prove that they characterize the superparticular ratios uniquely.

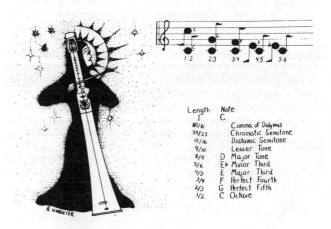

The first part of this problem can be answered by writing each whole number in a superparticular ratio in terms of its prime factors, i.e. $4=2^2$, $6=2\cdot 3$, $8=2^3$, $9=3^2$, $10=2\cdot 5$, $15=3\cdot 5$, $16=2^4$, $24=2^3\cdot 3$, $25=5^2$, $80=2^4\cdot 5$, $81=3^4$. One can infer that two properties characterize these ratios: (1) Each number is of the form $2^a 3^b 5^c$ where a,b,c 0, and (2) the numbers in each ratio differ by one. The difficult question is whether these two properties determine the superparticular ratios uniquely. Expressed another way, is it true that if a ratio of two whole numbers satisfies conditions (1) and (2) then it must be one of the 10 superparticular ratios? This statement is, in fact, true, and it was first proved by Stormer (1897). More recently it was re-examined by Halsey and Hewitt (1972). However, to understand the formal proof requires a considerable amount of mathematical expertize. In contrast, it is quite easy to write a computer program to generate successive numbers of the form $2^a 3^b 5^c$ from which a number of inferences can be made. Of course these inferences do not represent a strict proof but at least they increase one's understanding of the problem to the point where a proof may be easier to discover.

Program FIDDLE was written to do precisely that—fiddle around. It allows one to specify a set of primes $p_1$, $p_2$ ... $p_k$ from which consecutive whole numbers are generated which have these primes as their only factors. In the sample run the primes 2,3,5 are specified, and all numbers, up to 1000, which have these primes as their only factors and which differ by one are printed out. What we observe are precisely the numbers in the superparticular ratios.

### Conclusion

Our ability to make inferences in problem solving is strongly dependent upon our past experience as illustrated in the pigs and chickens problem. We can overcome this difficulty in many instances by using the computer to generate information to enrich our understanding of a variety of problems—even those which are not 'divinely inspired.'

### Post Script

What are the superparticular ratios for the primes 2,3,5 and 7? The answer may surprise you! Be patient; there are 23 ratios. Also, what can you infer about superparticular ratios relative to any set of primes that does not contain 2? Perhaps we have not gone far enough, and there exists two consecutive numbers beyond 1000 of the form $2^a 3^b 5^c$. Because a computer is limited to calculating a finite number of cases, it is impossible to absolutely rule this out. However, one can generate more evidence to weaken the case by observing the sequence of differences between successive numbers that are of the form $2^a 3^b 5^c$. This information is also shown in the sample run. Although the differences are not constantly increasing, at least they appear to be moving toward higher and higher values. This information prompted the authors to conjecture that for any specified distance d, successive terms of the form $2^a 3^b 5^c$ will eventually all differ by at least d. Shortly thereafter we found it had been proved mathematically by Stormer (1898) for any number of specified primes.

## FIDDLE PROGRAM

```
LIST
FIDDLE
10   PRINT "THIS PROGRAM CAN BE USED TO STUDY SUPERPARTICULAR"
20   PRINT "RATIOS RELATIVE TO ANY SPECIFIED SET OF PRIMES."
30   PRINT
40   PRINT "   HOW MANY PRIMES DO YOU WANT TO SPECIFY";
50   INPUT K
60   PRINT "WHICH ONES ARE THEY (SMALLEST ONE FIRST)";
70   MAT INPUT P[K]
80   PRINT "         HOW HIGH DO YOU WANT TO SEARCH ";
90   INPUT M
100  PRINT
110  PRINT "THE SUPERPARTICULAR RATIOS UP TO";M
120  PRINT "FOR THE SPECIFIED PRIMES ARE:"
130  PRINT
140  DIM D[1000]
150  J=0
160  Y=1
170  X=P[1]
180  N=X
190  FOR I=1 TO K
200  IF N/P[I]#INT(N/P[I]) THEN 230
210  N=N/P[I]
220  GOTO 200
230  NEXT I
240  IF N#1 THEN 300
250  J=J+1
260  D[J]=X-Y
270  IF X-Y#1 THEN 290
280  PRINT Y"/ "X
290  Y=X
300  X=X+1
310  IF X<M THEN 180
320  PRINT LIN(2)
330  PRINT "THE DIFFERENCES BETWEEN SUCCESSIVE INTEGERS  UP TO";M
340  PRINT "WITH THE GIVEN PRIMES AS THEIR ONLY FACTORS ARE:"
350  PRINT
360  FOR I=1 TO J
370  PRINT D[I];
380  NEXT I
390  PRINT LIN(2)
400  END
```

## SAMPLE RUN

```
THIS PROGRAM CAN BE USED TO STUDY SUPERPARTICULAR
RATIOS RELATIVE TO ANY SPECIFIED SET OF PRIMES.

   HOW MANY PRIMES DO YOU WANT TO SPECIFY?3
WHICH ONES ARE THEY (SMALLEST ONE FIRST)?2,3,5
         HOW HIGH DO YOU WANT TO SEARCH ?1000

THE SUPERPARTICULAR RATIOS UP TO 1000
FOR THE SPECIFIED PRIMES ARE:

1    /   2
2    /   3
3    /   4
4    /   5
5    /   6
8    /   9
9    /   10
15   /   16
24   /   25
80   /   81


THE DIFFERENCES BETWEEN SUCCESSIVE INTEGERS  UP TO 1000
WITH THE GIVEN PRIMES AS THEIR ONLY FACTORS ARE:

1    1    1    1    1    2    1    1    2    3    1    2
4    4    1    2    3    2    4    4    5    3    2    5
6    4    8    3    5    1    9    6    4    8    12   5
3    7    9    6    10   2    18   12   8    16   9    15
3    7    6    14   18   12   20   4    36   15   9    16
5    27   18   30   6    14   12   28   36   24   25   15
8    27   45   9    21   18   32   10   54   36   60   12

DONE
```

# FLOWCHART

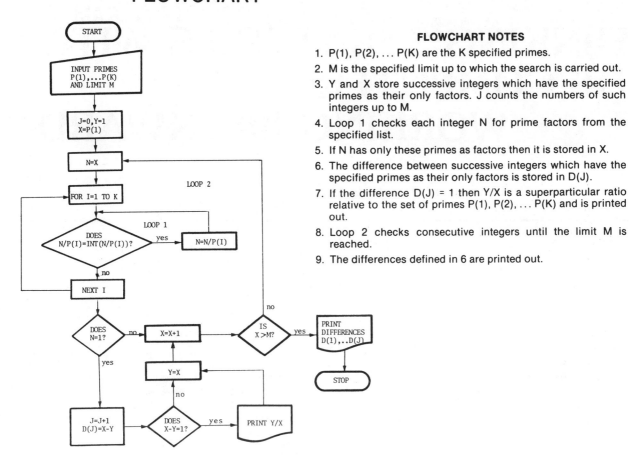

**FLOWCHART NOTES**

1. P(1), P(2), ... P(K) are the K specified primes.
2. M is the specified limit up to which the search is carried out.
3. Y and X store successive integers which have the specified primes as their only factors. J counts the numbers of such integers up to M.
4. Loop 1 checks each integer N for prime factors from the specified list.
5. If N has only these primes as factors then it is stored in X.
6. The difference between successive integers which have the specified primes as their only factors is stored in D(J).
7. If the difference D(J) = 1 then Y/X is a superparticular ratio relative to the set of primes P(1), P(2), ... P(K) and is printed out.
8. Loop 2 checks consecutive integers until the limit M is reached.
9. The differences defined in 6 are printed out.

## REFERENCES

Bronowski, J. *The Ascent of Man,* Little, Brown and Company, 1973.

Halsey, G.D., and Hewitt, Edwin. "More on the Superparticular Ratios in Music," *American Mathematical Monthly,* 79:1096-1100; December, 1972.

Helm, E.E. "The Vibrating String of the Pythagoreans," *Scientific American,* 217:92-103; December, 1967.

Polya, G. *How To Solve It.* Princeton University Press, 1945.

Silver, A.L. Leigh. "Musimatics or The Nun's Fiddle," *American Mathematical Monthly,* 78:351-57; April, 1971.

Stormer, Carl. "Quelques theoremes sur l'equation de Pell $x^2 - Dy^2 = 1$ et leurs applications." Skrifter Videnskabsselskabet (Christiania) I, Mat.-Naturv. K1., no. 2 (1898), 752-754.

Wickelgren, Wayne A. *How To Solve Problems.* W.H. Freeman and Company, San Francisco, 1974. (Available from *Creative Computing* Book service for $6.75 ppd.)

"About our prospects for that merger, it says: 'you have a snowball's chance in . . . .'"

# THINKING STRATEGIES WITH THE COMPUTER: WORKING BACKWARD

## D.T. Piele and L.E. Wood[*]

*"The so-called 'Treasury of Analysis' is, to put it shortly, a special body of doctrine for the use of those who, after having studied the ordinary Elements, are desirious of acquiring the ability to solve problems."*

Pappus, Book VII
Mathematical Collection

Pappus of Alexandria, who lived at the end of the third century A.D., wrote a comprehensive guidebook and commentary on the geometrical works of the great Greek mathematicians Pythagorus, Euclid, Archimedes, and Apollonius—to name a few. His *Mathematical Collection* consists of eight books describing the important developments of the classical Greek geometers and is punctuated with numerous original propositions, improvements, and historical original propositions, improvements, and historical comments of his own. Book VII is historically very important because it collects together the fundamental discoveries of Greek geometers into a "Treasury of Analysis" which, after Euclid's Elements, became essential reading for serious mathematicians of the day. The "Treasury" is also valuable as an early source for heuristic problem-solving strategies. The strategies of *analysis* and *synthesis* are particularly significant because together they constitute the earliest known description of the problem-solving strategy known today as working backward.

"...for in analysis we assume that which is sought as if it were already done, and we inquire what it is from which this results, and again what is the antecedent cause of the latter, and so on, until, by so retracing out steps, we come upon something already known or belonging to the class of first principles, and such a method we call analysis as being solution backwards.

"But in *synthesis*, reversing the process, we take as already done that which was last arrived at in the analysis and, by arranging in their natural order as consequences what before were antecedents, and successively connecting them one with another, we arrive finally at the construction of that which was sought; and this we call *synthesis*." (7)

## Working Backward

**In this second article on problem-solving, we will discuss strategy of** *working backward*. Any solution to a problem can be thought of as a path that leads from the given information to the goal. The point Pappus emphasized was that in cases where the goal is known or can be assumed known, it may be easier to start at the goal and work backward to the initial state (analysis). Once this is accomplished, the solution is simply the same series of steps in reverse (synthesis). As an example, consider the following problem.

### MATCHING COINS
Three men agree to match coins for money. They each flip a coin and the one who fails to match the other two is the loser. The loser must double the amount of money that each opponent has at that time. After three games, each player has lost once, and has $24. How much did each man begin with?

[*]University of Wisconsin-Parkside, Kenosha, Wisconsin 53140

The end result in this problem is known — all three players end up with $24. The initial state can be found by working backward one game at a time. For example, since each player had $24 after the 3rd game, the two winners of this game (who doubled their money) must have had $12 each at the end of the 2nd game. In order to pay each winner $12 and still end up with $24, the loser of this game must have had $48. Thus the distribution of money among the three players after the 2nd game has been determined. In a similar fashion one can continue working backward to reach the initial state.

If we let $P_1$, $P_2$ and $P_3$ represent the players who lost the first, second, and third games respectively, then Figure 1 shows the distribution of money between the three players at each stage constructed by working backward.

| States | Players | | |
|---|---|---|---|
| | $P_1$ | $P_2$ | $P_3$ |
| After 3rd game | $24 | $24 | $24 |
| After 2nd game | $12 | $12 | $48 |
| After 1st game | $ 6 | $42 | $24 |
| Initial State | $39 | $21 | $12 |

Figure 1. Solution to Matching Coins

Note that in this problem the path from the goal back to the initial state is uniquely determined; thus at each state in the solution, the previous state is forced upon us by the conditions of the problem. By working backward, we were able to arrive at the solution directly without any detours. This property is illustrated in Figure 2.

We turn now to a more complex problem where the strategy of working backward is not necessary but where it can be used very effectively in a computer program.

## FIVE SAILORS AND A MONKEY

Five sailors and a monkey were on an island. One evening the sailors rounded up all the coconuts they could find and put them in one large pile. Being exhausted from working so hard, they decided to wait and divide them up equally in the morning. During the night, a sailor awoke and

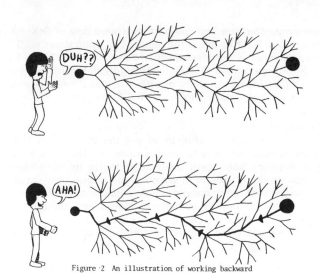

Figure 2. An illustration of working backward

separated the nuts into five equal piles, with one left over which he gave to the monkey. He took one pile, hid it, pushed the other four together and went back to his hammock. He was followed in turn by the other four sailors, each of whom did exactly the same thing. Next morning the remaining nuts were divided equally with one remaining nut going to the monkey. What is the least number of coconuts they could have begun with?

Philip W. Brashear (1) derived an elegant algebraic solution to this problem which solves it for any number of sailors. Unfortunately, to conceive such a solution requires a high level of mathematical maturity. But with a computer handy and an understanding of the strategy of working backward, a solution is relatively easy to find.

Consider the general problem where $S$ is the number of sailors on the island and $A$ is the number of coconuts that each sailor received in the final division of the pile. Since one coconut was given to the monkey at every division, the total number of coconuts left in the morning is $S \cdot A + 1$. But this pile came from pushing together $S - 1$ equal piles. Thus, the key condition that must hold is for $(S \cdot A + 1)(S - 1)$ to be an integer $K$, which represents the number of

coconuts that the last sailor stole from a pile of S•K + 1 coconuts. But this pile is the result of pushing together S - 1 equal piles by the previous theif so again (S•K + 1) (S - 1) is an integer as we move back through all S raids on the pile. This idea is explained further in the flowchart and notes which accompany the SAILOR program.

## Conclusion

From textbooks, it is easy to get the impression that there is only one way to solve a problem. The trouble is, our memory soon gets overloaded trying to remember which solution goes with which problem and vice-versa. On top of that, what should you do if classical algebraic or analytical techniques become awkward and difficult to solve? Quit? Never!!! Learn a few simple problem solving skills and start cracking some tough coconuts with the computer. ●

## Postscript

The algebraic solution to this problem is given by $S(S+1)-(S-1)$. Thus for S larger than 5, the program given here takes an appreciable amount of time to get an answer. Are there ways to make the program more efficient?

## References

1. Brashear, Philip W., "Five Sailors And A Monkey", *The Mathematics Teacher,* October 1967, pp 597-599.
2. Kordemsky, Boris A. *The Moscow Puzzles.* Charles Scribner's Sons. New York 1972.
3. Gardiner, M. *Mathematical Puzzles of Sam Loyd,* Vol. I. Dover Publications, Inc., New York, 1959.
4. Newell, A. and Simon, H.A. *Human Problem Solving.* Prentice-Hall, Inc., Englewood Cliffs, N.J. 1972.
5. Polya, G. *How To Solve It.* Princeton University Press, 1945.
6. Wickelgren, Wayne A. *How To Solve Problems.* W. H. Freeman and Company, San Francisco, 1974.
(7) Heath, Sir Thomas L., *The Thirteen Books of Euclid's Elements,* Dover publications, New York, 1956, p. 138.

Illustrations drawn by Robert Schroeter, a student at UW-Parkside.

## Flowchart Notes

1. S is the number of sailors on the island.
2. A = 1 is the initial value for the morning share.
3. K is an integer.
4. N is a counter for loop 2.
5. (S•K + 1) (S - 1) is the number of coconuts stolen by sailor number (S - N) the night before.
6. The value of A is increased by 1 in loop 1 until it reaches a number for the final share that could have come from a pile formed by pushing together S-1 equal shares.
7. Loop 2 checks to see when a number is reached for the final share that can survive being pushed back through S consecutive raids and regroupings and still give integers at each stage.
8. The print-out gives the number of sailors, the least number of coconuts they could have begun with, and the share each sailor received in the morning.

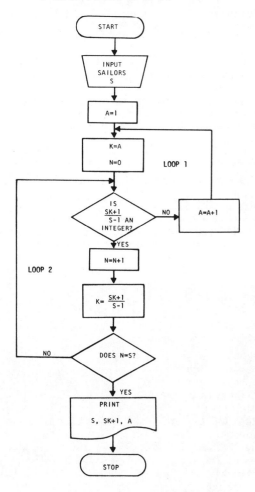

## SAILOR PROGRAM

```
LIST
SAILOR
100  PRINT "THIS PROGRAM SOLVES THE SAILORS AND"
110  PRINT "MONKEY PROBLEM BY WORKING BACKWARDS."
120  PRINT
130  PRINT "HOW MANY SAILORS ARE THERE ON THE ISLAND ";
140  INPUT S
150  PRINT
160  A=1
170  K=A
180  N=0
190  IF (S*K+1)/(S-1)=INT((S*K+1)/(S-1)) THEN 220
200  A=A+1
210  GOTO 170
220  N=N+1
230  K=(S*K+1)/(S-1)
240  IF N=S THEN 260
250  GOTO 190
260  PRINT "THE LEAST NUMBER OF COCONUTS THAT'S"
270  PRINT "SAILORS CAN BEGIN WITH IS"S*K+1
280  PRINT
290  PRINT "IN THE MORNING, EACH SAILOR GETS"A
300  END
```

## SAMPLE RUN

```
THIS PROGRAM SOLVES THE SAILORS AND
MONKEY PROBLEM BY WORKING BACKWARDS.

HOW MANY SAILORS ARE THERE ON THE ISLAND ?5

THE LEAST NUMBER OF COCONUTS THAT 5
SAILORS CAN BEGIN WITH IS 15621

IN THE MORNING, EACH SAILOR GETS 1023

DONE
```

Third in a series, this article shows how to break a problem into simpler subproblems.

# THINKING STRATEGIES WITH THE COMPUTER: SUBGOALS

## Donald T. Piele and Larry E. Wood*

*Nothing is more important than to see the sources of invention which are, in my opinion, more interesting than the inventions themselves.*

Leibnitz

One of the earliest and most famous problems in the field of topology (a branch of geometry) is the four-color problem. Conjectured by the English mathematician Francis Guthrie in 1850, it states that any map on a plane or a sphere can be colored with at most four colors so that any two countries that share a common boundary are colored differently. All attempts to prove this conjecture had been unsuccessful until last year when it was announced by kenneth Apple and Wolfgang Haken of the University of Illinois that it was indeed true. While listening to Professor Haken outline the proof at a recent colloquium, we were struck by his frequent use of clearly defined problem-solving strategies. Of paramount importance was the strategy of subgoals. After the problem had been represented in the rich language of graph theory, it was broken down into 1,930 subproblems each of which could be routinely solved on a computer. After 1,200 hours of computer time, the announcement was made, as anyone knows who has recently received a letter postmarked from the Mathematics Department at the University of Illinois.

### Subgoals

Basically, the method of subgoals consists of breaking a problem into simpler subproblems, solving each part, and regrouping the parts to solve the original problem. We often attack problems this way without thinking of it as a particular strategy since it seems so obvious. However, when we identify this strategy in a variety of problems, we learn how to use it much more effectively. As an example, consider the following balance problem (from Moscow Puzzles). How many glasses will balance a bottle?

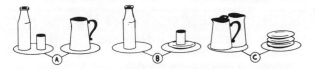

From the information given, it is apparent that the solution will require several step or subgoals. If we spend a few moments actively searching for appropriate subgoals, the solution can be obtained easily. From B, it is obvious that a bottle weighs as much as a glass plus a plate, so to solve the problem it is sufficient to replace the plate by its equivalent weight in glasses. Thus, obtaining a balance between glasses and one plate is a useful subgoal. This relationship is not given explicitly in A, B or C so it is necessary to establish a second subgoal. One possibility is to replace the two pitchers in balance C with glasses and plates. When this second subgoal is achieved, it is possible to reduce the number of plates on both sides until the first subgoal is achieved. The complete solution is:

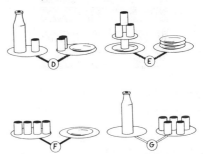

As a rule, subgoals are easier to attain than the entire goal, and this problem is no exception. Adding a glass to both sides of B yields the balance D. Combining this relationship with A shows that one plate and two glasses balances a pitcher. Hence, the two pitchers in C may be replaced with equivalent plates and glasses as shown in E. This solves the second subproblem which leads to the solution of the first subproblem F and the main problem G.

We next turn to an application of the subgoal strategy where recursive relationships can be used. Polya (1957) advises, "If you can't solve the problem posed, try to solve a simpler related problem." Many times the solutions to simpler problems may be combined and expanded in a recursive way to solve the original problem. As an example, consider the following AMAZE problem. A mouse enters a maze in search of a piece of cheese. There are infinitely many paths the mouse could follow but only a finite number will lead the mouse closer to the goal with every step. How many such paths are there?

*University of Wisconsin-Parkside, Kenosha, Wisconsin 53140

A poor way to attach this problem is to try to trace all the distinct routes and add them up. A better way is to build from simpler subproblems by placing the cheese at any one of the 14 other intersections. These are certainly related problems since any path that leads to the upper-right-hand corner must pass through a sequence of intersections. Also, the solutions for the simpler problems can be obtained through recursive relationships. There are two AMAZing things about solving the problem this way. The first is that it is really unnecessary to trace all of the paths to count them, and the second is that anyone could solve the problem this way in five minutes or less. For example, the number of paths that lead to intersection A, shown in the next figure, is the sum of the number of paths that lead to B, C and D because the only routes to A are through those intersections. The number of paths to each intersection is found recursively by starting in the lower left hand corner of the figure and moving to the upper right hand corner. As the problem is stated, there are 53 different paths the rat could take to the cheese.

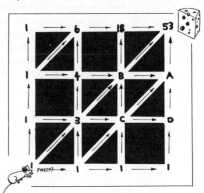

A classic example of the use of recursively defined subgoals appears in the solution to the Tower of Brahma (Hanoi) puzzle (see *Creative Computing*, January-February, 1976 and Wickelgren, 1974).

### Change for a Dollar

There are many ways, similar to those above, to apply the subgoal strategy to sole problems with the aid of a computer. A good example appears in the solution to the DOLLAR problem posed for *Creative Computing* by Brian Hess (1976): How many ways can you change a dollar bill?

Begin the solution by searching for ways to divide the problem into a set of smaller subproblems, each of which is easier to solve. One way to do this is illustrated by the tree diagram:

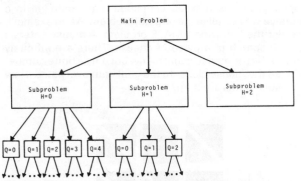

Since these two cases are mutually exclusive, we can infer that

$E_n = D_n + E_{n-50}$,
Similarly,
$D_n = C_n + D_{n-25}$
$C_n = B_n + C_{n-10}$
$B_n = A_n + B_{n-5}$

Now we begin with the simplest cases and build up to $E_{100}$. First of all, it is easy to understand why $E_0 = 1$. From above, when n = 50, $E_{50} = D_{50} + E_0$, and it is possible to make change for 50 cents only *one* more way if half dollars are allowed. Therefore $E_0 = 1$. Likewise, we can argue that $D_0 = $

The main problem is broken up into three subproblems: the number of ways of making change for one dollar using,

1. No half dollars (H = 0),
2. One half dollar (H = 1), or
3. Two half dollars (H = 2).

The last problem is trivial (only one way), while the other two need to be broken down further. This is done by dividing the remaining money into quarters and considering subproblems that specify the number of quarters (Q) used. Continuing on to lower denominations, subgoals are established that specify dimes (D), and nickels (N). As the number of subproblems is expanded, each one becomes easier to solve. In fact, with this problem, subgoals are reached which can be solved in only one way. For example, if H = 1, Q = 1, D = 2, and N = 0, then the pennies (P) must equal five in order to total up to one dollar.

While it is possible to continue the tree diagram in the figure by hand, it is very laborious to do so. However, it is a simple matter to program the computer to keep track of each subproblem with nothing more than nested loops. To demonstrate this, program DOLLAR was written so that each loop breaks the main problem down according to the scheme shown in the figure. Notice that at the quarter stage Q and thereafter, adjustments are made on the limits of the loops depending on how much money is left to change. For example, if H = 1 the only possible subgoals for quarters are Q = 0, 1, or 2 but not 3 or 4. Also, notice that there is no need to test combinations of coins to see whether they add up to $1.00. Simply counting the number of subgoals is sufficient since each one can be solved in only one way, (i.e., once H, Q, D, and N are specified then P must equal 100-50H-25Q-10D-5N).

### Recursion Relationships

Another way to attack this problem was suggested by Polya (1957) and uses recursion relationships in a similar way to that shown in the solution to AMAZE. Begin by defining quantities which represent the number of ways to make change for n cents using specified coins.

$A_n$ only pennies
$B_n$ nickels and pennies
$B_n$ nickels and pennies
$C_n$ dimes, nickels, and pennies
$D_n$ quarters, dimes, nickels, and pennies
$E_n$ half dollars, quarters, dimes, nickels, and pennies
The problem is to find $E_n$ for n = 100.

We can distinguish two cases in making change for n cents:

1. No half dollars are used, in which case $D_n$ is the number of ways to change n cents, or
2. One or more half dollars are used. After one half dollar is paid, there remains n - 50 cents to pay which can be done in $E_{n-50}$ ways.

$C_0 = B_0 = A_0 = 1$. It is also true that $A_n = 1$ for all values of n since there is only one way to make change using only pennies. We are now ready to apply the recursive relationships to solve the original problem. This is the strategy followed in program CHANGE which also has the added advantage that it can count the number of wasy of making change (with coins) for any specified number of cents, n.

## Conclusion

Forming subgoals is certainly one of the more common problem-solving strategies. Mathematical induction, recursion, and tree diagrams all contribute to its versatility. When used in conjuction with the computer, this strategy promises applications for solving old problems in new syas and for solving new problems in ways yet to be discovered.

## Postscript

The Dollar Problem has been around for some time and can be solved using analytical techniques. Kac and Ulam (1968) discuss a solution to this problem using power series. Specifically, if

$P(x) \times 1 + x + x^2 + x^3 + x^4 + \ldots$
$N(x) = 1 + x^5 + x^{10} + x^{15} + x^{20} \ldots$
$D(x) = 1 + x^{10} + x^{20} + x^{30} + x^{40} \ldots$
$Q(x) = 1 + x^{25} + x^{50} + x^{75} + \ldots$
$H(x) = 1 + x^{50} x^{100} + x^{150} + \ldots$

then the product series $\pi(x) = P(x)N(x)D(x)Q(x)H(x)$ is the key to finding the number of ways of making change for n cents. For example, the coefficient of the term $x^{100}$ in the product series $\pi(x)$ is the number of ways to make change for a dollar bill. Note, $1 \cdot x^5 \cdot x^{20} \cdot x^{25} \cdot x^{50} = x^{100}$ and this product corresponds to making change for one dollar using a half dollar ($x^{50}$), a quarter ($x^{25}$), two dimes ($x^{20}$), a nickel ($x^5$) and no pennies (1) and at the same time contributes 1 to the coefficient of $x^{100}$ in the product series $\pi(x)$. However, finding this coefficient by power-series analysis is very tedious and requires a high degree of mathematical sophistication. On the other hand, by making this connection between the coefficients of $\pi(x)$ and changing money, we can turn the table around and use program CHANGE to compute the coefficients for the product series $\pi(x)$ very quickly.

Suppose we expand AMAZE so that a computer program would be necessary for finding a solution quickly. Can you write a program that will handle any specified arrangement of blocks, some which have alleys? If a certain proportion of the blocks has alleys, how should be blocks be arranged so that the number of paths to the goal is maximized?

Russian coins made of copper and nickel come in denominations of 10, 15, 20, 50, and 100 kopecks (100 kopecks = one ruble). Copper-zinc coins come in denominations of 1, 2, 3, and 5 kopecks. How many ways are there to make change for one ruble?

## References

Hess, B. "How Many Ways Can You Change A Dollar?" *Creative Computing,* September-October 1976, p. 70.

Kac, M., and Ulam, S. *Mathematics and Logic,* Frederick Praeger, Publishers, New York, 1968, pp. 24-26.

Polya, G. *How to Solve It,* Princeton University Press, Princeton, New Jersey, 1957: 252-253.

"Tower of Brahma," *Creative Computing,* January-February, 1977, p. 25.

Wickelgreen, AW.A. *How to Solve Problems,* W.H. Freeman and Company, San Francisco, 1974, p. 103.

*Illustrations by Rodney Schroeter.*

## CHANGE FLOWCHART

Flowchart steps:
- START
- Input Money $ in Dollars. Cents
- Transform $ into Nickels M
- Initialize $A_0 = B_0 = C_0 = D_0 = E_0 = 1$ and $A_n = 1$ for all n.
- Increment on the Nickels J=1 to M. Apply recursion relationships
  $B_J = A_J + B_{J-1}$
  $C_J = B_J + C_{J-2}$
  $D_J = C_J + D_{J-5}$
  $E_J = D_J + E_{J-10}$
  where,
  $C_{J-2} = 0$ if $J < 2$ (less than a dime)
  $D_{J-5} = 0$ if $J < 5$ (less than a quarter)
  $E_{J-10} = 0$ if $J < 10$ (less than a half dollar)
- PRINT $E_M$
- STOP

## CHANGE PROGRAM

```
10  PRINT "PROGRAM CHANGE COMPUTES THE NUMBER OF WAYS OF MAKING"
20  PRINT "CHANGE IN COINS FOR ANY AMOUNT OF MONEY UP TO $5.00."
30  PRINT
40  DIM A[101],B[101],C[101],D[101],E[101]
50  PRINT "HOW MUCH DO YOU WANT TO CHANGE?"
60  PRINT "INPUT $ AS DOLLARS. CENTS "
70  INPUT C
80  M=INT(20*C)+1
90  A[1]=B[1]=C[1]=D[1]=E[1]=1
100 FOR J=2 TO M
110 A[J]=1
120 B[J]=A[J]+B[J-1]
130 C[J]=B[J]
140 IF J <= 2 THEN 160
150 C[J]=B[J]+C[J-2]
160 D[J]=C[J]
170 IF J <= 5 THEN 190
180 D[J]=C[J]+D[J-5]
190 E[J]=D[J]
200 IF J <= 10 THEN 220
210 E[J]=D[J]+E[J-10]
220 NEXT J
230 PRINT
240 PRINT "YOU CAN MAKE CHANGE FOR $ ";C
250 PRINT "IN ";E[M];"DIFFERENT WAYS."
260 END

PROGRAM CHANGE COMPUTES THE NUMBER OF WAYS OF MAKING
CHANGE IN COINS FOR ANY AMOUNT OF MONEY UP TO $5.00.

HOW MUCH DO YOU WANT TO CHANGE?
INPUT $ AS DOLLARS. CENTS ?5.00

YOU CAN MAKE CHANGE FOR $  5
IN      59576        DIFFERENT WAYS.

DONE
```

## DOLLAR PROGRAM

```
10  REM ***PROGRAM DOLLAR COMPUTES THE NUMBER OF WAYS OF
20  REM ***MAKING CHANGE FOR ONE DOLLAR.
30  C=0
40  FOR H=0 TO 2
50  FOR Q=0 TO 4-2*H
60  FOR D=0 TO 10-5*H-.5*Q
70  FOR N=0 TO 20-10*H-5*Q-2*D
80  C=C+1
90  NEXT N
100 NEXT D
110 NEXT Q
120 NEXT H
130 PRINT "THERE ARE";C;"DIFFERENT WAYS TO CHANGE A DOLLAR BILL."
140 END

THERE ARE 292  DIFFERENT WAYS TO CHANGE A DOLLAR BILL.

DONE
```

**Random, systematic and guided trial-and-error strategies are described in this fourth article in the series**

# Thinking Strategies with the Computer: Trial-and-Error

### Donald T. Piele and Larry E. Wood*

*What is the difference between a method and a device? A method is a device which you use twice."*
G. Polya

In the course of our formal education, we are taught a great variety of *devices* for solving problems that have already been neatly grouped together at the end of each chapter of a textbook. Typical examples of these devices are formulas, equations, rules, and theorems which are studied for the purpose of attacking certain types of problems. After a careful study of these specific techniques, exams are given to test our ability to recall them. But what do you do when you are faced with the more realistic situation of not being told what device is likely to solve a problem or, worse yet, of having forgotten how to use a technique altogether? Is all hope lost? Of course not, although many students, by the time they reach college, believe that it is.

At the beginning of each semester, we like to ask the students in our freshman and sophomore classes to try to solve a favorite problem of ours by any method they can. It is the Pigs and Chickens problem which appeared in our first article on Thinking Strategies (Piele & Wood, 1977).

#### PIGS AND CHICKENS

A boy and his sister visited a farm where they saw a pen filled with pigs and chickens. When they returned home, the boy observed that there were 18 animals in all, and his sister reported that she had counted a total of 50 legs. How many pigs were there in the pen?

Typically the response to this problem is as follows: approximately 40% of the students don't know how to begin because they haven't studied any specific methods for this problem, 25% recognize that the problem could be solved with two equations and two unknowns but have forgotten how to do it, 25% can set up the two equations and get a solution, and 10% quickly try a few numbers and get the answer by trial-and-error in two or three tries. On the other hand, when we give the same problem to elementary-school children who know nothing about two equations and two unknowns and again ask them to solve it using any method they choose, they turn to trial-and-error very naturally and a higher percentage answer it.

What does this all mean? To us it indicates that in the teaching of mathematical and scientific problem-solving in school we overemphasize the memorization of specific devices and techniques for attacking problems, and we underemphasize some very simple and powerful problem solving strategies useful for a variety of problems. In this article on general problem-solving strategies, we would like to turn the tables around and elaborate on the strategy of trial-and-error, which is always available but seldom used to its full potential. It is often frowned upon in school because it is thought to be a lazy approach which requires very little thinking. But with the computer available, trial-and-error takes on a whole new dimension which we will only begin to explore here.

## taxes problem

The strategy of trial-and-error can be used in a number of different ways, which we will illustrate with the following problem about income taxes.

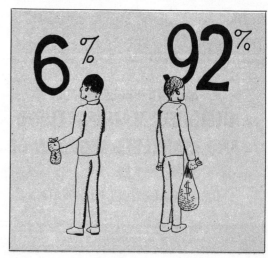

*University of Wisconsin-Parkside, Kenosha, Wisconsin 53141.

### TAXES IN TAXES???

Naturally, many people believe that rich people should pay more taxes than poor people, since the wealthier ones have more money. But sometimes this policy is carried to extremes. In one place I recently heard of, the tax rate was made the same as the number of thousands of dollars a person earns. For example, if a person earns $6,000, then his tax rate is 6% of that. But if a person earns $92,000, then his taxes are a whopping 92% of that.

What income between $1 and $100,000 would leave you the most money after taxes?

One way to solve this problem is *random* trial-and-error. As the term implies, random trial-and-error consists of arbitrarily choosing a series of values (gross salaries in this case) calculating results (net salaries), and then testing to see which one yields the highest value. This method takes little thought and only produces a solution if one happens to pick the correct value or values. In the case of the taxes problem the method would be extremely inefficient because each result must be compared to all previous ones to see if it is larger. It is this type of an approach to problem-solving that has given trial-and-error a bad name.

An improvement over random trial-and-error that requires additional thought and substantially improves its utility is *systematic* trial-and-error. Here a rule is devised to make certain all the reasonable alternatives will be systematically considered and evaluated until a problem is solved or shown to be unsolvable. With the tax problem this might consist of beginning with $1,000 and trying successive values in increments of $1,000 until a solution is reached. The result is that the net pay would continue to increase up to a gross income of $50,000 and then begin to decrease. This result implies, of course, that the optimum income is $50,000. While this method may be tedious and time-consuming, it will usually produce a solution, and therefore is a substantial improvement over random trial-and-error.

A further refinement of trial-and-error, and one that makes it much more respectable as a general problem-solving strategy, is *guided* trial-and-error. The key to its success lies in the fact that the results from each trial are used to guide the choice of a value for the next trial that will produce a result closer to the correct solution. This process is continued until the solution is finally attained, and is usually much more efficient than systematic trial-and-error. As an example of guided trial-and-error, let us return to the taxes problem. As a starting point, we might try both values suggested in the problem ($6,000 and $92,000). Because the results show that $92,000 yields a higher income than $6,000, it seems reasonable to choose a value higher than $92,000. As it turns out, however, values higher than $92,000 provide less net income than does $92,000. Therefore, it is logical to choose values less than $92,000 in large increments ($10,000) as long as they continue to result in larger net incomes than values chosen previously. Because the correct answer is $50,000, values closer to $50,000 will produce larger and larger net incomes, and values less than $50,000 will produce decreasing net incomes. Thus, the correct value can be determined quite efficiently. As mentioned earlier, the key to success is to carefully examine the result from each trial to guide the selection of values for the next trial in a way that will guarantee a movement closer to the correct solution.

## systematic trial-and-error

Now we turn to the computer and write a program to find the solution using systematic trial-and-error. If X represents the gross income then .01X is the tax rate and f(X) = (1−.01X)X is the net pay. Program 1STMAX uses a systematic procedure to find the value of X for which the net pay achieves a maximum value on the interval 0 to 100. The system is based on the following principle: Let X=0 be the starting point and increment to the right by I=10, comparing f(X) with f(X+I). If f(X) < f(X+I) then move up one step (X=X+I) and compare f(X) and f(X+I) again. Continue this procedure until f(X) > f(X+I). Now move back one step (X=X−I), reduce the step size by a factor of 2 and continue as before. As soon as the step size falls below the specified level of accuracy D (I < D), print out the value of X that corresponds to the first maximum value of f(X). For functions that have more than one relative maximum, the systematic procedure used in 1STMAX can be easily extended to find all relative maximum and minimum points for a given function on a specified interval. Can you do it?

## 1STMAX PROGRAM

```
LIST
1STMAX

10  PRINT "***THIS PROGRAM FINDS THE X VALUE WHERE THE FUNCTION X-.01*X*X"
20  PRINT "***IS A MAXIMUM ON THE INTERVAL 0 TO 100 BY SYSTEMATIC TRIAL"
30  PRINT "***AND ERROR."
40  PRINT
50  PRINT "INPUT THE DESIRED DEGREE OF ACCURACY.";
60  INPUT D
70  DEF FNF(X)=X-.01*X*X
80  PRINT
90  X=0
100 I=10
110 PRINT " I","X AT MAX"
130 Y1=FNF(X)
140 Y2=FNF(X+I)
150 IF Y1>Y2 THEN 180
160 X=X+I
170 GOTO 130
180 PRINT I,X
190 IF I<D THEN 230
200 X=X-I
210 I=I/2
220 GOTO 130
230 PRINT
240 PRINT "THE X VALUE WHERE X-.01*X*X IS MAXIMUM IS";X
250 END
```

## SAMPLE RUN

```
RUN
1STMAX

***THIS PROGRAM FINDS THE X VALUE WHERE THE FUNCTION X-.01*X*X
***IS A MAXIMUM ON THE INTERVAL 0 TO 100 BY SYSTEMATIC TRIAL
***AND ERROR.

INPUT THE DESIRED DEGREE OF ACCURACY.?.1

 I          X AT MAX
 10         50
 5          50
 2.5        50
 1.25       50
 .625       50
 .3125      50
 .15625     50
 .078125    50

THE X VALUE WHERE X-.01*X*X IS MAXIMUM IS 50
```

# x·log(x)=100

Although the taxes problem can be solved analytically using calculus or even more easily with the properties of quadratic functions, it is more likely that the majority of non-textbook problems one encounters will not have a nice closed-form solution. One such example is the following: *Find a value of X such that $X \cdot log(X) = 100$.* (We assume here that log(X) is the natural logarithm.)

Systematic trial-and-error, which is frequently used to search for solutions with a computer, could be applied to this problem in much the same way it was applied to the taxes problem. But for variety, we will use a different type of systematic trial-and-error.

Clearly, the solution to $X \cdot log(X) = 100$ lies somewhere between 1 and 100 since $1 \cdot log(1) = 0$, $100 \cdot log(100) > 400$, and $X \cdot log(X)$ increases with increasing values of X. If we let $X_l$ be the guess that is low ($X_l \cdot log(X_l) < 100$) and let $X_h$ be the guess that is high ($X_h \cdot log(X_h) > 100$) then we can use a systematic procedure which generates new trials by dividing the search area in half at each step as follows: Let the new trial $X_n$ be the average of the last two trials $X_n = (X_l + X_h)/2$ and then test $X_n$ to see whether it is high, low, or within the accuracy desired. If $X_n$ is high ($X_n \cdot log(X_n) > 100$) then replace the last high guess with $X_n$ ($X_h = X_n$) or if $X_n$ is low ($X_n \cdot log(X_n) < 100$) replace the last low guess with $X_n$ ($X_l = X_n$) and repeat the process of taking averages. Since the distance between $X_l$ and $X_h$ is cut in half with each new trial, $X_l$ and $X_h$ will both approach the desired solution within any pre-set degree of accuracy given a sufficient number of iterations. Program SLICE solves $X \cdot log(X) = 100$ for X using this method. The sample run following the program lists the upper and lower bounds at each halving of the search area to illustrate the approach. When this printout is suppressed the answer is computed immediately.

## SLICE PROGRAM

```
SLICE
10  PRINT "*****THIS PROGRAM SOLVES X*LOG(X)=100 FOR X"
20  PRINT "*****BY USING SYSTEMATIC TRIAL AND ERROR."
25  PRINT
30  PRINT "INPUT A LOWER AND UPPER GUESS AND THE DESIRED ACCURACY";
40  INPUT X1,X2,D
50  N=1
60  Y1=X1*LOG(X1)
70  Y2=X2*LOG(X2)
80  PRINT
90  PRINT " N","X-LOWER","X-UPPER"
100 PRINT N,X1,X2
110 X3=(X1+X2)/2
120 Y3=X3*LOG(X3)
130 IF ABS(Y3-100) <= D THEN 210
140 IF Y1<100 AND Y3>100 THEN 180
150 X1=X3
160 N=N+1
170 GOTO 100
180 X2=X3
190 N=N+1
200 GOTO 100
210 PRINT LIN(1)"THE ANSWER IS "X3
220 END
```

## SAMPLE RUN

```
*****THIS PROGRAM SOLVES X*LOG(X)=100 FOR X
*****BY USING SYSTEMATIC TRIAL AND ERROR.

INPUT A LOWER AND UPPER GUESS AND THE DESIRED ACCURACY?1,100,.001

 N         X-LOWER         X-UPPER
 1         1               100
 2         1               50.5
 3         25.75           50.5
 4         25.75           38.125
 5         25.75           31.9375
 6         28.8437         31.9375
 7         28.8437         30.3906
 8         28.8437         29.6172
 9         29.2305         29.6172
 10        29.4238         29.6172
 11        29.5205         29.6172
 12        29.5205         29.5688
 13        29.5205         29.5447
 14        29.5326         29.5447
 15        29.5326         29.5386
 16        29.5356         29.5386
 17        29.5356         29.5371
 18        29.5364         29.5371

THE ANSWER IS  29.5367
```

---

## guided trial-and-error

The systematic trial-and-error algorithm for solving $X \cdot log(X) = 100$ given above does not take into account all the information available after each trial. For example, this method takes the same amount of time to reach a solution whether the first guess is close to the solution already or not and is independent of the problem being solved. On the other hand, guided trial-and-error uses more of the information available from each trial (such as how close a particular trial is to a solution) to make a more educated next trial. This technique is used in program GUIDE to solve the problem $X \cdot log(X) = 100$ and is based on the following principle: Let $X_n$ be a given trial and $Y_n = X_n \cdot log(X_n)$ be the corresponding value of the function. If $Y_n$ is too large ($Y_n > 100$) then the exact trial is decreased by the factor $100/Y_n$ which is less than one. If $Y_n$ is too small ($Y_n < 100$) then the next guess is increased by the factor $100/Y_n$ which is greater than one. Thus the new trial is guided by the outcome of the previous trial as follows:

$$X_{n+1} = X_n * 100/Y_n.$$

Notice that this algorithm has the following important properties: Trials that are far from the correct value are changed by a bigger factor than those that are close. All trials oscillate above and below the desired solution. When $Y_n$ reaches 100, all subsequent trials remain the same.

## GUIDE PROGRAM

```
GUIDE
10  PRINT "*****THIS PROGRAM SOLVES X*LOG(X)=100 FOR X"
20  PRINT "*****BY USING GUIDED TRIAL AND ERROR."
30  PRINT
40  PRINT "INPUT AN INITIAL GUESS AND THE DESIRED ACCURACY";
50  INPUT X,D
60  PRINT
70  PRINT " n"," Xn"
80  N=1
90  Y=X*LOG(X)
100 PRINT N,X
110 IF ABS(Y-100)<D THEN 150
120 X=X*100/Y
130 N=N+1
140 GOTO 90
150 PRINT LIN(1)"THE ANSWER IS"X
160 END
```

## SAMPLE RUN

```
RUN
GUIDE

*****THIS PROGRAM SOLVES X*LOG(X)=100 FOR X
*****BY USING GUIDED TRIAL AND ERROR.

INPUT AN INITIAL GUESS AND THE DESIRED ACCURACY ?100,.001

 n         Xn
 1         100
 2         21.7147
 3         32.4887
 4         28.7283
 5         29.7807
 6         29.465
 7         29.5578
 8         29.5303
 9         29.5384
 10        29.536
 11        29.5368

THE ANSWER IS 29.5368

DONE
```

# newton's method

For completeness, we conclude our search for solutions to the X•log(X) = 100 problem with Newton's method which is a form of guided trial-and-error. This method is fully explained in almost every calculus book so we will not repeat it here. In comparison with the other algorithms given above, Newton's method is a very efficient approach as shown in the sample run. However, there is a price to pay for this efficiency — a knowledge of calculus and the ability to remember the method — which may not be worth the time and effort in many cases. For example, even though Newton's method converges to the solution over four times faster than the systematic trial and error method of program SLICE, the difference in the response time at the terminal is unnoticeable. Besides, there are some problems where Newton's method will fail if one makes an unlucky first guess, whereas the systematic trial-and-error technique never fails.

## NEWTON PROGRAM

```
NEWTON
10   PRINT "*****THIS PROGRAM SOLVES X*LOG(X)=100 FOR X"
20   PRINT "*****BY USING NEWTONS METHOD FOR GUIDED TRIAL AND ERROR."
30   PRINT
40   PRINT "INPUT AN INITIAL GUESS AND THE DESIRED ACCURACY ";
50   INPUT X,D
60   N=1
70   PRINT
80   PRINT " n "," Xn"
90   PRINT N,X
100  X1=X-(X*LOG(X)-100)/(LOG(X)+1)
110  IF ABS(X-X1)<D THEN 150
120  X=X1
130  N=N+1
140  GOTO 90
150  PRINT LIN(1)"THE SOLUTION IS "X1
160  END
```

## SAMPLE RUN

```
RUN
NEWTON

*****THIS PROGRAM SOLVES X*LOG(X)=100 FOR X
*****BY USING NEWTONS METHOD FOR GUIDED TRIAL AND ERROR.

INPUT AN INITIAL GUESS AND THE DESIRED ACCURACY ?100,.001

  n              Xn
  1              100
  2              35.6813
  3              29.6595
  4              29.5367

THE SOLUTION IS  29.5366

DONE
```

# turkey puzzle

A completely different application of guided trial-and-error is illustrated in the solution to the *Turkey Puzzle*.

**TURKEY PUZZLE**

Two people, Jane and Mary, went to the butcher shop to buy turkeys for Thanksgiving dinner. Since Mary had a larger family than Jane, she wanted a larger turkey. The butcher just happend to have two turkeys left, a small one and a large one. "Together these two turkeys weigh twenty pounds," he said. "The little one sells for two cents a pound more than the large one." Jane purchased the little one for 82¢ and Mary paid $2.96 for the big turkey. How much did each gobbler weigh?

Story problems like this one are the bane of most beginning algebra students. They ask "Where do I begin? What method do I use? Did I set it up right?" Frequently, students are primarily concerned about setting up the machinery for a problem so the answer will drop out like an egg into a basket and forget the most fundamental property of any solution: *A solution is an answer that works!* Of course, the Turkey problem can be solved using algebra and the reader may want to try solving it this way. But we will eschew any algebraic devices for attacking this problem to illustrate the power of guided trial and error.

The easiest way to begin the Turkey problem is to try a few numbers. Let's assume, as a first guess, that the small turkey weighs 8 lbs. What implications does this have, given the condition stated in the problem? First, the big turkey must weigh 20 - 8 or 12 lbs. Next, since the big turkey cost $2.96, the price per lb is $2.96/12, about $.25. The small bird cost 2 cents more per lb so its price is $.27. But $.82 was spent for the small bird which means it must have weighed .82/.27, about 3 lbs.

We have come full circle and our results are conflicting. We started out with a small bird weighing 8 lbs and ended up with the same bird weighing 3 lbs. If the first guess and the outcome had agreed for the weight of the small bird, the system would have been consistent and we would have had a solution simply because it worked.

The process of guided trial-and-error is based on the idea of adjusting the next trial depending upon the results of the previous trial. It is a feedback control system similar to the control of a guided missile to its target. From the first trial of 8 lbs the feedback told us we were too high so our next guess should be smaller. In 3 or 4 trials anyone should be able to narrow in on the target of 4 lbs for the small turkey and 16 for the large one.

Would it be possible to set up an automatic trial-and-error or feedback system that is simple enough to be programmed by anyone familiar with the BASIC language and solve the turkey puzzle with only an initial guess? The answer is yes, and program TURKEY is one example of how to do it. The feedback systems of this program follows very closely the first discussion of the Turkey problem given above. If we let L be the weight of the small turkey, B the weight of the large turkey, P the price per lb for the large turkey, then the conditions in the problem can be summarized as follows:

1. $B = 20 - L$
2. $P = 296/B$
3. $L = 82/(P+2)$

Begin with an initial guess of L1 for the weight of the small turkey L and follow through the consequences of this guess in 1, 2, and 3 above. If L1 and L agree, the system is consistent and we have a solution. If not, then use the outcome L in 3 as the next trial in 1 and loop back through the system. This algorithm was used in program TURKEY, and within 7 iterations of the algorithm the system has narrowed in on a solution as shown in the sample run. The program terminates when the difference between successive trials falls below .001.

## TURKEY PROGRAM

```
TURKEY
10  PRINT "SOLUTION TO THE TURKEY PUZZLE BY GUIDED TRIAL AND ERROR"
20  PRINT "MAKE ANY GUESS FOR THE WEIGHT OF THE SMALL TURKEY";
30  INPUT L1
40  PRINT
50  PRINT " SMALL"," BIG"
60  L=L1
70  B=20-L
80  P=296/B
90  L=82/(P+2)
100 IF ABS(L-L1)<.001 THEN 140
110 PRINT L1,B
120 L1=L
130 GOTO 70
140 PRINT
150 PRINT USING 160;L+.001,P+2.001
160 IMAGE "THE SMALL TURKEY WEIGHS",2D.2D," LBS AND THE PRICE PER LB IS ",2D.2D
170 PRINT USING 180;B+.001,P+.001
180 IMAGE "THE BIG TURKEY WEIGHS ",2D.2D," LBS AND THE PRICE PER LB IS ",2D.2D
190 END
```

## SAMPLE RUN

```
RUN
TURKEY
SOLUTION TO THE TURKEY PUZZLE BY GUIDED TRIAL AND ERROR
MAKE ANY GUESS FOR THE WEIGHT OF THE SMALL TURKEY?8

 SMALL        BIG
 8            12
 3.075        16.925
 4.20752      15.7925
 3.95312      16.0469
 4.01057      15.9894
 3.99761      16.0024

THE SMALL TURKEY WEIGHS  4.00 LBS AND THE PRICE PER LB IS 20.50
THE BIG TURKEY WEIGHS   16.00 LBS AND THE PRICE PER LB IS 18.50

DONE
```

## conclusion

The development of digital computers has revolutionized the methods available for solving problems. In many areas of modern science, computer-oriented numerical methods have been developed to solve problems that have been impossible to solve analytically. The diversity of fields being affected includes planetary astrodynamics, wave diffraction, weather prediction, thermodynamics, electrostatics and gravitational potential, molecular interaction, quantum theory, and relativistic collapse (Greenspan, 1974).

The guided trial-and-error methods discussed above are but a small sample of the growing number of computer techniques being used to find solutions or approximate solutions to given mathematical problems. The immense power of the digital computer to perform arithmetical operations with exceptional speed has added a quantum jump in the number of techniques now available to solve problems.

## postscript

The iteration technique used in the TURKEY program will converge to the solution, given any initial guess. This of course is the desirable outcome, but it is only one of three typical outcomes. An iteration can diverge away from any solution, getting increasingly larger with each iteration or getting locked into an endless loop that oscillates between two numbers that are not solutions. It is remarkable that the outcome that does occur depends only upon the way the conditions are expressed. For example, if you try writing a program to solve the Turkey puzzle using the following equivalent set of conditions:

1. $L = 20 - B$
2. $P = 82/L - 2$
3. $B = 296/P$

the sequence of iterations diverge and no solution is reached. Try it! The above equations were obtained by inverting the original three equations. Therefore, if you use a set of equations that give a diverging outcome, simply invert them and try again.

### REFERENCES

Piele, D. T., and Wood, L. E. "Thinking Strategies with the Computer: Inference", *Creative Computing Magazine*, Vol. 3, No. 2, Mar-Apr., 1977, p. 96-99.

Greenspan, Donald. *Discrete Numerical Methods in Physics.* Academic Press, Inc., New York, N.Y. 10003, 1974.

**This fifth article in the series shows how indirect reasoning can lead to a contradiction and thus to the truth.**

# Thinking Strategies with the Computer: Contradiction

**Donald T. Piele and Larry E. Wood**

*Let craft, ambition, spite
Be quenched in Reason's night,
Till weakness turn to might
Till what is dark be light
Till what is wrong be right!*
    Lewis Carroll

One of the first and possibly most elegant applications of the method of contradiction was given by Euclid in the 3rd century B.C. He proved that among the natural numbers 1, 2, 3, 4, ... there exists an infinite number of primes. (A prime number is one which has no factors other than itself and 1; for example, 2, 3, 5, 7 are prime, while 9 is not.) Euclid's idea is delightfully simple and illustrates the problem-solving strategy of contradiction.

To begin with, there are two possible outcomes:
1. A finite number of natural numbers are prime or
2. An infinite number of natural numbers are prime.

No one has been able to show directly that the second alternative is true. However, by reasoning indirectly, Euclid showed that if you assume the first alternative is true, you can make a sequence of logical inferences that lead to a contradiction. Thus, the first alternative is untenable and the second must be true.

This indirect method called *reductio ad absurdum* is used throughout mathematics and represents one form of the strategy of contradiction. Interested readers may want to study Euclid's clever sequence of logical inferences where he shows that the assumption of only finitely many primes leads to a contradiction (Eves, 1976).

The following problem will be used to illustrate the method of proof by contradiction. The reader is encouraged to try the problem before reading the solution.

## TRUTH AND FALSEHOOD

In a faraway land there dwelt two races. The Ananias were inveterate liars, while the Diogenes were unfailingly veracious. Once upon a time, a stranger visited the land, and on meeting a party of three inhabitants inquired to what race they belonged. The first murmured something that the stranger did not catch. The second remarked, "He said he was an Anania." The third said to the second, "You're a liar!" Now the question is, of what race was this third man?

As in Euclid's problem, there are only two possibilities. The third person is either an Anania (liar) or a Diogene (truar). Let us assume one of the alternatives, he/she is an Anania, and see if we can reach a contradiction. Since the third person said "You're a liar" to the second person and we are assuming the third person is an Anania (liar), it must follow that the second person is really telling the truth (a Diogene). But if the second person is a truar then the statement made about the first person, "He said he was an Anania," is true. But if we examine this statement closely we find that neither a liar or a truar could make it. A liar could not admit to being a liar and a truar would have to say he/she was a Diogene. Thus, we have arrived, by a sequence of logical inferences, to a contradiction based on the original assumption that the third person was a Diogene. Since there are only two alternatives, the third person must be a Diogene and the problem is solved. The reader may want to check that the problem is well-posed and that indeed, if the third person is a truar, the statements made by the first and second persons are non-contradictory.

Now we turn to an application of the method of contradiction where the choice of alternative is more than just two. Whenever the set of alternatives is small enough, it is still feasible to systematically examine each of them and derive a contradiction to all but one. As an example, consider the following problem:

---

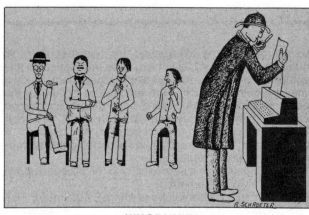

# whodunit?

### WHODUNIT?

Four men, one of whom is known to have committed a certain crime, said the following when questioned by an inspector from Scotland Yard.

    Growley:   "Snavely did it."
    Snavely:   "Gaston did it."
    Gus:   "I didn't do it."
    Gaston:   "Snavely lied when he said I did it."

If only one of the four statements is true, whodunit?

Brain teasers like this one often lead the novice, who is likely to attack the problem directly, into an endless loop. The experienced problem-solver recognizes that by first assuming a particular suspect committed the crime, it is an easy matter to check whether this assumption is consistent with the given information or leads to a contradiction. For example, if we assume that Growley did it, we can determine the truth or falsity of the statements given in the problem as follows:

    "Snavely did it" is false.
    "Gaston did it" is false.
    "I didn't do it" is true.
    "Snavely lied when he said I did it" is true.

Since two of the suspects are telling the truth and we are given that only one of the first four statements is true, the assumption that Growley did it has lead, through a sequence of logical inferences, to a contradiction. Thus, Growley is innocent. Each suspect, in turn, can be checked out in a similar way. Whenever a problem can be reduced to a bookkeeping chore like this one, it is natural to call in the computer. Even though the work involved in this problem is small and can be easily done by hand, the ideas learned in programming the computer to do the job will be useful when we are faced with a more difficult and time-consuming problem.

We begin the WHODUNIT problem by assigning variables to statements as follows:

    $P(1) \longleftrightarrow$ "Growley did it."
    $P(2) \longleftrightarrow$ "Snavely did it."
    $P(3) \longleftrightarrow$ "Gus did it."
    $P(4) \longleftrightarrow$ "Gaston did it."

A statement is designated as true by setting the corresponding variable equal to one, $P(I) = 1$. If we assume it is false, then $P(I) = 0$. For example, if Growley is the culprit, then the statement "Growley did it" is true and $P(1) = 1$, otherwise $P(1) = 0$. Using this new representation, it is a simple matter to express all the statements (facts of the case) in one equation:

$$P(2) + P(4) + \text{NOT } P(3) + \text{NOT } P(4) = 1.$$

This expression embodies the condition that only one of the statements; "Snavely did it" $P(2)$, "Gaston did it" $P(4)$, "Gus didn't do it" NOT $P(3)$, "Gaston didn't do it" NOT $P(4)$ is true and hence the expression adds to 1. (Recall that in the BASIC language, if $A = 1$ then NOT $A = 0$ and vice versa if $A = 0$ then NOT $A = 1$.) It is a simple matter for the computer to systematically assume each suspect, in turn, is guilty (for I = 1 to 4, P(I) = 1, and P(J) = 0, J ≠ 1) and check this assumption for any contradiction with the facts of the case, (P(2) + P(4) + NOT P(3) + NOT (P(4) = 1).

This is the technique employed in program CRIME in lines 80 to 170 to crack the case. The program also checks to see if the problem as posed has a unique solution, no solution, or perhaps many solutions. By changing line 120 (the facts of the case) a new problem can be examined with the same program. The reader is invited to make up new circumstances and test them for solutions. For example, suppose that the statements given by the suspects are: "Growley did it" P(1), "Snavely did it" P(2), "Gaston didn't do it" NOT P(4), "Snavely didn't do it" NOT P(2) and we know that only one of the suspects is lying. Then the facts of the case would be expressed as

P(1) + P(2) + NOT P(4) + NOT P(2) = 3.

Under this assumption, whodunit?

## CRIME PROGRAM

```
10  PRINT "THIS PROGRAM SOLVES THE WHODUNIT PROBLEM AS ORGINALLY WRITTEN."
15  PRINT
16  PRINT "FOUR SUSPECTS - GROWLEY, SNAVELY, GUS AND GASTON - ARE QUESTIONED"
17  PRINT "ABOUT A CERTAIN CRIME. THE FOLLOWING STATEMENTS ARE GIVEN AND ONLY"
18  PRINT "ONE IS TRUE."
19  PRINT
20  PRINT "    SNAVELY DID IT."
22  PRINT "    GASTON DID IT."
24  PRINT "    GUS DIDN'T DO IT."
26  PRINT "    GASTON DIDN'T DO IT."
28  PRINT
30  PRINT "THIS IS EXPRESSED IN LINE 120 AS P(2)+P(4)+NOTP(3)+NOTP(4)=1."
32  PRINT
34  PRINT "(YOU CAN MAKE UP YOUR OWN MYSTERY BY CHANGING THE STATEMENTS"
36  PRINT "AND THE CORRESPONDING EXPRESSION IN LINE 120.)"
40  PRINT LIN(2)
50  J=1
60  DIM P[4],A$[72]
70  A$="GROWLEY DID IT.SNAVELY DID IT.GUS DID IT.    GASTON DID IT. "
80  FOR I=1 TO 4
90  MAT P=ZER
100 P[I]=1
110 REM ***********************************************
120 IF P[2]+P[4]+ NOT P[3]+ NOT P[4]=1 THEN 150
130 REM ***********************************************
140 GOTO 170
150 J=J+1
160 K=I
170 NEXT I
180 GOTO J OF 190,210,230
190 PRINT "SORRY THERE IS NO SOLUTION TO THIS CASE. MORE INFO IS NEEDED."
200 GOTO 250
210 PRINT "THE INESCAPABLE CONCLUSION IS THAT "A$[15*K-14,15*K]
220 GOTO 250
230 PRINT "THERE IS NO UNIQUE SOLUTION TO THIS CASE. MORE THAN ONE SUSPECT"
240 PRINT "IS IMPLICATED BY THE INFORMATION GIVEN."
250 END
```

## SAMPLE RUN

```
RUN
CRIME

THIS PROGRAM SOLVES THE WHODUNIT PROBLEM AS ORGINALLY WRITTEN.

FOUR SUSPECTS - GROWLEY, SNAVELY, GUS AND GASTON - ARE QUESTIONED
ABOUT A CERTAIN CRIME. THE FOLLOWING STATEMENTS ARE GIVEN AND ONLY
ONE IS TRUE.

    SNAVELY DID IT.
    GASTON DID IT.
    GUS DIDN'T DO IT.
    GASTON DIDN'T DO IT.

THIS IS EXPRESSED IN LINE 120 AS P(2)+P(4)+NOTP(3)+NOTP(4)=1.

(YOU CAN MAKE UP YOUR OWN MYSTERY BY CHANGING THE STATEMENTS
AND THE CORRESPONDING EXPRESSION IN LINE 120.)


THE INESCAPABLE CONCLUSION IS THAT GUS DID IT.

DONE
```

# cryptarithmetic

In contrast to the previous examples, where the number of alternatives were relatively small, cryptarithmetic problems usually leave a large number of possible assignments of digits to letters to be examined. For example, consider the following problem.

**WIRE MONEY**

A college student sent the above message to his father. If each letter represents a unique number, how much should his dad send?

In the jargon of computer science, the *search space* — the set of possible alternatives for considerations as solutions — is 8! or 40,320. It would be impractical in this problem to blindly try each possible assignment one at a time. It would be much more efficient to divide up the search space into large classes, according to a common property shared by members of each class, and then attempt to eliminate entire classes by the method of contradiction. Wickelgren (1974) has labeled this technique *classificatory contradiction*.

We can illustrate this approach with the problem above by the following argument.

Consider all possible solutions where E = 0. Since E + E = Y, Y also equals 0, contradicting the fact that Y and E must be different digits. Thus, the entire class of solutions where E = 0 is ruled out. For another, less trivial example, consider E = 3. Now Y equals 6 and there is no carry to the next column. Thus in the record column we have R + R = E or perhaps R + R = E + 10 if a carry is involved. But in either case, E would be an even number, since 2R is always even, and this contradicts the assumption that E = 3. Thus the entire class of solutions with E = 3 is ruled out.

By following this same type of classificatory contradiction strategy for each of the digits in order of E, R, I, O, and N, program CRYPT search out all five solutions to the WIRE + MORE = MONEY problem. Can you write your own program to solve the cryptarithmetic problem.

```
      DONALD
    + GERALD
    --------
      ROBERT    ?
```

This problem has been extensively studied by Newell and Simon (1972) using a program called General Problem Solver (GPS) which was written to demonstrate that general problem-solving strategies exist and may be discussed at the very concrete level of computer programming.

## CRYPT PROGRAM

```
10  PRINT "THIS PROGRAM SEARCHES OUT ALL SOLUTIONS"
20  PRINT "TO THE CRYPTARITHMETIC PROBLEM"
30  PRINT
40  PRINT "      W    I    R    E"
50  PRINT "  +   M    O    R    E"
60  PRINT "---------------------------"
70  PRINT "  M   O    N    E    Y"
80  PRINT LIN(2)
90  M=1
100 FOR E=2 TO 9
110 Y=E+E
120 IF Y >= 10 THEN 150
130 C1=0
140 GOTO 170
150 C1=1
160 Y=Y-10
170 FOR R=0 TO 9
180 IF R=M OR R=E OR R=Y THEN 470
190 IF R+R+C1=E THEN 220
200 IF R+R+C1=E+10 THEN 240
210 GOTO 470
220 C2=0
230 GOTO 250
240 C2=1
250 FOR I=0 TO 9
260 IF I=M OR I=E OR I=Y OR I=R THEN 460
270 FOR O=0 TO 9
280 IF O=M OR O=E OR O=Y OR O=R OR O=I THEN 450
290 N=I+O+C2
300 IF N >= 10 THEN 330
310 C3=0
320 GOTO 350
330 C3=1
340 N=N-10
350 IF N=M OR N=E OR N=Y OR N=R OR N=I OR N=O THEN 450
360 FOR W=0 TO 9
370 IF W=M OR W=E OR W=Y OR W=R OR W=I OR W=O OR W=N THEN 440
380 IF O+10*W+M+C3 THEN 440
390 PRINT "      ";W;I;R;E
400 PRINT "    ";M;O;R;E
410 PRINT "---------------------------"
420 PRINT M;O;N;E;Y
430 PRINT LIN(2)
440 NEXT W
450 NEXT O
460 NEXT I
470 NEXT R
480 NEXT E
490 END
```

## SAMPLE RUN

```
RUN
CRYPT

THIS PROGRAM SEARCHES OUT ALL SOLUTIONS
TO THE CRYPTARITHMETIC PROBLEM

        W    I    R    E
   +    M    O    R    E
   ---------------------
   M    O    N    E    Y


        9    7    6    2
   1    0    6    2
   ---------------------
   1    0    8    2    4


        9    2    7    4
   1    0    7    4
   ---------------------
   1    0    3    4    8


        9    5    7    4
   1    0    7    4
   ---------------------
   1    0    6    4    8


        9    2    8    7
   1    0    8    7
   ---------------------
   1    0    3    7    4


        9    5    8    7
   1    0    8    7
   ---------------------
   1    0    6    7    4

DONE
```

# conclusion

When it comes to problem solving, one of the most significant advantages that people have over computers is the ease in which humans can make inferences about a problem and quickly reduce the search space. Programming a computer to make similar inferences is at best extremely difficult and at worst, impossible. People can be very unpredictable in ways to survey a problem looking for the easy inferences to attack first, while the computer is easiest to program to attack each problem in a very predetermined and algorithmic way. Polga (1954) contrasts the algorithmic method with the *heuristic* approach in which the nature of the solution is guessed and then proved to be correct. These two approaches have been combined into "heuristic programming" which is a new way of thinking about what a computer program should do. The idea is to find a good set of rules for generating guesses and then prove they are correct. This technique evolved from the work of Newell and Simon (1972) working on the program Logic Theorist (LT) and General Problem Solver (GPS).

The reader can appreciate the difficulty of heuristic programming by imagining how one would write a program to solve general cryptarithmetic problems in which the simplest inferences in a problem are first made similar to the way a human would approach it. In contrast a general algorithmic program can be written to solve any cryptarithmetic problem in a way similar to program CRYPT. Can you do it?

### References

Carroll, Lewis. *The Humorous Verse of Lewis Carroll,* Dover Publications, New York, NY, 1960.

Eves, Howard. *An Introduction to the History of Mathematics,* Holt, Rinehart, and Winston, New York, NY, 1976, p. 120.

Newell, A., and Simon, H. *Human Problem Solving,* Prentice Hall, Englewood Cliffs, NJ, 1972.

Wickelgren, Wayne A. *How to Solve Problems,* W.H. Freeman and Company, San Francisco, CA, 1974.

### Acknowledgement

We are extremely grateful to Rodney Schroeter for the drawings which he provided throughout the series.

# On Solving Alphametrics

John Beidler

## 1. Introduction.
There are those who frown upon mathematicians spending time studying various esoteric mathematical games and pastimes. There are a variety of good responses one can give in defense of these pastimes. Personally, we believe they need no defense.

Occasionally, one finds a correlation between these games and pastimes and other important fields of endeavor. About 1966 we became interested in solving alphametics as a pastime. About the same time we also became interested in computing. A natural outgrowth of this was the writing of computer programs to verify the solutions to alphametics.

After a while one begins to wonder: Rather than writing a program to solve each alphametic, why not write a single program which accepts as input an alphametic and then solves the alphametic? We accomplished this about 1970. We do not claim this to be the only program around which solves alphametics. However, we did receive several inquiries about this program after we indicated its existence in (4). That program was written by Mr. Tabor when he was a sophomore computer science major.

From the response we received we felt a description of the program would be appropriate. For those who also have an interest in computer programming, this program also serves as an example of the static use of pointer variables. Pointer variables (pointers) are variables used in programming which do not directly contain the data the program is manipulating but indicate or point to data. Indices into arrays are examples of pointers. However, the concept goes far beyond the use of indices and many times the use of pointers is at the heart of a sophisticated use of computing. Further, we refer to this as a "static" example because once the pointers are established, their values do not change.

John Beider, Computer Science, University of Scranton, Scranton, PA 18510.

To the best of our knowledge, the FORTRAN programs which appear here are not dependent on our compiler. However, if the programs are run on a WATFOR or WATFIV compiler, they should be compiled with the execute time diagnostics turned off. Otherwise, an error message might be produced. Also, for the sake of readability we have taken some liberties with forming indices. For this reason, some obvious modifications will have to be made if the program is run on a standard IBM 1130 FORTRAN compiler or any FORTRAN compiler which follows the strict ASA standards on allowable forms of indices.

## 2. Alphametics.
Alphametics are arithmetic expressions in which the digits are replaced by letters of the alphabets. Each digit associates to a distinct letter and the corresponding alphabetic statement should be of some interest. For example,

$$\begin{array}{r} \text{SEND} \\ \text{MORE} \\ \hline \text{MONEY} \end{array}$$

becomes

$$\begin{array}{r} 9567 \\ 1085 \\ \hline 10652 \end{array}$$

and this is the only possible solution.

Many examples of alphametics can be found in the Problems Section of the *Mathematics Magazine* as well as in the *Journal of Recreational Mathematics*. Several examples are listed in figure 1. These examples are solvable in many bases. The base establishes the number of degrees of freedom. Hence if a problem is solvable in one base, it will have solutions in higher bases.

```
    THE          DOUR
   EARTH         DONS
   VENUS         DONT
   SATURN        STOP
   URANUS        DROP
   ──────        ────
   NEPTUNE       OUTS

VIOLIN+VIOLIN+VIOLA+CELLO=QUARTET

    THREE+NINE=EIGHT+FOUR

     A+GO+GO+GAL=LOOK
```

*Figure 1. Some alphametics*

Given an alphametic, how does one find the solution? In the example above, there are eight letters, D, E, Y, N, R, O, S, and M. If an exhaustive attempt is made to solve this problem it would require the testing of 8! combinations. A program must use the relationships which hold between the digits in order to reduce the number of combinations attempted. For example, if we try replacing D by 2 and E by 5 in
 SEND+MORE=MONEY,
then we must replace Y by 7.

Figure 2 is a program which solves this alphametic. The function DIFF determines if the number associated to a particular letter differs from the values associated to other letters. DIFF is 1 if the value associated to a letter is different from the values already associated to other letters, otherwise DIFF = 2. The EQUIVALENCE statement shows the order in which letters have values associated to them as the program executes. Basically, this order is the order in which multi-digit numbers are added together. That is, the letters associated to low order digits are processed, then the tens column, then hundreds, etc.

Just as a value for Y is forced because of the values established for D and E, a value is forced for R because of the values determined for N and E. An analysis of the alphametic reveals that values for Y, R, O and M are forced once values are established for the other letters. Hence it is necessary only to exhaustively try all combinations for D, E, N, and S. This translates into 4 nested loops for the program. These loops begin at lines 11, 12, 20, and 34. With only 4 nested loops, the execution time for this program reduces to .48 seconds on a Xerox Sigma 6 computer.

## 3. A General Additive Alphametic Solver.
Once you observe the techniques employed in solving one alphametic, it is not difficult to write programs to solve others. The real challenge then is to write a single program which solves all alphametics. What follows is a description of a simplified version of an additive alphametic solver. A faster version exists but a description of it would get more wrapped up in minor details rather than, as we wish to do here, emphasize the fundamentals of solving alphametics and the use of pointer variables.

There are three types of structures used by the program, one dimensional arrays, two dimensional arrays, and a two dimensional array of pointers. First an input is translated from an input string into a two dimensional structure,

SEND
MORE
MONEY

Next, this two dimensional structure is scanned a column at a time, starting with the low order digits column and the letters are placed into a one dimensional array and their positions in the two dimensional array are replaced by pointers to the positions of the letters in the one dimensional array of letters (see figure 3).

While this occurs, we establish a second two dimensional array. If a position in the original two dimensional array had been blank, the corresponding position in the ACTION array contains a 1. While scanning the characters and placing them into the one dimensional array, a 3 goes into the corresponding position in the ACTION array if it is the first time that particular character has been scanned, otherwise a 2 is placed into that position in the ACTION array. For example, in the ACTION array described in figure 3, the position in the right most column of the ACTION array corresponding to the letter "E" contains a 3 while all other positions corresponding to "E"s contain 2s.

Three arrays are used to assist in solving an alphametic. These are arrays to contain the values associated to each character, VALUE, to hold the carry from the summation of the previous column, CARRY, and an array of logical values which indicate if the corresponding letter represents a leading digit and hence cannot be zero. Figure 3 shows all the arrays and their contents when the alphametic

SEND+MORE=MONEY

has been solved.

**4. The Program.** This program is written modularly with several subprograms which provide the tools for decoding the alphametic, setting up the structures, solving the alphametic, and printing the solution. In addition, output has been inserted into several routines so that the ACTION and POINT arrays can be seen and also there is a procedure to print values so that the solution can be easily verified.

We start our description with the main program, figure 4. It reads the alphametic and the base in which it is to be solved. It determines the number of rows and columns in the alphametic (lines 9-15), locates the space for the various arrays (lines 17-21), and calls the routine SETUP (lines 22-23).

SETUP appears in figure 5. It takes the various arrays and the alphametic and sets up the necessary information in the arrays to solve the alphametic. First, the characters are taken from the input image and placed into the POINT array (lines 9-22). In doing this DELIM is used to check for the three allowable delimiters, "+," "=," and "(blank space)." Next (lines 24-38), the characters which are now in the POINT array are placed into the array CHAR and the corresponding position in POINT is replaced by a pointer to the position in CHAR which now holds the character.

The function APPEAR is used (line 28) to see if the character under consideration was seen before. If it had not, a new entry is made in the POINT and ACTION arrays (lines 29-33). Otherwise, if the character had been seen before, entries are made only in the POINT and ACTION arrays (lines 34-35) but not in the array CHAR.

Next, the logical array LEAD is initialized. This is done to guarantee that no solutions are created with leading zeros (lines 39-44). For informational purposes, the ACTION and POINT arrays are printed, then the procedure SOLVE is called (lines 60-61).

The procedure SOLVE, (figure 6), now tries to find a solution to the alphametic. The key to what this procedure does is the variable DIR. When DIR = 1 the procedure is successfully proceeding towards a solution. When DIR = 2, the procedure has come upon a relationship it cannot resolve and is in the process of backtracking in order to change the value or values associated to some letters.

```
1.            IMPLICIT INTEGER (A-Z)
2.            DIMENSION A(10)
3.            EQUIVALENCE (A(1),D),(A(2),E),(A(3),Y),(A(4),N)
4.           2  ,(A(5),R),(A(6),0),(A(7),S),(A(8),M)
5.            READ(105,5)BASE
6.       5    FORMAT(1G)
7.            WRITE(108,910)
8.     910    FORMAT('1     S E N D',/,'     M O R E',
9.           2  /,'     M O N E Y',/)
10.           D = 0
11.    10     E = 0
12.    20     IF(DIFF(A,2) .EQ. -2) GO TO 100
13.             C1 = 0
14.             Y = D + E
15.             IF( Y .LT. BASE) GO TO 30
16.             Y = Y - BASE
17.             C1 = 1
18.    30     IF(DIFF(A,3) .EQ. 2) GO TO 100
19.             N = 0
20.    40     IF(DIFF(A,4) .EQ. 2) GO TO 90
21.             C2 = 0
22.             R = E - (N+C1)
23.             IF(R .GE. 0) GO TO 50
24.             R = R + BASE
25.             C2 = 1
26.    50     IF(DIFF(A,5) .EQ. 2) GO TO 90
27.             C3 = 0
28.             O = N - (E+C2)
29.             IF(O .GE. 0) GO TO 60
30.             O = O + BASE
31.             C3 = 1
32.    60     IF(DIFF(A,6) .EQ. 2) GO TO 90
33.             S = 1
34.    70     IF(DIFF(A,7) .EQ. 2) GO TO 80
35.             M = O - S
36.             IF( M .GE. 0) GO TO 80
37.             M = M + BASE
38.             IF(M*BASE+O .NE. M+S+C3 .OR. M.EQ. 0)GO TO 80
39.             IF(DIFF(A,8) .EQ. 2)GO TO 80
40.             WRITE(108,970)S,E,N,D,M,O,R,E,M,O,N,E,Y
41.    970    FORMAT(2(/,' ',3X,4I3),/,' ',15('='),/,' ',5I3)
42.    80     S = S+1
43.             IF(S .LT. BASE) GO TO 70
44.    90     N = N+1
45.             IF(N .LT. BASE) GO TO 40
46.    100    E = E+1
47.             IF(E.LT.BASE) GO TO 20
48.             D = D+1
49.             IF(D .LT. BASE) GO TO 10
50.           STOP
51.           END
```

```
1.            INTEGER FUNCTION DIFF(A,N)
2.            IMPLICIT INTEGER(A-Z)
3.            DIMENSION A(N)
4.            X = A(N)
5.            MX = N-1
6.            DO 20 I = 1,MX
7.             IF( A(I) .NE. X) GO TO 20
8.             DIFF = 2
9.             RETURN
10.    20     CONTINUE
11.           DIFF = 1
12.           RETURN
13.           END
```

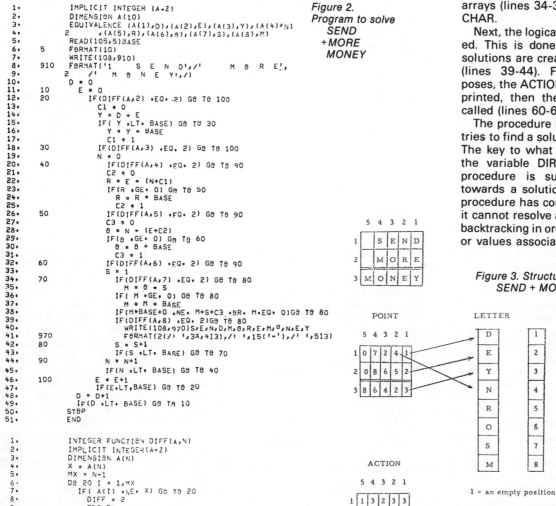

Figure 2. Program to solve
SEND
+MORE
MONEY

Figure 3. Structure for solving
SEND + MORE = MONEY

1 = an empty position
2 = this letter occurred before
3 = first occurrence of this letter

*Figure 4. The main program.*

```
1.              IMPLICIT INTEGER (A-Z)
2.              DIMENSION IMAGE(100)
3.              LOGICAL LEAD(2000)
4.              DIMENSION SPACE(2000)
5.              COMMON SPACE
6.              EQUIVALENCE (LEAD,SPACE)
7.       5      READ(105,9,END=100) BASE,IMAGE
8.       9      FORMAT(1G,/100A1)
9.              LAST = 0
10.             COLS = 0
11.             DO 20 ROWS = 1,40
12.                NEXT = DELIM(IMAGE,LAST)
13.                IF( NEXT .EQ. LAST+1 ) GO TO 30
14.                IF( NEXT-LAST-1 .GT. COLS) COLS = NEXT-LAST-1
15.      20        LAST = NEXT
16.      30     ROWS = ROWS - 1
17.             SECOND = ROWS*COLS + 1
18.             THIRD = SECOND + ROWS*COLS
19.             FOURTH = THIRD + BASE
20.             FIFTH = FOURTH + BASE
21.             SIXTH = FIFTH + BASE
22.             CALL SETUP(SPACE(1),SPACE(SECOND),SPACE(THIRD),SPACE(FOURTH)
23.          *        ,LEAD(FIFTH),SPACE(SIXTH),COLS+1,ROWS,COLS,BASE,IMAGE)
24.             GO TO 5
25.      100    STOP
26.             END
```

*Figure 5. The procedure SETUP.*

```
1.              SUBROUTINE SETUP(POINT,ACTION,CHAR,VALUE,LEAD,CARRY,NCP1
2.           *        ,ROWS,COLS,BASE,IMAGE)
3.              IMPLICIT INTEGER(A-Z)
4.              DIMENSION POINT(ROWS,COLS),ACTION(ROWS,COLS),CARRY(NCP1)
5.           *        ,CHAR(BASE),VALUE(BASE),IMAGE(100)
6.              LOGICAL LEAD(BASE)
7.              DATA EQUAL/'='/
8.              LAST = 0
9.              DO 50 ROW = 1,ROWS
10.                DO 10 COL = 1,COLS
11.                   POINT(ROW,COL) = 0
12.      10           ACTION(ROW,COL) = 1
13.                END = DELIM(IMAGE,LAST)
14.                IF(IMAGE(END) .EQ. EQUAL ) LINE = ROW
15.                THIS = END - 1
16.                COL=COLS
17.      20        IF(THIS.EQ.LAST) GO TO 50
18.                POINT(ROW,COL)=IMAGE(THIS)
19.                THIS=THIS-1
20.                COL=COL-1
21.                GO TO 20
22.      50        LAST = END
23.             NCHARS = 0
24.             COL = COLS
25.      60     DO 100 ROW = 1, ROWS
26.                IF(POINT(ROW,COL) .EQ. 0) GO TO 100
27.                THIS = APPEAR(POINT(ROW,COL),CHAR,NCHARS)
28.                IF(THIS .GT. 0) GO TO 70
29.                   NCHARS=NCHARS + 1
30.                   CHAR(NCHARS) = POINT(ROW,COL)
31.                   POINT(ROW,COL) = NCHARS
32.                   ACTION(ROW,COL) = 3
33.                   GO TO 100
34.      70        POINT(ROW,COL) = THIS
35.                ACTION(ROW,COL) = 2
36.      100    CONTINUE
37.             COL = COL - 1
38.             IF(COL .GT. 0) GO TO 60
39.             DO 110 C = 1,NCHARS
40.      110    LEAD(C) = .FALSE.
41.             DO 200 ROW = 1, ROWS
42.                DO 150 COL = 1,COLS
43.      150         IF(POINT(ROW,COL) .NE. 0)GO TO 200
44.      200         LEAD(POINT(ROW,COL)) = .TRUE.
45.             VALUE(1) = -1
46.             IF(LEAD(1)) VALUE(1) = 0
47.             CARRY(COLS+1) = 0
48.      C      WRITE(108,209)
49.      C209   FORMAT('1        TABLES')
50.      C      DO 350 ROW = 1,ROWS
51.      C      WRITE(108,349)(ACTION(ROW,COL),POINT(ROW,COL),COL=1,COLS)
52.      C349   FORMAT(/1X,15(I1,I3,4X))
53.      C350   IF(ROW .EQ. LINE) WRITE(108,359) COLS
54.      C359   FORMAT(1X,N('!-----'))
55.      C      WRITE(108,369)(I,CHAR(I),LEAD(I),I=1,NCHARS)
56.      C369   FORMAT(//(1X,I2,1X,A1,2X,L4))
57.             WRITE(108,409)IMAGE,BASE,(CHAR(I),I=1,NCHARS)
58.      409    FORMAT('1',T10,'CRYPTORYTHM SOLVER',//' SOLVE ',100A1
59.          X        ,/' IN BASE ',I4
60.          X        ,(T27,12(A1,2X)))
61.             CALL SOLVE(POINT,ACTION,CHAR,VALUE,LEAD,CARRY,NCP1
62.          X        ,ROWS,COLS,BASE,LINE,NCHARS)
63.             RETURN
64.             END
```

*Figure 6. The procedure SOLVE.*

```
1.              SUBROUTINE SOLVE(POINT,ACTION,CHAR,VALUE,LEAD,CARRY,NC
2.           *        ,ROWS,COLS,BASE,LINE,NCHARS)
3.              IMPLICIT INTEGER (A-Z)
4.              LOGICAL LEAD(BASE),DIFF
5.              DIMENSION POINT(ROWS,COLS),ACTION(ROWS,COLS),CARRY(NC)
6.           *        ,VALUE(BASE),CHAR(BASE)
7.              DIR = 1
8.              CHANGE=0
9.              ROW = 1
10.             COL = COLS
11.      5      IF(COL.LT.1 .OR. COL.GT.COLS) GO TO 150
12.      10     IF(ROW.LT.1 .OR. ROW.GT.ROWS)GO TO 60
13.             OP=ACTION(ROW,COL)
14.             IF(DIR.EQ.2 .AND. OP.EQ.3) DIR = 1
15.             IF(DIR.EQ.2 .OR. OP.LE.2) GO TO 30
16.             I=POINT(ROW,COL)
17.      20     VALUE(I)=VALUE(I)+1
18.             IF(.NOT.DIFF(VALUE(I),VALUE,I-1))GO TO 20
19.             IF(VALUE(I).GE.BASE) DIR = 2
20.             IF(I+1.GT.NCHARS)GO TO 30
21.             VALUE(I+1)=-1
22.             IF(LEAD(I+1))VALUE(I+1)=0
23.      30     IF(DIR.EQ.1)GO TO 40
24.                ROW=ROW-1
25.                GO TO 50
26.      40        ROW = ROW+1
27.      50     GO TO 10
28.      60     IF(DIR.EQ.1) GO TO 70
29.                COL = COL+1
30.                ROW=ROWS
31.                GO TO 120
32.      70     SUM=CARRY(COL+1)
33.             CARRY(COL)=0
34.             DO 75 I=1,ROWS
35.                IF(POINT(I,COL).EQ.0) GO TO 75
36.                IF(I.LE.LINE) SUM=SUM+VALUE(POINT(I,COL))
37.                IF(I.GT.LINE) SUM=SUM-VALUE(POINT(I,COL))
38.      75     CONTINUE
39.             IF(SUM.GE.0) GO TO 80
40.      77        SUM=SUM+BASE
41.                CARRY(COL)=CARRY(COL)-1
42.                IF(SUM.LT.0) GO TO 77
43.                GO TO 100
44.      80     IF(SUM.EQ.0) GO TO 100
45.      85        SUM = SUM-BASE
46.                CARRY(COL)=CARRY(COL)+1
47.                IF(SUM.GT.0)GO TO 85
48.      100    IF(SUM.EQ.0)GO TO 110
49.                DIR=2
50.                ROW=ROWS
51.                GO TO 120
52.      110    COL=COL-1
53.             ROW=1
54.      120    GO TO 5
55.      150    IF(DIR.EQ.2)RETURN
56.             IF(CARRY(1).EQ.0)CALL ANSWER(POINT,VALUE,CHAR,ROWS
57.          *        ,COLS,NCHARS,LINE)
58.             DIR=2
59.             COL=1
60.             GO TO 5
61.             END
```

*Figure 7a. Numeric logic function.*

```
1.              LOGICAL FUNCTION DIFF(THIS,THAT,SIZE)
2.              IMPLICIT INTEGER(A-Z)
3.              DIMENSION THAT(SIZE)
4.              DIFF = .FALSE.
5.              IF( SIZE .EQ. 0 ) GO TO 30
6.              DO 10 I = 1,SIZE
7.                 IF( THIS .EQ. THAT(I) ) RETURN
8.      10      CONTINUE
9.      30      DIFF = .TRUE.
10.             RETURN
11.             END
```

*Figure 7b. Alphabetic logic function.*

```
1.              FUNCTION APPEAR(THIS,THAT,SIZE)
2.              IMPLICIT INTEGER(A-Z)
3.              DIMENSION THAT(SIZE)
4.              IF( SIZE .EQ. 0 ) GO TO 30
5.              DO 10 APPEAR = 1, SIZE
6.                 IF(THIS .EQ. THAT(APPEAR)) RETURN
7.      10      CONTINUE
8.      30      APPEAR = -1
9.              RETURN
10.             END
```

*Figure 7c. Numeric operation logic function.*

```
1.              INTEGER FUNCTION DELIM(IMAGE,START)
2.              IMPLICIT INTEGER(A-Z)
3.              DIMENSION IMAGE(100),LIMITR(3)
4.              DATA LIMITR/' ','=','+'/
5.              MIN = START + 1
6.              DO 20 DELIM = MIN, 100
7.                 DO 20 I = 1,3
8.                    IF( IMAGE(DELIM) .EQ. LIMITR(I) ) RETURN
9.      20      CONTINUE
10.             DELIM = -1
11.             RETURN
12.             END
```

*Figure 7d. Subroutine to print answer.*

```
1.              SUBROUTINE ANSWER(POINT,VALUE,CHAR,ROWS,COLS,NCHARS,LINE)
2.              IMPLICIT INTEGER (A-Z)
3.              DIMENSION POINT(ROWS,COLS),VALUE(NCHARS),CHAR(NCHARS)
4.              WRITE(108,9) (         VALUE(I),I=1,NCHARS)
5.      9       FORMAT(/(T25,12I3))
6.      C       WRITE(108,69)
7.              DO 50 I=1,ROWS
8.                 DO 10 MIN=1,COLS
9.                    IF(POINT(I,MIN) .NE. 0) GO TO 20
10.     10         CONTINUE
11.     20         CONTINUE
12.                WRITE(108,29)MIN,(VALUE(POINT(I,J)),J=MIN,COLS)
13.     29         FORMAT(N(3X),20I3)
14.     50         IF(I .EQ. LINE)WRITE(108,59) COLS
15.     59      FORMAT(3X,N('---'))
16.     69      FORMAT(' VERIFY ')
17.             RETURN
18.             END
```

*FORTRAN program and subroutine for solving alphametrics.*

If the alphametic processes successfully, the solution is printed (lines 56-57). The variables ROW and COL determine the position in the alphametic; which is being processed. The action to be taken, ACTION(ROW, COL), is placed in the variable OP (line 13) and used to determine in conjunction with the variable DIR the appropriate action that is to be taken (lines 14-15).

While DIR is 1, the program executes as follows: If OP is 3 (lines 16-22), an attempt is made to associate a value to the letter. If the attempt is unsuccessful DIR is reset to 2 (line 19). Once a column is processed successfully, it is summed (lines 32-38) and the carry to the next column is formed (cards 39-47).

The backtracking process (lines 24-25, 29-31) simply backtracks until ACTION (ROW, COL) is 3 (line 14). Figure 7 presents several of the additional procedures used by SOLVE and SETUP, and the main program.

**5. Concluding Remarks.** As you can see, one can learn some of the intricacies in the use of pointer variables in an attempt to write a general alphametic solver. For those interested in pursuing a similar venture, we can suggest two exercises. The first would be to write a program which solves multiplicative alphametics. For example, solve

TWO * SIX = TWELVE

and

ZERO * TWO = NOTHING.

A second exercise would be to modify the program to consider secondary conditions. For example, solve

THREE + FOUR = SEVEN

where

1. 3 divides THREE;
2. 4 divides FOUR;
3. 7 divides SEVEN.

Neither exercise is trivial. The second can be more difficult, especially if you allow for such things as simultaneous alphametics. In either case, one will readily see the importance of the concept of pointer variables and its use in achieving the logical structure of information while the information is physically in another form.

### BIBLIOGRAPHY

1. Hunter, J.A.H., Problem 768, *Math. Mag.* (Sept. 1970).
2. McCravig, E.P., Problem 789, *Math. Mag.* (March 1971).
3. Suer, B. and Demir, H., Problem 859, *Math. Mag.* (March 1973).
4. Tabor, J., and Beidler, J., Solution to 859, *Math. Mag.* (Jan. 1974).
5. Tiner, J.H., Problem 761, *Math. Mag.* (May 1970).
6. Usiskin, Z., Problem 810, *Math. Mag.* (Nov. 1971).

■

*Another new game from Creative Computing....*

# BRAIN TEASER

### Hal Knippenberg

```
100 REM     BRAINT      10 APRIL 79     HAL KNIPPENBERG
105 REM
110 REM     PATTERNED AFTER A MACHINE LANGUAGE PROGRAM USED TO
115 REM     DEMONSTRATE DIGITAL GROUP EQUIPMENT
120 REM
125 REM     VARIABLES:  A$.....ANSWERS TO COMPUTER QUESTIONS
130 REM                 B()....ARRAY TO HOLD BOARD
135 REM                 E......ERROR FLAG
140 REM                 I......INDEX COUNTER
145 REM                 M......MOVE COUNTER
150 REM                 S......SUMMING REGISTER
155 REM                 S$.....STORAGE STRING
160 REM                 T......TEMPORARY STORAGE REGISTER
165 REM                 W$.....WIN/LOSE FLAG
170 REM                 X......POINTS TO PLAYERS MOVE
175 REM                 Z......CALLS MACHINE LANGUAGE SUBROUTINES
180 REM
185             DIM B(9)
190             DIM S$(45)
195 REM
200 REM *******************************
205 REM         START OF MAIN ROUTINE
210 REM
215         LET M = 0                   :REM   INITIAL MOVE COUNT
220         GOSUB 625                   :REM   SET UP BOARD
225 REM
230         Z = CALL(12762)             :REM   CLEAR SCREEN
235 REM
240         PRINT""
245         PRINT"    B R A I N   T E A S E R"
250         PRINT""
255 REM
260         GOSUB 700                   :REM   PRINT BOARD
265 REM
270         INPUT"DO YOU WANT INSTRUCTIONS ?   ",A$
275         IF A$(1,1) <> "Y" THEN 450
280 REM
285         Z = CALL(12762)             :REM   CLEAR SCREEN
290 REM
295         PRINT"    THE OBJECT OF THIS PUZZLE IS TO CHANGE THE"
300         PRINT"PATTERNS OF 0'S AND 1'S UNTIL THE BOARD HAS A 0"
305         PRINT"IN THE CENTER AND 1'S IN ALL OTHER POSITIONS."
310         PRINT"    TO CHANGE THE BOARD PATTERN, ENTER THE "
315         PRINT"NUMBER OF A SQUARE THAT CONTAINS A 1.  ENTER THE"
320         PRINT"SQUARES POSITION NUMBER AS FOLLOWS:"
325         PRINT""
330         PRINT"              1   2   3"
335         PRINT"              4   5   6"
340         PRINT"              7   8   9"
345         PRINT""
350         PRINT"    CHOOSING A SQUARE IN THE CENTER OF AN EDGE"
355         PRINT"(2,4,6,8) CAUSES ALL POSITIONS ALONG THE EDGE TO"
360         PRINT"CHANGE STATE.(0'S BECOME 1'S AND 1'S BECOME 0'S)"
365         PRINT""
370         INPUT"PRESS RETURN TO CONTINUE ",A$
375 REM
380         Z = CALL(12762)             :REM   CLEAR SCREEN
385 REM
390         PRINT""
395         PRINT"    CHOOSING A CORNER SQUARE (1,3,7,9) CAUSES"
400         PRINT"THE CORNER SQUARE AND THE THREE ADJACENT SQUARES"
405         PRINT"TO CHANGE STATE."
410         PRINT""
415         PRINT"    FINALLY IF YOU CHOOSE THE CENTER SQUARE (5),"
420         PRINT"ALL BUT THE CORNER SQUARES WILL CHANGE STATE."
425         PRINT""
430         PRINT"    TO END THE GAME, ENTER A MOVE OF 0."
435         PRINT""
440         INPUT"(PRESS RETURN TO BEGIN THE GAME) ",A$
445 REM
450         GOSUB 690                   :REM   PRINT BOARD
455 REM
460         GOSUB 780                   :REM   WIN CHECK
465         IF W$ <> "" THEN 535
470 REM
475         INPUT"YOUR MOVE ? ",X
480         IF X = 0 THEN 535
485 REM
```

```
490     GOSUB 845                       :REM  MOVE CHECK
495     IF E = 1 THEN 475
500 REM
505     GOSUB 310                       :REM  REVISE BOARD
510 REM
515     LET M = M + 1
520 REM
525     GOTO 450
530 REM
535     PRINT""
540     IF W$ = "WON" THEN PRINT"    Y O U    W O N  !"
545     IF W$ = "LOST" THEN PRINT"    Y O U    L O S T !"
550     PRINT""
555     PRINT"IT TOOK YOU ";M;" MOVES."
560     PRINT""
565 REM
570     INPUT"WOULD YOU LIKE TO TRY AGAIN ? ",A$
575     IF A$(1,1) = "Y" THEN 215
580 REM
585     PRINT""                         :REM  END IF PLAYER QUITS
590     PRINT""
595     END
600 REM
605 REM          END OF MAIN ROUTINE
610 REM * * * * * * * * * * * * * * * * * * * * * * * *
615 REM  SUBROUTINE:    SET UP BOARD
620 REM
625     LET S = 0
630     FOR I = 1 TO 9
635        LET T = RND(0)
640        IF T <= .9 THEN B(I) = 0
645        IF T > .9 THEN B(I) = 1
650        LET S = S + B(I)
655     NEXT I
660     IF S = 0 THEN 630
665     RETURN
670 REM
675 REM - - - - - - - - - - - - - - - - - - - - - - - -
680 REM  SUBROUTINE:    PRINT BOARD
685 REM
690     Z = CALL(12762)                 :REM  CLEAR SCREEN
695 REM
700     PRINT"THE BOARD AFTER MOVE ";M
705     PRINT""
710     PRINT""
715     FOR I = 0 TO 6 STEP 3
720        FOR J = 1 TO 3
725           LET T = I + J
730           PRINT"  ";B(T);
735        NEXT J
740        PRINT""
745     NEXT I
750     PRINT""
755     RETURN
760 REM
765 REM - - - - - - - - - - - - - - - - - - - - - - - -
770 REM  SUBROUTINE:    WIN CHECK
775 REM
780     LET S = 0
785     LET W$ = ""
790     FOR I = 1 TO 9
795        LET S = S + B(I)
800     NEXT I
805     IF S = 0 THEN W$ = "LOST"
810     IF S <> 8 THEN 820
815        IF B(5) = 0 THEN W$ ="WON"
820     RETURN
825 REM
830 REM - - - - - - - - - - - - - - - - - - - - - - - -
835 REM  SUBROUTINE:    MOVE CHECK
840 REM
845     LET E = 0
850     IF X > 9 THEN 870
855     IF B(X) = 0 THEN 870
860     RETURN
865 REM
870     PRINT"  ILLEGAL MOVE, RE-ENTER."
875     PRINT""
880     LET E = 1
885     RETURN
890 REM
895 REM - - - - - - - - - - - - - - - - - - - - - - - -
900 REM  SUBROUTINE:    REVISE BOARD
905 REM
910     S$ ="124501230023560147002456336900457807390058890"
915     LET B(0) = 5
920     FOR I = (5*X - 4) TO 5*X
925        LET T = VAL(S$(I,I))
930        IF B(T) = 0 THEN B(T) = 1 ELSE B(T) = 0
935     NEXT I
940     RETURN
945 REM
950 REM - - - - - - - - - - - - - - - - - - - - - - - -
```

At first glance, "BRAIN TEASER" appears deceptively simple but, unless you are very lucky or very smart, you will use far more than the six maximum moves needed to solve this puzzle.

The program begins by setting up a 3 x 3 playing board. The board is randomly filled with 0's and one or more 1's.

The object of the puzzle is to change the patterns of 0's and 1's until the board has a 0 in the center and 1's in all other positions.

To make a board change, select a square that contains a 1. You tell the computer which square you are pointing to by entering the square's address number.

A possible starting board and its addresses are as follows:

```
0 1 0      1 2 3
0 1 0      4 5 6
0 0 1      7 8 9
```

There are three ways the board will change as you select different addresses. If you print to the center square (Address 5), all but the corner squares will change state (1's will change to 0's and 0's will change to 1's). For example:

```
0 1 0                      0 0 0
0 1 0  will change to  1 0 1
0 0 1                      0 1 1
```

If you select a corner square (Addresses 1, 3, 7 or 9), that corner and the three adjacent squares will change state. For example, if you point to Square 9:

```
0 1 0                      0 1 0
0 1 0  will change to  0 0 1
0 0 1                      0 1 0
```

Finally, if you choose a square at the center of an edge (Addresses 2, 4, 6 or 8), the three squares along that edge will change state. For example, choose Square 2:

```
0 1 0                      1 0 1
0 1 0  will change to  0 1 0
0 0 1                      0 0 1
```

One last fact, if you happen to end up with all 0's, you lose! (I've been able to lose several times in fewer than six moves.)

A "WIN" looks like this:

```
1 1 1
1 0 1
1 1 1
```

Try it — You'll hate it!

The program was written in MAXI-BASIC and should work with few

---

Hal Knippenberg, 2514 Blueberry Drive, Augusta, Georgia 30906.

modifications in most other BASIC's. However, there are several lines of code that might best be explained.

Z = CALL(12762) is a call to a machine language subroutine that instantly clears the TV monitor. (MAXI-BASIC has no specific command that will clear the screen.) If you have no CLEAR SCREEN command, try a BASIC subroutine to do the job, i.e.:  FOR I = 1 to 16
    PRINT " "
   NEXT I
   RETURN

The BASIC subroutine will be slower than its machine language counterpart, but it will get the job done.

Try programming a simple graphics version of BRAIN TEASER with light squares representing the 1's and dark squares the 0's. Try to incorporate sound effects too. □

# Computer Simulations

# How to write a computer simulation

## David H. Ahl

This article demonstrates how to write a computer simulation to be used for classroom instruction in some other subject. The object is not to teach about computers or computer programming; indeed very little programming knowledge is required to write a useful and effective simulation.

Nor is the purpose of this article to set down guidelines for determining instructional objectives or choosing subjects appropriate for simulation. The booklet *Designing Classroom Simulations* by Glenn Pate and Hugh Parker, Jr., fulfills these needs nicely. It should be noted, however, that not all subjects suitable for a classroom simulation or game are equally suitable for one on the computer.

For the most part, simulations and games have been found to be an extremely effective way of teaching a wide variety of subjects. Several studies have described the learning effectiveness of games (Allen, Allen, and Ross, 1970; Boochock and Schild, 1968; Fletcher, 1971). Learning games also generally create an intense and often enjoyable interpersonal experience. This is due in part to the interdependent task structure that requires interaction among the players (Inbar, 1968). The rationale for computer simulations is very compelling given the unique ability of such models to compress time, duplicate expensive, massive, delicate, or dangerous equipment, and produce large sample sizes (Braun, 1972).

However, lest we get overly ardent in using simulations, it should also be noted that certain things can be learned perfectly well by reading about them in a book, and to write a simulation is probably a waste of time to everyone but the writer, who will inevitably have learned more about the subject than when he or she started.

---

Presented at CCUC/5 (Conference on Computers in the Undergraduate Curricula) in Pullman, Washington, June 1974.

## The Subject

In response to current events, the subject for our computer simulation is electrical energy generation and usage. The objectives of this simulation are:

1. To develop within the student the ability to recognize the need for advance planning and construction of electrical generation capacity.

2. To develop within the student the ability to plan an economically and environmentally sound electrical generation policy using five major fuel sources—coal, oil, gas, nuclear, and water.

## Background Data and Assumptions

After poring over innumerable reference sources, you will quickly discover that:

1. There are far more facts and data available than you can use.

2. Several vital facts that you need are unavailable.

Here you can either quit before you've wasted too much time, or fearlessly plunge ahead. With a surge of optimism, we chose the latter course.

Ultimately, the following facts were obtained:

1. The demand for electricity is increasing 5.4% per year. In 1971, usage was 1.466 trillion kwh.

2. Some mathematical gyrations indicate that to produce 1.466 trillion kwh requires approximately 308 million kw of generation capacity plus a 20% margin of reserve (for extra-hot days, generator failures, emergencies, etc.) or a total capacity of 367 million kw.

3. The amount of generating capacity operational in 1971 is as follows:

| Fuel | Generating Capacity | Annual Growth |
|---|---|---|
| Coal | 167 million kw | 6.7% |
| Fuel Oil | 50 | 7.4 |
| Gas | 87 | 8.4 |
| Nuclear | 7 | 54.8 |
| Hydro | 56 | 5.2 |
| Total | 367 | |

4. Relationships between generation, fuels, and their supply is as shown in Table I.

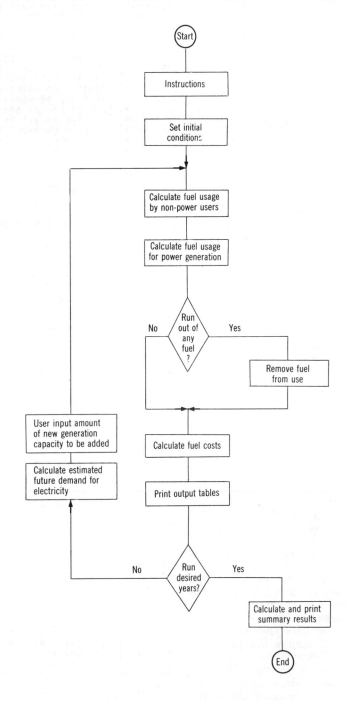

Figure 1

### TABLE I

| Fuel | Amount to Generate 1 Kwh | Fuel Cost | Proven Reserves, United States |
|---|---|---|---|
| Coal | .92 lb | $8.63/ton | 830 billion tons |
| Fuel Oil | .00183 lb | $9.00/bbl | 48 billion barrels |
| Gas | 10.6 cu ft | $0.44/1000 cu ft | 286 trillion cu ft |
| Nuclear | .0000075 lb | $530/lb | 1.05 million tons |
| Hydro | n/a | $0.50/1000 kwh | 190 million kw |

Naturally, few of these numbers are constant over time. Since the simulation to be written will span a period of 50 to 100 years, some of the trends will have to be taken into account. This leads into the uncharted area of assumptions. For example, increasing efficiency of power plants has reduced the amount of coal required to produce one kilowatthour of electricity from 1.29 lb. in 1946 to 0.92 in 1971; an average reduction of 0.015 lb per year or a compound rate of 1.3% per year. Will this continue? Who knows? Let's assume it will but certainly not to using zero fuel. As the model is formulated, one must be careful of these assumptions and either put limits on the time period covered by the model or alter the trends and assumptions over time.

Any computer model or simulation requires a fair number of assumptions if it is not to get completely out of hand. Here is a summary of the major assumptions used in this model.

1. The demand for electricity will continue to increase at an annual rate of 5.4% per year.

2. Efficiency of generators will continue to improve as discussed above at a rate of 1.3% per year.

3. The major cost variable in generating electricity using different fuels is the cost of the fuel itself. In other words, the capital cost for new generating equipment is about the same per kilowatt hour for a coal, oil, gas, or nuclear plant. (In fact, the costs range from $240 to $280 per kwh of new capacity which is remarkably close).

4. Cost for additional hydro capacity depends greatly upon the cost of the dam which are generally built for irrigation as well as for power. Furthermore, most dams are government-built and the cost figures are not comparable with privately-owned utilities. However, due to environmental and other considerations, it is unlikely that new hydro capacity can be added at a rate of much over 5.5% per year. For the model, this rate is assumed.

5. Fuel reserves will not change significantly over time. This assumption is subject to the most debate. Discoveries of new natural gas sources, for example, have been keeping pace with usage. As fuel prices rise, new sources become economically feasible to exploit; deep coal mines or ones with thinner veins, for examples. Also, new recovery methods may be discovered or become cost effective; for example, the recovery of shale oil. In the computer program, it is wise to

```
10 PRINT "PROGRAM 'POWER'"
20 PRINT "SIMULATION OF POWER GENERATION, FUEL REQUIREMENTS, AND COSTS"
30 PRINT "(VERSION 1-A, JAN 1974, WRITTEN BY DAVID H. AHL)"
40 INPUT "WANT INSTRUCTIONS (Y OR N)";A$ \ PRINT \ IF A$="N" THEN 230
100 !        INSTRUCTIONS
110 PRINT "YOU ARE THE DIRECTOR OF THE U.S. FEDERAL POWER COMMISSION,"
120 PRINT "IT IS UP TO YOU TO APPROVE THE BUILDING OF NEW POWER GEN-"
130 PRINT "ERATION FACILITIES FOR BOTH INVESTOR-OWNED AND GOVERNMENT"
140 PRINT "CONTROLLED UTILITIES.  THEREFORE, AN IMPORTANT PART OF"
150 PRINT "YOUR JOB IS KNOWING HOW MUCH ADDITIONAL POWER IS REQUIRED,"
160 PRINT "HOW IT CAN BE OBTAINED (I.E., USING WHAT TYPE OF FUEL),"
170 PRINT "AND THE IMPACT ON FUEL RESERVES.  SINCE YOUR ACTIONS WILL"
180 PRINT "HAVE AN ENORMOUS IMPACT ON FUTURE GENERATIONS (WHO ALSO"
190 PRINT "NEED FUEL), YOU HAVE AT YOUR DISPOSAL A COMPUTER SIMULA-"
200 PRINT "TION OR MODEL OF THE U.S. POWER GENERATION SYSTEM,"\PRINT
210 PRINT "THE MODEL WILL ASK FOR YOUR DECISIONS ON NEW GENERATING"
220 PRINT "CAPACITY EVERY 5 YEARS AND THEN GIVE YOU THE RESULTS OF"
225 PRINT "YOUR DECISIONS." \ PRINT
230 PRINT "YOU MAY RUN THE MODEL FOR 25 TO 100 YEARS."
240 INPUT "HOW MANY YEARS";Y3 \ IF Y3<25 OR Y3>100 THEN 230
250 !        INITIAL CONDITIONS AND VALUES
260 F$(1)="COAL" \ F$(2)="OIL" \ F$(3)="GAS" \ F$(4)="NUCLEAR"
265 F$(5)="HYDRO"
270 G$(1)="BILLION TONS" \ G$(2)="BILLION BARRELS" \ G$(3)="TRILLION CU FT"
275 G$(4)="MILLION TONS" \ G$(5)="MILLION KW"
280 H(1)=832 \ H(2)=50 \ H(3)=700 \ H(4)=1.05 \ H(5)=190   ! 1966 RESERVES
290 S(1)=1.16 \ S(2)=3.66 \ S(3)=45
295 S(4)=34.8E-6                        ! OTHER FUEL USAGE
310 Y,Y2=1966 \ W=1466/1.3 \ C(6)=367   ! YEAR, USAGE, GEN CAPACITY
320 C(1)=167 \ U(1)=613                 ! COAL CURRENT CAPACITY, 1966 USAGE
330 C(2)=50 \ U(2)=76
340 C(3)=87 \ U(3)=251
350 C(4)=7 \ U(4)=5.5
360 C(5)=56 \ U(5)=200
370 R2(I)=0 FOR I=1 TO 5                ! USAGE OF RESERVES
380 B$="\         \#####        ######"
390 C$="             ##.#"
400 D$="\         \###.##        #######        ###.##"
410 !        FUEL USAGE BY NON-POWER GENERATION SOURCES
420 FOR I=1 TO 4
425 IF R9(I)=1 THEN 440
430 S(I)=S(I)*1.3 \ R2(I)=R2(I)+S(I)
440 NEXT I
500 !        CAPACITY AND USAGE
510 Y=Y+5 \ W=W*1.3 \ K=W/C(6)          ! YEAR, DEMAND, CONSTANT
520 P=W/(C(6)*.05) \ W1=W1+5.75*W       ! % FUEL USAGE, CUM GENERATION
530 P1=P1+P \ FOR I=1 TO 5
535 IF R9(I)=1 THEN U(I)=0
540 U2(I)=U(I) \ U(I)=C(I)*K            ! ALLOCATE GENERATION BY FUEL
550 NEXT I
560 !        FUEL USAGE (5 YEARS)
590 E9=.9867^(Y-1968.5)                 ! 5-YEAR EFFICIENCY FACTOR
600 PRINT \ PRINT "NOW CALCULATING 5-YEAR FUEL USAGE" \ PRINT
610 R1(1)=2.5*(U(1)+U2(1))*.92*E9/2000  ! COAL BILLION TONS
620 R1(2)=2.5*(U(2)+U2(2))*.00183*E9               ! OIL BILLION BBL
630 R1(3)=2.5*(U(3)+U2(3))*10.6*E9/1000            ! GAS TRILLION CU FT
640 R1(4)=2.5*(U(4)+U2(4))*3.75                    ! NUCLEAR TONS
650 FOR I=1 TO 3
655 IF R9(I)=1 THEN 710
660 R2(I)=R2(I)+R1(I)
670 H1(I)=(H(I)-R2(I))*100/H(I)         ! % CUM FUEL USAGE
680 IF H1(I)>10 THEN 710
690 PRINT "THE U.S. HAS VIRTUALLY RUN OUT OF "F$(I)" !" \ R9(I)=1
710 NEXT I
715 IF R9(4)=1 THEN 760
720 R2(4)=R2(4)+R1(4)*1E-6
730 H1(4)=(H(4)-R2(4))*100/H(4)         ! % CUM FUEL USAGE
740 IF H1(4)>10 THEN 760
750 PRINT "THE U.S. HAS VIRTUALLY RUN OUT OF NUCLEAR FUEL !" \ R9(4)=1
760 R(5)=C(5)*100/H(5)
770 IF R(5)>=100 THEN PRINT "YOU HAVE NO REMAINING HYDRO CAPACITY !"
780 FOR I=1 TO 4
785 IF R7(I)=1 THEN 810
790 IF R9(I)=0 THEN 810
800 C(6)=C(6)-C(I) \ C(I)=0 \ R7(I)=1
810 NEXT I
900 !        COST OF FUEL PER KW AND TOTAL
```

```
910 E9=.9867^(Y-Y2)*1000
920 D(1)=.92*E9*8.63/2000              ! COST=FUEL/KW + COST/QUAN
930 D(2)=.00183*E9*9
940 D(3)=.0106*E9*.44
950 D(4)=7.5E-6*E9*530
960 D(5)=.5
970 T(6)=0
980 FOR I=1 TO 5
990 T(I)=U(I)*D(I) \ T(6)=T(6)+T(I)    ! TOTAL FUEL COSTS
1000 NEXT I
1010 T1=T1+T(6) \ D(6)=T(6)/W
1290 !
     5-YEAR RESULTS
1300 PRINT \ PRINT "RESULTS FOR"Y \ PRINT
1310 PRINT "","CAPACITY","USAGE","USAGE (%"
1320 PRINT "FUEL","(MILLION KW)","(BILLION KWH)","OF CAPACITY)"
1330 FOR I=1 TO 5 \ IF R8(I)=1 THEN 1350
1340 PRINT USING B$,F$(I),C(I),U(I)
1350 NEXT I
1360 PRINT USING B$+C$,"TOTAL",C(6),W,P
1361 IF P>80 THEN 1362 ELSE IF P<70 THEN 1366 ELSE GOTO 1370
1362 PRINT \ PRINT "YOUR MARGIN OF RESERVE CAPACITY IS ONLY"INT(100-P)"%"
1363 IF P>90 THEN 1364 ELSE PRINT "LEADING TO NUMEROUS BROWNOUTS."\GOTO 1368
1364 PRINT "LEADING TO MAJOR PROBLEMS IN INDUSTRIAL PRODUCTION AND URBAN"
1365 PRINT "TRANSPORTATION SYSTEMS." \ GOTO 1368
1366 PRINT "YOUR RESERVE CAPACITY MARGIN OF"INT(100-P)"% IS TOO GREAT."
1367 PRINT "THE INVESTMENT FOR NEW PLANTS WAS NEEDED MORE IN OTHER AREAS."
1368 PRINT "YOUR SUCCESSOR IS ABOUT TO BE APPOINTED UNLESS THERE IS A"
1369 PRINT "DRAMATIC IMPROVEMENT IN THE NEXT 5 YEARS."
1370 PRINT \ PRINT "","FUEL COST","TOTAL FUEL $","REMAINING FUEL"
1380 PRINT "FUEL","PER 1000 KWH","(MILLION $)","(% OF 1966)"
1385 H1(5)=(H(5)-C(5))*100/H(5)
1390 FOR I=1 TO 5 \ IF R8(I)=1 THEN 1420
1400 PRINT USING D$,F$(I),D(I),T(I),H1(I)
1420 NEXT I
1430 PRINT USING D$,"TOTAL",D(6),T(6)
1440 IF Y-Y2>=Y3 THEN 1700
1450 PRINT \ PRINT "EST. DEMAND FOR ELECTRICITY IN "Y+5"IS"W*1.3"BILLION KWH"
1460 PRINT "GENERATING CAPACITY IN"Y"IS"C(6)"BILLION KW."
1470 PRINT "DESIRABLE GENERATING CAPACITY IN"Y+5"IS"W*.325"BILLION KW."
1480 PRINT
1490 R8(I)=R9(I) FOR I=1 TO 4
1500 !
     INPUT NEW VALUES
1510 PRINT "AMOUNT OF NEW CAPACITY TO BE ADDED IN NEXT 5 YEARS (BILLION KW)"
1520 C1(6)=0
1530 FOR I=1 TO 5 \ IF R9(I)=1 THEN 1560
1540 PRINT F$(I), \ INPUT N(I)
1542 IF I<>5 THEN 1550
1543 IF N(5)<=.3*C(5) THEN 1544 ELSE A=.3*C(5) \ GOTO 1546
1544 IF C(5)+N(5)<190 THEN 1550 ELSE A=190-C(5)
1546 PRINT "THAT'S TOO MUCH ADDITIONAL HYDRO CAPACITY." A"IS THE LIMIT."
1548 GOTO 1540
1550 C1(I)=C(I)+N(I) \ C1(6)=C1(6)+C1(I)
1560 NEXT I
1570 IF C1(6)>W*.26 THEN 1600
1580 PRINT "THAT'S NOT ENOUGH ADDITIONAL CAPACITY TO MEET THE DEMAND."
1590 PRINT "YOU NEED A MINIMUM OF"INT(W*.26-C1(6))"BILLION KW OVER"
1595 PRINT "THE CURRENT CAPACITY OF"C(6)". PLEASE TRY AGAIN." \ GOTO 1510
1600 !
     UPDATE 5 YEAR AGO FIGURES
1610 FOR I=1 TO 6
1620 C2(I)=C(I) \ C(I)=C1(I)
1630 NEXT I
1640 GOTO 500
1700 !
     END SUMMARY
1710 PRINT \ PRINT \ PRINT "SUMMARY RESULTS -- 1966 TO"Y \ PRINT
1715 Y1=Y-1966 \ H1(5)=H(5)-C(5)        ! TOTAL YEARS, % HYDRO USAGE
1720 PRINT "TOTAL ELECTRICITY GENERATED ="INT(W1)"BILLION KWH."
1725 PRINT "AVERAGE PERCENT OF CAPACITY IN USE ="P1*5/Y1 \ PRINT
1730 Y1=Y-1966 \ PRINT Y1"YEAR FUEL COST SUMMARY:"
1740 PRINT "     TOTAL COST = $"5*T1"MILLION"
1745 PRINT "     AVERAGE COST PER YEAR = $"5*T1/Y1
1750 PRINT "     AVERAGE COST PER 1000 KWH = $"5*T1/W1
1760 PRINT \ PRINT "","RESERVES","RESERVES"
1770 PRINT "FUEL","IN 1966","IN"Y \ PRINT
1790 FOR I=1 TO 5
1795 A=H(I)-R2(I) \ IF A<0 THEN A=0
1800 PRINT F$(I),H(I),H(I)-R2(I),G$(I)
1810 NEXT I
1820 PRINT \ PRINT \ PRINT "FUEL","% REMAINING OF 1966 RESERVE" \ PRINT
1830 FOR I=1 TO 5
1835 IF H1(I)<0 THEN H1(I)=0
1840 PRINT F$(I),H1(I)
1850 NEXT I
2000 END
```

leave all these factors as variables so that you, or an enterprising student, can try out other alternatives or "improve" the simulation.

6. The model will assume a self-sufficient energy policy; that is, only fuel reserves within the United States can be called upon. If you're an optimistic internationalist, you can easily change this assumption in the one line (280) where these values are set. Again, the general principle is: leave everything a variable.

7. All available fuel cannot be used for the generation of electricity. We've assumed the following percentages are available: coal 55%, oil 25%, gas 25%, nuclear 90%.

You'll discover other assumptions that have to be made or facts that have to be gathered as you actually write the model but if you have most of it beforehand, you'll have an easier time.

### Writing the Model

The temptation, of course, is to plunge into the computer program. DON'T! Write a flowchart first. Virtually all interactive models and simulations for teaching follow these steps, exclusive of instructions.

1. Set initial conditions.
2. Perform the calculations for one period, cycle, steady-state at the start, etc. Output this.
3. Accept input from the user.
4. Perform calculations for the next period using inputted data. Output this.
5. Go back to Step 3 until the model has run the desired number of steps, periods, etc. At this point go to Step 6.
6. Calculate and output final results. These steps have been transformed into the simplified flowchart shown in Figure 1.

It's probably also a good idea to state any formulas used in the model. Other than percentage calculations, the electrical generation model didn't use many formulas. Here are the most notable ones:

Five year fuel usage = 5x (kw Generation Year 1 + kw Generation Year 5) ÷ 2x Efficiency improvement factor x fuel per kw.

Cost of fuel per yr. = kw Generation x Efficiency improvement factor x fuel per kw x cost per unit of fuel

It's also worthwhile to state the format of the output and assign letter variables right on the mocked-up output. For example:

```
        Capacity      Usage        Usage (%
Fuel   (Million kw) (Billion k wh) of Capacity)
F$(1)    c(1)         u(1)           p
F$(2)    c(2)         u(2)           p
etc.
```

Now, finally, you're ready to write the program. If you've done the previous steps thoroughly, the writing of the program will be trivial (well, perhaps not trivial, but at least straightforward). It's probably good practice to write the program in sections. Worry about the calculations section first. If you can hack through that, the other sections will fall nicely into place. While you're doing the calculations section, jot down the variables that will have to be set in other sections like initial conditions or user input.

Use comments liberally throughout your program. You'll be glad you took the time when you start to debug the program or even look at it a few weeks later.

Add some elements to keep the program interesting to the user. If a value gets out of control, don't just print a message but try to describe the impact of that value. For example, instead of printing:

RESERVE POWER MARGIN = 12%

the user will learn more and be more interested if you print something like this:

YOUR MARGIN OF RESERVE POWER IS ONLY 12%. THERE HAVE BEEN NUMEROUS BROWNOUTS AND YOUR SUCCESSOR IS ABOUT TO BE APPOINTED UNLESS THERE IS A DRASTIC IMPROVEMENT.

Look over the power-generation program. Not for the specifics of how it works, but rather note the liberal use of comments, its division into sections, and explanations to the user.

### Writing The User's Guide

Most simulations are of sufficient complexity that all the necessary background information cannot be included in the computer instructions in the program. Even with the simplest simulations, additional background information and suggested readings for further pursuit of the topic are desirable for the student to have in hand. The user's guide need not be long but it should include the following:

A. *User Section*

1. A brief description of the event, system, or apparatus being simulated.

2. Instructions for running the program and a sample run.

3. Questions, exercises, and/or activities to do with the program output. These should be as open-ended as possible, consistent with the student fulfilling your learning objectives.

```
RUNNH
PROGRAM 'POWER'
SIMULATION OF POWER GENERATION, FUEL REQUIREMENTS, AND COSTS
(VERSION 1-A, JAN 1974, WRITTEN BY DAVID H. AHL)
WANT INSTRUCTIONS (Y OR N)? Y

YOU ARE THE DIRECTOR OF THE U.S. FEDERAL POWER COMMISSION.
IT IS UP TO YOU TO APPROVE THE BUILDING OF NEW POWER GEN-
ERATION FACILITIES FOR BOTH INVESTOR-OWNED AND GOVERNMENT
CONTROLLED UTILITIES.   THEREFORE, AN IMPORTANT PART OF
YOUR JOB IS KNOWING HOW MUCH ADDITIONAL POWER IS REQUIRED,
HOW IT CAN BE OBTAINED (I.E., USING WHAT TYPE OF FUEL),
AND THE IMPACT ON FUEL RESERVES.   SINCE YOUR ACTIONS WILL
HAVE AN ENORMOUS IMPACT ON FUTURE GENERATIONS (WHO ALSO
NEED FUEL), YOU HAVE AT YOUR DISPOSAL A COMPUTER SIMULA-
TION OR MODEL OF THE U.S. POWER GENERATION SYSTEM.

THE MODEL WILL ASK FOR YOUR DECISIONS ON NEW GENERATING
CAPACITY EVERY 5 YEARS AND THEN GIVE YOU THE RESULTS OF
YOUR DECISIONS.

YOU MAY RUN THE MODEL FOR 25 TO 100 YEARS.
HOW MANY YEARS? 50

NOW CALCULATING 5-YEAR FUEL USAGE

RESULTS FOR 1971

              CAPACITY       USAGE          USAGE (%
FUEL          (MILLION KW)   (BILLION KWH)  OF CAPACITY)
COAL          167            667
OIL           50             200
GAS           87             348
NUCLEAR       7              28
HYDRO         56             224
TOTAL         367            1466           79.9

              FUEL COST      TOTAL FUEL $   REMAINING FUEL
FUEL          PER 1000 KWH   (MILLION $)    (% OF 1966)
COAL          3.71           2477           99.65
OIL           15.40          3077           88.04
GAS           4.36           1516           89.45
NUCLEAR       3.72           104            99.97
HYDRO         0.50           112            70.53
TOTAL         4.97           7285

EST. DEMAND FOR ELECTRICITY IN 1976 IS 1905.8 BILLION KWH
GENERATING CAPACITY IN 1971 IS 367 BILLION KW.
DESIRABLE GENERATING CAPACITY IN 1976 IS 476.45 BILLION KW.

AMOUNT OF NEW CAPACITY TO BE ADDED IN NEXT 5 YEARS (BILLION KW)
COAL           ? 50
OIL            ? 0
GAS            ? 0
NUCLEAR        ? 50
HYDRO          ? 10

NOW CALCULATING 5-YEAR FUEL USAGE
RESULTS FOR 1976

              CAPACITY       USAGE          USAGE (%
FUEL          (MILLION KW)   (BILLION KWH)  OF CAPACITY)
COAL          217            867
OIL           50             200
GAS           87             348
NUCLEAR       57             228
HYDRO         66             264
TOTAL         477            1906           79.9

              FUEL COST      TOTAL FUEL $   REMAINING FUEL
FUEL          PER 1000 KWH   (MILLION $)    (% OF 1966)
COAL          3.47           3011           99.46
OIL           14.41          2878           84.74
GAS           4.08           1418           87.07
NUCLEAR       3.48           792            99.74
HYDRO         0.50           132            65.26
TOTAL         4.32           8230

EST. DEMAND FOR ELECTRICITY IN 1981 IS 2477.54 BILLION KWH
GENERATING CAPACITY IN 1976 IS 477 BILLION KW.
DESIRABLE GENERATING CAPACITY IN 1981 IS 619.385 BILLION KW.

AMOUNT OF NEW CAPACITY TO BE ADDED IN NEXT 5 YEARS (BILLION KW)
COAL           ? 60
OIL            ? 0
```

```
     GAS            ? 0
     NUCLEAR        ? 50
     HYDRO          ? 15

NOW CALCULATING 5-YEAR FUEL USAGE

RESULTS FOR 1981

              CAPACITY        USAGE           USAGE (%
     FUEL     (MILLION KW)    (BILLION KWH)   OF CAPACITY)
     COAL     277             1140
     OIL       50              206
     GAS       87              358
     NUCLEAR  107              440
     HYDRO     81              333
     TOTAL    602             2478             82.3

YOUR MARGIN OF RESERVE CAPACITY IS ONLY 17 %
LEADING TO NUMEROUS BROWNOUTS.
YOUR SUCCESSOR IS ABOUT TO BE APPOINTED UNLESS THERE IS A
DRAMATIC IMPROVEMENT IN THE NEXT 5 YEARS.

              FUEL COST       TOTAL FUEL $    REMAINING FUEL
     FUEL     PER 1000 KWH    (MILLION $)     (% OF 1966)
     COAL      3.25           3702             99.22
     OIL      13.47           2772             81.60
     GAS       3.82           1366             84.81
     NUCLEAR   3.25           1432             99.14
     HYDRO     0.50            167             57.37
     TOTAL     3.81           9439

EST. DEMAND FOR ELECTRICITY IN 1986 IS 3220.8 BILLION KWH.
GENERATING CAPACITY IN 1981 IS 602 BILLION KW.
DESIRABLE GENERATING CAPACITY IN 1986 IS 805.201 BILLION KW.


(Output for Period 1986-2011 is not shown)


RESULTS FOR 2016

              CAPACITY        USAGE           USAGE (%
     FUEL     (MILLION KW)    (BILLION KWH)   OF CAPACITY)
     COAL     3274            13095
     OIL        50              200
     GAS        87              348
     NUCLEAR   287             1148
     HYDRO     189              756
     TOTAL    3887            15546            80.0

              FUEL COST       TOTAL FUEL $    REMAINING FUEL
     FUEL     PER 1000 KWH    (MILLION $)     (% OF 1966)
     COAL      2.03           26614            93.44
     OIL       8.43           1686             64.80
     GAS       2.39            831             72.72
     NUCLEAR   2.04           2336             86.04
     HYDRO     0.50            378              0.53
     TOTAL     2.05           31845

SUMMARY RESULTS -- 1966 TO 2016

TOTAL ELECTRICITY GENERATED = 359261 BILLION KWH.
AVERAGE PERCENT OF CAPACITY IN USE = 80.3055

   50 YEAR FUEL COST SUMMARY:
       TOTAL COST = $ 809391 MILLION
       AVERAGE COST PER YEAR = $ 16187.8
       AVERAGE COST PER 1000 KWH = $ 2.25293

              RESERVES        RESERVES
     FUEL     IN 1966         IN 2016

     COAL     832             777.45          BILLION TONS
     OIL       50              32.3996        BILLION BARRELS
     GAS      700             509.022         TRILLION CU FT
     NUCLEAR    1.05             .903427      MILLION TONS
     HYDRO    190             190             MILLION KW


     FUEL     % REMAINING OF 1966 RESERVE

     COAL     93.4436
     OIL      64.7993
     GAS      72.7175
     NUCLEAR  86.0407
     HYDRO     1

     READY
```

B. *Description Section*
 1. Reprints or summaries of articles or information about the vent being simulated.
 2. Background data and its source.
 3. Key assumptions and their rationale.
 4. Short bibliography for further reference.
C. *Computer Section* (optional)
 1. Program listing.
 2. Flowchart.

**References — Computer Simulations**

1. Ahl, D.H., (Ed.), *Basic Computer Games*, Morristown, NJ: Creative Computing Press, 1978.
2. Allen, L.E., Allen, R.W., & Ross, Jr., "The Virtues of Nonsimulation Games." *Simulation and Games*, 1970, 1, 319-326.
3. Boocock, S.S., & Schild, E.O. (Eds.), *Simulation Games in Learning*. Beverly Hills, CA: Sage. 1968.
4. Braun, L., "Digital Simulation in Education." *J. Educational Technology Systems*, 1972, 1, 1, 5-27.
5. DeVries, D.S. & Edwards, K.J., "Learning Games and Student Teams: Their Effects on Classroom Process." *American Education Research Journal*, 1973, 10, 307-318.
6. Edwards, K.J., DeVries, D.L., & Snyder, J.P., "Games and Teams: A Winning Combination." *Simulation and Games*, 1972, 3, 247-269.
7. Fletcher, J.L., "The Effectiveness of Simulation Games as Learning Environments." *Simulation and Games*, 1971, 2, 425-454.
8. Inbar, M., "Individual and Group Effects on Enjoyment and Learning in a Game Simulating a Community Disaster." In S.S. Boocock & E.O. Schild (Eds.), *Simulation Games in Learning*. Beverly Hills, CA: Sage, 1968, pp. 169-190.
9. Pate, G.S. & Parker, H.A., Jr., *Designing Classroom Simulations*, Belmont, CA: Fearon, 1973.
10. Spencer, D.D., *Game Playing With Computers*. New York: Spartan Books, 1968.

**References — Electric Power**

1. "Power Industry Statistics for 1971." *Electrical World*, Sept. 15, 1972.
2. *Questions and Answers About the Electric Utility Industry*, Edison Electric Institute, 1973. ■

# GREED: A Game Playing Program With Adjustable Skill Level

## Ronald G. Ragsdale

In the Nov.-Dec. 74 issue of *Creative Computing*, the game of NOT ONE was described and readers were invited to create programs that played the game. In the intervening months a program to play that game has evolved at this institution, but under the name of GREED, which seems to be well suited to the game's characteristics. The rules of the game, as given to the players that request instructions, are shown below:

GREED IS A GAME FOR TWO PLAYERS. ON EACH OF THEIR TEN TURNS, THE PLAYERS THROW THE DICE AND ADD THE RESULT TO THEIR SCORES. PLAYERS MAY THROW THE DICE AS MANY TIMES AS THEY LIKE BUT IF THE FIRST RESULT OF THE TURN IS REPEATED, THEIR SCORE FOR THAT TURN IS ZERO.
THIS PROGRAM PLAYS AT VARYING SKILL LEVELS. IT PLAYS THE BEST AT LEVEL 100 (AVERAGE SCORE ABOUT 300) AND THE WORST AT LEVEL 0 (AVERAGE SCORE ABOUT 150).
GOOD LUCK!

The basic strategy of the GREED program is based on the comparison of the expected gain (probability of not repeating the first roll times the average of the "non-losing" rolls) from a particular roll of the dice with the expected loss (probability of repeating the first roll times the points already accumulated on this turn). As long as the expected gain is larger the program continues "rolling" the dice. A summary of this strategy appears in Table 1.

| If the first roll is a | 2 | 3 | 4 | 5 | 6 | 7 | 8 | 9 | 10 | 11 | 12 |
|---|---|---|---|---|---|---|---|---|---|---|---|
| Keep rolling until the points for the turn exceed | 252 | 123 | 80 | 58 | 44 | 35 | 42 | 54 | 74 | 115 | 240 |

Table 1. The basic GREED Strategy.

In an effort to manipulate the success rate of the program a "skill level" was introduced so that at skill level=100 the program followed the strategy of Table 1, but at skill levels=N, it altered its goal (the lower line of Table 1) by (100-N)%. Thus at skill level=0, the program either played an extremely conservative (all goals=0) or extremely reckless (all goals doubled) game (randomly determined on a per-game basis).

The straightforward application of this basic strategy led to situations where the program behaved in a silly manner, so that the following three sub-strategies were added:

1. The program takes advantage of the fact that it plays last and, on the last turn, stops rolling when it reaches a winning score.

2. In order to temper the reckless play at low skill levels, the program stops "rolling" if it is at least 100 points ahead and has accumulated at least 100 points (250 if the first roll was a 2 or 12) on this turn.

3. In order to adjust to the progress of the game, the setting of the goal for a particular turn is modified if either player is sufficiently far ahead (defined as 50 times the number of remaining turns). If the program is ahead by that amount it halves the calculated goal and if it is behind it sets a lower limit of 50 on the goal.

The additions of the sub-strategies led to an obvious improvement, but it also attenuated the effect of "skill level" by improving the lower levels more than the higher ones.

Because of the uncertainty about the effects of the sub-strategies and because the recently installed DECsystem-10 was not heavily loaded, the GREED program was evaluated against another program. The program used in the comparison was NOT ONE, which appeared in the Mar.-Apr. '75 issue as one of the best programs to be received in response to the previously mentioned invitation.

The strategy of NOT ONE is similar to the basic strategy for GREED, but is based on the number of rolls in a turn rather than the number of points. The strategy is summarized in Table 2.

| If the first roll is a | 2 | 3 | 4 | 5 | 6 | 7 | 8 | 9 | 10 | 11 | 12 |
|---|---|---|---|---|---|---|---|---|---|---|---|
| Keep rolling until the number of rolls = | 18 | 18 | 9 | 9 | 6 | 6 | 6 | 9 | 9 | 18 | 18 |

Table 2. The basic strategy for NOT ONE.

Except for initial rolls of 2 and 12 the two strategies are quite similar; the expected value for 18 rolls is 126, for 9 rolls 63 and for 6 rolls 42. Since 2 and 12 each occur with probability=1/36, the results for the two strategies are essentially the same.

Comparisons were made between the NOT ONE program and eight versions of the GREED program. The eight versions consisted of the basic strategy (shown in Table 1) and all possible combinations of the three sub-strategies. In each comparison, 1000 games were played at each of 11 different GREED "skill levels" ranging from 0 to 100 in multiples of 10.

The results of these comparisons are summarized in Table 3. It can be seen that the basic strategy wins from 34 to 50.5% of the games, while the additions of the sub-strategies can alter the range so that GREED wins a minimum of 50% and a maximum of about 61% depending on the skill level. In any case the difference between the percentage of games won at the highest and lowest skill levels is relatively constant at about 15-20%.

Ronald G. Ragsdale, The Ontario Institute for Studies in Education, 252 Bloor Street, Toronto, Ontario, Canada M5S 1V6.

| Strategies | skill level | | |
|---|---|---|---|
| | 0 | 50 | 100 |
| basic plus 1, 2 & 3 | 50 | 56 | 61 |
| basic plus 1 & 2 | 50 | 56 | 64 |
| basic plus 1 & 3 | 42 | 54 | 60 |
| basic plus 1 | 43 | 51 | 60 |
| basic plus 2 & 3 | 46 | 48 | 50 |
| basic plus 2 | 45 | 47 | 53 |
| basic plus 3 | 36 | 44 | 51 |
| basic | 34 | 48 | 47 |

Table 3. Games won by GREED at various strategy and skill level combinations when playing against NOT ONE. Scores, as percentages, are based on 1000 games at each combination.

The initial reaction to these results was to try and redefine "skill level" in terms of various combinations of sub-strategies, in order to increase the "ability range." Fortunately, serendipity reared its lovely head before this could be accomplished and the solution became much simpler.

The fortuitous event occurred when the program to compare all strategy GREED with NOT ONE was being recoded into regular DECsystem-10 BASIC. During the recoding a test within sub-strategy three was inverted so that 50 points was set as a maximum (rather than minimum) goal when the GREED program was losing by a sufficient amount. The result of the comparison between this version and NOT ONE is shown in Table 4. In this comparison, based on 2000 games at each skill level, the percentage of games won by GREED varies from 31 to 58%.

| Greed skill level | 0 | 10 | 20 | 30 | 40 | 50 | 60 | 70 | 80 | 90 | 100 |
|---|---|---|---|---|---|---|---|---|---|---|---|
| Percentage of games won by GREED | 31 | 34 | 40 | 45 | 50 | 51 | 55 | 55 | 56 | 58 | 55 |

Table 4. Games won by GREED (where strategy 3 is modified) at various skill levels when playing against NOT ONE. Scores, as percentages, are based on 2000 games at each combination.

The final choice was to use the inverted strategy for the lower skill levels (0-59) and the regular strategy for the higher levels (60-100). The hope that this would result in an increased range of ability for GREED was tested by playing 10,000 games against NOT ONE at each of levels 0, 10, 90, and 100. The results indicated a range of about 29 percentage points, as shown in Table 5.

| GREED skill level | 0 | 10 | ... | 90 | 100 |
|---|---|---|---|---|---|
| Percentage of games won by GREED | 30.0 | 34.1 | ... | 59.3 | 59.2 |

Table 5. Games won by the final version of GREED playing against NOT ONE at skill levels of 0, 10, 90, and 100. Scores, as percentages, are based on 10,000 games at each combination.

Other improvements could be made in the GREED strategies, but the variability of the results makes the improvements difficult to detect. One possibility is to let the player insert his strategy as a subroutine and give the results of playing 1000 games against a particular skill level of GREED. This would reduce variability but might reduce interest in the game, as well as limiting the number of possible players (and increasing CPU time). Therefore, the strategies of GREED are unlikely to change in the foreseeable future. □

```
010 DIMENSION S(2)
020 RANDOMIZE
030 PRINT "DO YOU WANT INSTRUCTIONS ?"
040 INPUT N$
050 IF LEFT$(N$,1)="N" GOTO 100
```

Program lines 60-99 print the instructions given in the second paragraph.

```
100 PRINT "WHAT SKILL LEVEL WOULD YOU LIKE ?"
105 INPUT N
122 IF N<100 THEN 124
123 LET N=100
124 LET N=100-N
125 LET I=RND
130 IF I<.5 THEN 140
135 LET N=-N
140 LET S(1)=S(2)=0
141 I=1
145 PRINT "YOUR TURN #" I;
150 GOSUB 400
160 LET T=R1=R9
162 PRINT "FIRST THROW IS " R9;
164 PRINT "ROLL AGAIN?";
166 INPUT N$
168 IF LEFT$(N$,1)="N" GOTO 200
180 GOSUB 400
185 PRINT "ROLL =" R9;
190 IF R1=R9 THEN 195
191 LET T=T+R9
192 PRINT "TOTAL= " T;
193 GOTO 164
195 LET T=0
200 LET S(1)=S(2)+T
202 PRINT
203 PRINT "THIS TURN " T " YOUR TOTAL IS " S(1)
206 PRINT "MY TURN #" I;
210 GOSUB 400
215 PRINT "FIRST THROW IS "R9;
220 LET R1=T=R9
230 LET N1=R1-1
240 LET G=252/N1-R1
241 G=(N*G)/100+G
242 IF (S(2)-S(1))<((10-I)*50) THEN 244
243 LET G=G/2
244 IF (S(1)-S(2))<((10-I)*50) THEN 250
245 IF G<50 THEN 248
246 IF N<60 THEN 249
247 GOTO 250
248 IF N<60 THEN 250
249 LET G=50
250 IF T>=G THEN 285
255 IF I<>10 THEN 257
256 IF S(2)+T>S(1) THEN 290
257 IF T<100 THEN 261
258 IF (T+S(2)-S(1))<100 THEN 261
259 IF 2=R1 THEN 261
260 IF R1<12 THEN 290
261 IF T<250 THEN 265
262 IF (T+S(2)-S(1))>100 THEN 290
265 GOSUB 400
267 PRINT "ROLL=" R9;
270 IF R9=R1 THEN 280
271 LET T=T+R9
273 PRINT "TOTAL=" T
275 GOTO 250
280 LET T=0
281 GOTO 290
285 IF L<>10 THEN 290
286 IF S(2)+T<S(1) THEN 255
290 LET S(2)=S(2)+T
292 PRINT
295 PRINT "THIS TURN " T, "MY TOTAL IS " S(2)
300 LET I=I+1
301 IF I=11 THEN 310
302 GOTO 145
310 IF S(1)<=S(2)+T THEN 320
315 PRINT "YOU WIN"
316 GOTO 330
320 IF S(1)<S(2) THEN 324
321 PRINT "WE TIED"
322 GOTO 330
324 PRINT "SORRY, YOU LOST"
330 PRINT "PLAY AGAIN?"
332 INPUT N$
334 IF LEFT$(N$,1)="N" GOTO 343
336 GOTO 100
343 STOP
400 LET D3=INT(6*RND)+1
410 LET D2=INT(6*RND)+1
420 LET R9=D3+D2
421 RETURN
999 END
```

*Problems for Creative Computing....*

# Seeing is Believing but Simulating is Convincing

by Walter Koetke

Simulation represents one of the more promising areas in which the computer can be used to provide pertinent data on the options facing our society. Many computer related texts, however, pursue the subject no further than Buffon's needle, playing craps or dealing cards. While these are indeed valid examples of simulation, they are not sufficient because: they do not clearly demonstrate a potential connection between simulation results and human decision making; they seem to associate simulation with theoretical problems of mathematics as opposed to real problems of society; and they are too simple to give a feeling for the true complexity of societal simulations. More elaborate simulations are available (such as the variety of material from the Huntington Project), but these are intended for students to execute rather than write. The following two problems are neither outstanding problems of society nor outstanding problems of mathematics. They are offered because they provide the student with the opportunity to write simulations that go just a little further than the standard examples.

**Problem 1: Horse Ranch.** Suppose you are a rancher and own 200 horses. All of your stock is healthy, and that's very important because you expect a buyer to arrive unannounced sometime in the next 10 days. The buyer is looking for 160 healthy horses. You know from experience that if he finds fewer than 160 healthy animals, he won't buy any at all. Clearly your finances could not absorb the complete loss of the entire sale.

Just as you're ready to celebrate the pending sale, you learn that one of your horses has "day cough", a terrible sounding disease of short duration. Day cough lasts exactly one day and immunity to re-exposure results. You know that each sick horse will contact five other horses each day, and each contact has 0.6 chance of transmitting the sickness if the contacted horse is healthy and not immune. Although your situation appears bleak, do you really have such a serious problem? Which of the following three alternatives is your best course of action and why is it best?

a) Stop worrying. There's really very little chance that fewer than 160 horses will be healthy during the next 10 days.

b) Solicit the expensive support of modern medical science. "Cough shots" are available that are supposed to provide instant immunity when given to a healthy horse. However, only 13 horses per day can be inoculated by the local veterinarian, and neither he nor you can tell a healthy horse from a horse already immune. Thus you may innoculate an already immune horse. The only problem in doing this is that one of the cough shots is wasted. Actually, the cough shots have a 90% chance of providing instant immunity and a 10% chance of instantly giving day cough to a healthy horse. For all his expertise and advice, the veterinarian will charge you one healthy horse for each day of service he provides.

c) By calling right away, you can probably convince the buyer to delay his unannounced visit for 10 days. He will still come unannounced, but during the period 10 to 20 days from now rather than during the next 10 days.

**Problem 2: Fish Pond.** As part of a conservation and ecology project, a group of biology students has designed the following controlled experiment: A small pond is polluted with several common types of waste material. Exactly 100 male fish are then introduced to the previously fish free pond. Each day the students carefully net exactly 10 fish. All 10 fish are caught simultaneously and at the same time each day. The netted fish are examined for signs of gill disease, their tails dyed, and they are returned to the pond. The dye used is harmless and completely disappears after 13 full days in the water. If a netted fish is already dyed, it is dyed again so that it too will remain dyed for the next 13 days. The experiment continues until each of the fish netted on any one day all have dye on their tails.

How many days should the students allow to permit "a reasonable chance" for successful completion of the experiment?

A complete solution to this problem, as in many problems of society for which computer simulations might be useful, requires clarification of some human values. The definition of "a reasonable chance" is a personal one, and as such it is based on a wide variety of factors. Discussing the point alone makes this a very worthwhile problem in a classroom setting.

A closely related simulation problem can be described by not stating the number of fish in the pond, but instead specifying the number of days that pass before the 10 fish netted all have dye on their tails. The question then becomes "How many fish were originally in the pond?" For instance, if the experiment described was completed in 32 days, how many fish were originally placed in the pond?

**Problem 3: Superspy.** This one is included just for fun. The problem is really another disguise of the two dimensional random walk.

Each night IBF, the superspy, leaves his daytime refuge and emerges from a secret manhole cover in the center of a city. Being an exceptionally tricky spy, IBF is likely to sneak forward, backward, left or right at every intersection. If IBF happens to accidentally stray 8 blocks from the manhole (in any direction) he is captured by the arch enemies of society, the TUVEFOUT. If IBF happens to return to his manhole, he is safe for another night. What is the probability that IBF will return safely and thus complete a successful mission? What is the average number of blocks traveled during a mission?

# Probability

# COMPUTER SIMULATIONS AND PROBLEM-SOLVING IN PROBABILITY

**John S. Camp**

Probability is a subject that is used in a wide variety of disciplines. Examples of applications can be found in the study of marketing, population planning, system reliability, and even mathematics, itself. The purpose of this paper is to present problems (and solutions) from these areas to show how a computer simulation can be used as a problem-solving strategy in probability.

In probability, problem-solving often involves the use of known theory [P(A|B)=P(ANB)÷P(B)] and/or the study of actual experiments that are designed to suggest or give answers to questions of interest. For this discussion, is it the experimental aspect of probability that will be emphasized, for experiments are at the heart of probability and are what simulations are all about.

## PROBABILITY AS A MOTIVATOR

Probability is an almost guaranteed motivator. People enjoy predicting the outcome of elections, estimating the chance that a particular team will win a world series, or applying the subject to games of chance. As other examples, consider the following:

*Population Planning*

Suppose that you have decided that you want exactly four children in your family. What are the chances that the four children will be boys?

*Marketing*

Assume that you are responsible for marketing packages of bubblegum and to increase sales you enclose a picture of a famous football player in each package. If there are 25 pictures, what is the expected number of packages of bubblegum an individual would have to purchase to acquire a complete set?

*System Reliability*

The figure below is an electrical system that was built by using five components arranged in parallel and two small systems, A and B, arranged in a series, If each component has a 60% chance of lasting 1000 hours, what is the chance that the entire system lasts 1000 hours?

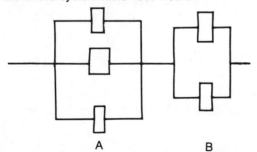

---

Paper delivered at NAUCAL '77, Dearborn, Michigan, Nov. 3-5, 1977. The author is a professor at Wayne State University, College of Education, Detroit, Michigan.

Although the problems could be presented "as is," if they are to be used in the classroom they should probably be introduced with a little flair. In the population example, you might ask, "Why is this an important question to some people? Is there anything wrong with all boys?" Students are usually quite willing to argue the *pros* and *cons* of this issue especially when there are boys and girls in the class.

The bubblegum problem is especially interesting to those students who collect cards. Ask if there are any collectors in your class and ask them "How hard is it to acquire the last card of a set?" You might ask the students for a show of hands for how many think it would take more than 10,000 packs of gum, how many think less than 200, and how many think between 200 and 10,000.

In introducing the exercise on system reliability, you might say that the component is an integral part of say a VOYAGER spacecraft and it is important to increase its reliability.

For each problem, a good strategy is to ask students to guess at the answer *before* attempting to solve it. If there are a wide range of guesses, this will cause students to want to find a solution to determine whose guesses are correct. For these examples, most students will be surprised at the answers.

## EXPERIMENTS AND SIMULATIONS

Probability tells us something about the "long run." For a fair die, we know that on a single toss of the die,

P(3 showing) = 1/6

and so in the "long run" (i.e. many tosses of the die), I expect to see 3 appear about 1/6th of the time. This "long run" aspect of probability can be used to approximate probabilities simply by collecting data on many trials of an experiment.

Actual experiments, however, may be costly as well as time consuming. For example, one could locate 4-child families and determine the ratio of the number that were all boys to the total. In the case of the electrical system, one could build many, turn them on for 1000 hours and determine the rate of success.

An alternative to an actual experiment is a simulation (representation) of the experiment. When a simulation can be conducted by studying arrangements of random numbers, then the computer becomes a powerful problem-solving tool.

## SOME SIMULATIONS

The heart of the simulation process is generating numbers at random. The following two methods are rather standard; BASIC is the language that is used.

Method 1:   Using a String

```
10  DIM A$(10)
20  A$= "0123456789"          The digit string
30  FOR N = 1 TO 100          Generate 100 numbers
40  FOR F = 1 TO 4            Each number has 4 digits
50  Z = INT(10*RND(8)) + 1    An integer between
                                1 and 10 inclusive
60  PRINT A$ (Z,Z);           Print the Zth digit
70  NEXT F
80  PRINT "    ";
90  NEXT N
100 END
```

Note how A$ contains the possible digits of the 4-digit numbers that are generated and printed in lines 30-90. The output is 100 4-digit numbers like:
   1257        9843        0016    ....
where *each* digit of each number has been generated at random. Another method that will generate *individually* produced digits is:

```
10  FOR N=1 TO 100             100 numbers
20  X = 0                      X is the number;
                                 initialize to 0
30  FOR F = 1 TO 4             4 digits in X
40  X = 10*X + INT(10*RND(8))  Successive passes
50  NEXT F                       through F loop "fills" X
60  PRINT X;
70  PRINT " ";
80  NEXT N
90  END
```

Method 2: Generating a number between *a* and *b*

In some systems, $0 \leq RND(8) < 1$ and so
$$0 \leq (b-a) * RND(8) < b-a$$
$$a \leq (b-a) * RND(8) + a < b$$
To generate 100 4-digit numbers, run

```
10  FOR I = 1 TO 100
20  Z = INT (1000*RND(8) + 2000)
30  PRINT Z;
40  NEXT I
50  END
```

Note that $1000 \leq Z < 3000$. The output is 100 4- digit numbers like
   1257        2639        2411    ...
where *each number* has been generated at random (rather than each digit of each number).

### Population Planning Simulation

Depending on the students and their backgrounds, 4-child families can be simulated in a number of ways. One method is to generate 4-digit numbers, as in Method 1 of this section, and for each number, let an even digit represent a boy, an odd digit a girl. Students can count the results. In the run of 100 numbers that follows, there are exactly 6 all even digit numbers and so
$$\frac{6}{100} = .06$$
is an approximation of the probability of having an all boy 4-child family (the exact probability is .0625). Although our approximation is fairly good, in practice one would simulate many more trials to increase the chances that the approximation is close to the true probability.

```
LIST
RNDDIG
10  DIM A$[10]
20  A$="0123456789"           (SAMPLE RUN AT
30  FOR A=1 TO 100             BOTTOM OF PAGE)
40  FOR F=1 TO 4
50  Z=INT(10*RND(8))+1
60  PRINT A$[Z,Z];
70  NEXT F
80  PRINT "  ";
90  NEXT A
100 END
```

If it is not important to display the intermediate results then run a program like the one which follows to *simulate* 1000 families (Here 0 = girl, 1 = boy) and determine if a family contains all boys (product of digits will be 1). The only information that is printed is the approximated probability.

```
10  C = 0                           Start counter
20  FOR I = 1 TO 1000               1000 trials
30  FOR J = 1 TO 4                  4 per family
40  F(J) = INT (2*RND(8))           Generate 0 or 1
50  NEXT J
60  IF F(1)*F(2)*F(3)*F(4)=0 THEN 80  If product 0, then at
                                        least 1 girl. Don't count.
70  C = C+1                         Count number of all
                                      boy families
80  NEXT I
90  PRINT "P(4 BOYS) IS APPROXIMATELY"; C/1000
100 END
```

```
RUN
RNDDIG
 0819   7981   0598   3150   5916   3600✓  3729   9761   6806✓  9971   6710   7968
 7883   1559   6670   6883   3864   0731   5821   6334   0080✓  7868   8275   7807
 9579   3696   0531   5335   7636   4959   5006   4957   0773   0945   2748   8443
 7189   3392   3545   3404   2667   2427   1546   0818   3242   5763   8450   7857
 1347   1806   9215   3326   4755   1135   4575   8989   0309   6394   3465   9619
 9726   1687   5042   0673   4341   7069   4729   2959   6568   4547   6118   3077
 6021   3417   9999   1263   5372   9399   8319   8487   9455   2019   5125   8993
 2866✓  3752   5297   6324   5962   2534   1671   3751   4805   4605   7660   6488✓
 6642✓  9119   2770   5394
```

# PROBABILITY con't...

### Bubblegum Simulation

This particular problem is a good one for computer simulation, for few people know how to calculate the answer directly. Do you?

Here's how we will proceed. Simulate the purchases made by 100 people (Line 50) in the following way:

(1) Initialize a 25-element matrix A to zero to represent the 25 pictures. 0 = picture not purchased; 1 = picture purchased.
(2) Start buying (Line 70). For each purchase, randomly generate an integer Z (Line 80) from 1 to 25 inclusive. Set A(Z,1) = 1 to show that the picture of star Z has been purchased. Check to see if set has been completed (Line 120).
(3) Compute average number of purchases required to complete set.

Here's the program:

```
10  DIM A(25,1),C(1,25),P(1,1),B(100,1),D(1,100)
20  MAT B = ZER
30  MAT C = CON
40  MAT D = CON
50  FOR I = 1 TO 100            100 trials
60  MAT A = ZER
70  FOR J = 1 TO 500            Allow for at most 500 purchases
80  Z = INT (25*RND(8)+1)       Generate integers 1 to 25
90  A(Z,1)=1                    Star Z purchased
100 MAT P = C*A                 Product is the sum of the elements of A
110 B(I,1) = B(I,1) + 1
120 IF P(1,1) = 25 THEN 140     If sum is 25 you have entire collection
130 NEXT J
140 NEXT I
150 MAT P = D*B
160 PRINT "EXPECTED NUMBER OF PURCHASES IS ABOUT"; P(1,1)/100
170 END
```

### System Reliability

The following program, when run, simulates an experiment to approximate the probability that the system described earlier works for 1000 trials. The program uses these facts:

(1) A system made up of components arranged in series, will work if and only if all components work.
(2) A system made up of components arranged in parallel will work when at least one component works.

```
5    S=0
10   FOR I = 1 TO 500             Try 500 systems
20   FOR I = 1 TO 5               5 components
30   C(J) = INT(10*RND(8))+1      Integer between 1 and 10 inclusive
40   IF C(J)> 6 THEN 70           If C(J)=7,8,9, or 10, then C(J) fails
50   C(J) = 1                     Replace C(J) by 1 to mean it works
60   GO TO 80
70   C(J) = 0                     Replace C(J) by 0 to mean it fails
80   NEXT J
90   IF C(1)=1 or C(2)=1 or C(3)=1 THEN 110   Does subsystem A work?
100  GO TO 140
110  IF C(4)=1 or C(5)=1 THEN 130              Does subsystem B work?
120  GO TO 140
130  S = S+1                                   If both work, count as a success
140  NEXT I
150  PRINT "RELIABILITY ABOUT" S/500
160  END
```

The reader should note that there are other ways to accomplish the test above. For example, lines 90-130 could be replaced with

```
90  IF (C(1) + C(2) + C(3)) * (C(4) + C(5)) = 0 THEN 110
100 S = S+1
110 NEXT I
```

## CONCLUSIONS

The three examples presented in this paper illustrate how computer simulations can be used to "problem-solve" in probability.

Teachers need not delay the study of probability just because their students lack theory. The foundation of probability is experiments and young children can be introduced to questions about chance events and can conduct experiments to suggest answers. Upper elementary school children can study coins, dice, cards, and other objects by actually experimenting with them. As the children get older and the experiments become more complex, simulations become a welcomed relief. Introduce simulations gently and with much practice so that the concept is understood. One approach is to devise simulations using tables of random numbers and then lead to the computer when appropriate. Good luck. ∎

"Null Character"

*Problems For Creative Computing....*

# Probability

by David C. Johnson, University of Minnesota

**PREREQUISITES**
Basic notions of probability including P(E) = 1 - P(E').

**DISCUSSION**
The CAMP project, University of Minnesota, has conducted research and development activities on the use of the computer as a problem-solving tool in school mathematics grades 7 - 12. The following problem while a "take-off" on the classical Birthday Problem has a number of real applications relative to expected occurrence of given events:

*from the everyday:* What is the probability of at least two girls wearing the same style and color outfit at a party with say 30 girls invited (assuming some given number of basic styles, say g, and number of colors, c, or gxc different outfits -- e.g., if g=8 and c=10 then gxc=80.)

*to a problem in manufacturing and sales:* How many different styles and colors are needed to give a low probability (p<.10) to the event that two or more families in the same neighborhood (of 100 families) will purchase identical automobiles (if, on the average, 10% of the families purchase a new XXX each year.) Note: the problem is actually a little more complex than this, but the statement should provide a general idea -- the assumption is also made that people like their cars to be different.

**PROBLEM**
The situations posed above can be stated in purely mathematical terms. The three problems posed below appear in the CAMP exercises in the book *Elements of Probability* by Robert J. Wisner, Scott, Foresman and Company, 1973, appropriate for a high school course in probability.

1. First, to warm up — write a computer program to calculate the probability that at least two people in a group of n people will have the same birthday. (Hint: since the $365^n$ may become very large, you will have to design a procedure to calculate 365/365 x 364/365 x 363/365 x 362/365 . . .)
2. Now for *the* problem:
   a. Write a general program which considers n people selecting an alternative (or having a characteristic) from m equally likely possibilities. What is the probability that at least two will select the same alternative? You might think of this as n people each picking a number between 1 and m and writing it down -- what is the probability that at least two will pick the same number? (Of course, m>n or the probability is 1.)
   b. Use your program to determine how many numbers you will need to use at a party with 12 people to give yourself better than a 50-50 chance of having two pick the same number (you might like a probability of about .75). Do you see the similarity between this and the manufacturing problem? Actually conduct the number experiment with some groups of friends -- how well do the experimental results agree with the mathematical? Note that the experiment can be done by asking your friends to pick a favorite color or object from a list with m items -- but, you have to be cautious here; not all of the items may be equally liked by your friends -- what does this do to your computation?

# Compounding

by Charles A. Reeves, Florida State University

▶ Try to fold a sheet of paper onto itself as many times as you can (i.e., fold it in half, then in half again, then again, etc.). What is the largest number of folds you can make? Someone has claimed that it is impossible to make more than 8 folds, no matter what size you start with!

But imagine for a moment that it *is* possible to fold it over onto itself a large number of times. The thickness of one sheet of notebook paper is about .004 inches. If you could fold it 50 times, how high would the stack be?

▶ Your rich uncle deposited $1000 in a savings account for you the day you were born. The account draws 6% simple interest, and the earnings are added back into the account each year. But your uncle didn't tell you about this — you found out when his will was read. He died when you were forty years old — how much did you get?

For those who want more: Same problem as above, but the interest rate is ½% each month instead of 6% per year. How much more money, if any, would you get this way?

▶ Consumer prices rose an average of 8.8% during 1973. Let's round this off to 9%, and assume that prices continue to go up this much *every* year.

Pick out an item that you think you might like to buy when you're an adult, and for which you know the present price. Write a program that will report to you how much the item will cost in the year 2000 AD.

▶ Your father gives you a penny as a gift on your first birthday. He promises to double the amount of the gift each year until you reach your 21st birthday. How much will you get from him on this birthday?

For those who want more: Have the computer print the amount you will receive on the 21st birthday, and also the *total* amount you will have gotten through the years.

▶ Erie County in upstate New York is one of the most heavily polluted areas in the United States. In a study of the residents of the county it was found that the number of people dying from respiratory diseases is *doubling* every five years. In 1950 there were 263 deaths attributed to respiratory diseases. How many deaths will there be in the year 2050 AD, assuming this same rate of increase every 5 years?

▶ The population of the world increases almost 2% each year over what it was the year before. In 1970, the world population was about 3.6 billion, or 3,600,000,000.

Have the computer calculate what the world population will be in the year 2000 AD.

▶ A salmon starts a 100 mile journey upstream to the placid lake where she was born. Each day she is able to swim 3 miles upstream, but each night when she sleeps she is pushed 2 miles back downstream. Exactly how many days will it take her to reach the quiet spawning grounds?

▶ The bristleworm can reproduce by splitting itself into 24 segments, each of which grows a new head and a new tail. What is the maximum number of bristleworms that could be obtained in this fashion, starting with only 1 worm, after ten "splittings"?

# WHAT WILL HAPPEN IF...?

*Glenda Lappan and M.J. Winter*

*This article describes a method of teaching the concept of expected value through an experiment - conjecture - explain mode which uses the computer to simulate repetitions of games. The games, the rationale for choosing those particular games, and the computer programs to simulate the games are given in detail.*

### WHAT WILL HAPPEN IF...?

In the 1600's a French gambler, the Chevalier de Mere, had a run of "bad luck". The dice turned against him and he suddenly found himself in the uncomfortable position of losing money! For some time the Chevalier had been systematically winning by betting (at even odds) that in four tosses of a die he would obtain a six at least once. When his source of takers dried up, he changed the bet. He now bet that in 24 tosses of a pair of dice, he would obtain a double-six at least once. Since 4/6 = 24/36, he assumed he had an equally profitable bet. To his surprise he began to lose. In his search to find out why, he consulted Blaise Pascal (1623-1662) who in turn discussed the problem with Pierre de Fermat (1601-1665), two of the greatest mathematicians of all time. From the analysis of de Mere's problem a new branch of mathematics was born. Born of a desire to understand the behavior of systems that cannot be entirely controlled.

Just as the Chevalier started his study of probability with a misunderstanding about the probabilistic basis of his new bet, many students start their study of probability with misconceptions about the nature of a probabilistic statement. Their prior study of mathematics has dealt (at least in their minds) with exact answers. Students expect probability to deliver an exact description of what will happen. If a fair coin is tossed 10 times you will get exactly 5 heads. Consequently, (the students' reasoning goes) if you have tossed it 9 times and have gotten h,t,h,h,t,h,t,h,t, then of course the next toss has a high probability of being a t. If you ask a class on day one which outcome is more likely to happen (h,h,h,h,h,h,h,h,h,h) or (h,t,t,h,h,t,h,h,t,t), many of the students will choose the latter because 5 t's and 5 h's are more likely than 10 t's. True, but not relevant to the question asked. Each of the outcomes has probability $1/2^{10} = 1/1024$ of happening. Let us give another example from a game situation. If the probability of winning a game which pays even money is 3/7, the student expects to be exactly $1 behind at the end of 7 games. The student assumes that you win exactly 3 and lose 4. So you win $3 and lose $4, for a total loss of $1. Theoretically he is correct. It is in the belief that this theoretical expected value determines what will happen for each 7 games that the student is in error. The authors feel that students need to have their "faith" in what probabilistic statements mean shaken. To accomplish this, students need to do a great deal of experimentation before being exposed to a theoretical discussion of certain aspects of probability. That is, they need to explore a situation until they have some feeling for what is happening or until, as de Mere, they are puzzled enough to understand how the system works. The computer is an excellent tool for this type of experimentation. In the time it takes students to repeat an experiment 100 times, a computer may simulate the experiment 10,000 times. The thinking required to set up a computer simulation is in itself useful in understanding the experiment. A useful sequence of activities for students might look like this:

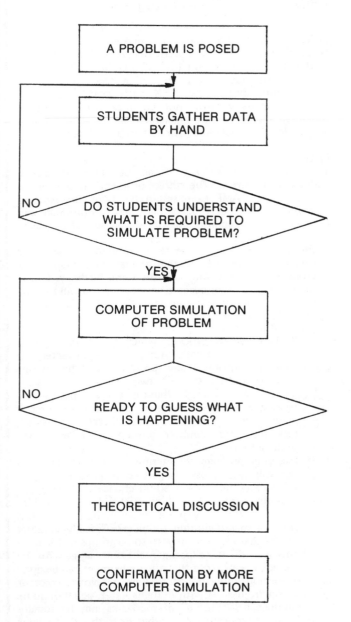

---

Paper first delivered at AEDS Convention, May 16-19, 1978, Atlanta, GA.

The rest of this paper describes a sequence of laboratory and hands-on computer activities designed to assist in the teaching and learning of one important aspect of the probability of game situations — expected value. These activities are divided into two groups.

The first group comprises several games of "chance". The games, with names such as Gambler's Choice, Over and Under, Coins in a Pocket, have been programmed in BASIC, using the RND function. The outcomes depend on a roll of dice, a random sample, or a coin toss. Students play the games several times, as many as they think necessary in order to predict the result if each game were to be played 10,000 times. The games are designed so that for some, a prediction can easily be made, for others, not. A classroom discussion of expected value enables the students to verify the accuracy of some of their predictions. Extended discussion of the expected value of independent variables leads to successful analysis of the remaining games.

The second group develops intuitively the expected number of trials until first success, preparing the way for a classroom derivation of the formula 1/p. Simulation of completing a collection of baseball cards allows the students to compare a complicated example with the results predicted by theory.

## ACTIVITIES

The unit is introduced by simulating **COINS IN A POCKET**. The story is that a newspaper costs 5¢. A customer has 5 pennies and a dime in his pocket and offers to pay for the paper by letting you, the vendor, select, at random, 2 of the 6 coins. Using marked chips, the students repeat the selection of coins 10 times and calculate the total value of the coins selected. One student might obtain 2 pennies 8 times and a dime and penny 2 times for a total of 38¢. Based on these 10 trials, the student would predict that after 100 trials she would have $3.80; in other words the average of 3.8¢ per trial would hold.

To actually repeat the game 100 times would be time-consuming, so we look for a way to do it by computer. On our machines, Tektronix 4051's, using BASIC, RND (-1) produces a random variable between 0 and 1, (the (-1) ensures that repetitions will produce different results). 6*RND(-1) lies between 0 and 6; so that X = INT(6*RND(-1)) produces a variable equally likely to take on the values 0,1,2,3,4,5. We identify selecting a coin with generating X; if X = 0 we'll say we selected the dime. The program we use is:

```
100 PRINT "TAKING TWO COINS AT RANDOM FROM A
    DIME AND 5 PENNIES"
110 PRINT
120 PRINT "HOW MANY TIMES SHALL WE REPEAT?"
130 INPUT N
140 U=O
150 FOR K=1 TO N
160 X=INT(6*RND(-1))
170 IF X=O THEN 220
180 Y=INT(5*RND(-1))
190 IF Y=O THEN 220
200 U=U+2
210 GO TO 240
220 U=U+11
240 NEXT K
250 PRINT "AVERAGE VALUE OF THE TWO COINS IS
    ";U/N;" CENTS"
260 END
```

Running this program for N = 100 leads most students to predict that after 1000 times the total amount would be close to 5000 cents. Coins in a Pocket is a good game situation to start with because the average value becomes apparent after a few runs of 100 trials. In the next sequence of games it is much harder to decide on the average value.

**GAMBLER'S CHOICE:** A gambler has a choice of two games. The first costs $10 to play; 3 dice are rolled and the player receives the sum of the numbers rolled in dollars. The second game costs $12 to play; 2 dice are rolled and the player receives the product of the numbers rolled in dollars. The program plays both games the same number of times.

```
100 PRINT "GAMBLERS CHOICE"
110 PRINT
120 PRINT "GAME 1 COSTS $10 TO PLAY. 3 DICE ARE
    ROLLED."
130 PRINT "YOU GET THE NUMBER OF DOLLARS
    EQUAL TO THE SUM OF THE DICE"
140 PRINT
150 PRINT "GAME 2 COSTS $12 TO PLAY. TWO DICE
    ARE ROLLED. YOU GET THE"
160 PRINT "NUMBER OF DOLLARS EQUAL TO THE
    PRODUCT OF THE DICE"
170 PRINT
180 PRINT "HOW MANY TIMES DO YOU WANT TO PLAY
    EACH GAME?"
190 INPUT N
200 I=0
210 J=0
220 FOR K=1 TO N
230 X=INT(6*RND(-1))+1
240 Y=INT(6*RND(-1))+1
250 Z=INT(6*RND(-1))+1
260 I=I+X+Y+Z-10
270 J=J+X*Y-12
280 NEXT K
290 PRINT "IN GAME 1 YOUR WINNINGS TOTAL ";I;"
    AVERAGE PER GAME=";I/N
300 PRINT
310 PRINT "IN GAME 2 YOUR WINNINGS TOTAL ";J;"
    AVERAGE PER GAME =";J/N
320 END
```

Typical outputs are:

*AVERAGE WINNINGS PER GAME*

| N | Game 1 | Game 2 |
|---|---|---|
| 1000 | .58 | .41 |
| 1000 | .45 | .20 |
| 5000 | .54 | .35 |

For the game one the students might say, "There appears to be an average value and it appears to be in the low 50's." For game two the data presents a much more confusing picture.

**OVER AND UNDER:** This game has been popular at fundraising events. Two dice are rolled; the player can bet that the sum of the numbers showing will be under 7, equal to 7, or over 7. Bets on over and under each pay even money; 7 pays four times the bet. If you always bet over (or under) what will your winnings be after 100 games? After 10,000 games? Will you do better if you always bet on 7? (Losing less *is*, mathematically, doing better!) For a series of 100 games, the outcome is clear, but the amount of the average loss is not clear.

```
100 DIM W(3)
110 PRINT "OVER AND UNDER"
120 PRINT "HOW MANY GAMES DO YOU WANT TO
    PLAY?"
130 INPUT N
140 PRINT "HOW DO YOU BET? FOR UNDER, ENTER
    -1; FOR 7 EXACTLY, ENTER 0;"
150 PRINT "FOR OVER, ENTER 1"
160 INPUT B
170 W(1)=O
171 W(2)=O
173 W(3)=O
180 FOR J=1 TO N
190 C=INT(6*RND(-1))+1
200 D=INT(6*RND(-1))+1
210 IF C+D  7 THEN 250
220 IF C+D=7 THEN 270
230 W(3)=W(3)+1
240 GO TO 300
250 W(1)=W(1)+1
260 GO TO 300
270 W(2)=W(2)+1
300 NEXT J
310 IF B=-1 THEN 360
320 IF B=O THEN 390
330 G=W(3)-W(2)-W(1)
340 GO TO 430
360 G=W(1)-W(2)-W(3)
370 GO TO 430
390 G=4*W(2)-W(1)-W(3)
430 PRINT "YOUR TOTAL WINNINGS ARE $";G
440 PRINT "AVERAGE WINNINGS PER GAME ARE
    $";G/N
450 END
```

| BETTING ON 7 | | BETTING ON OVER | |
|---|---|---|---|
| Number of Games | Average Winnings | Number of Games | Average Winnings |
| 100 | −.25 | 100 | −.22 |
| 100 | −.35 | 100 | −.20 |
| 100 | −.20 | 100 | −.08 |
| 100 | 0 | 100 | −.12 |
| 100 | −.5 | 100 | −.20 |

After the students have played these games and have tried to predict the outcome for 10,000 or 100,000 trials, we emphasize that their predictions are made on the basis of average winnings and then consider the following situation: Let 2 coins be tossed. If they both show heads, you will win $5. If one is heads and one tails, you will win $2, but if they both show tails, you will lose $20. Suppose the game has been played 1000 times with the outcomes:

Heads, Heads 200 times
Heads, Tails 721 times
Tails, Tails 79 times

Your winnings will be

$$\text{winnings} = 5 \times 200 + 2 \times 721 + (-20) \times 79$$

Your average winnings per game will be

$$\text{average winnings} = \frac{\text{winnings}}{1000} =$$

$$5 \times \left[\frac{200}{1000}\right] + 2 \times \left[\frac{721}{1000}\right] + (-20) \times \left[\frac{79}{1000}\right] = .862.$$

The bracketed terms, $\left[\frac{200}{1000}\right]$, $\left[\frac{721}{1000}\right]$ and $\left[\frac{79}{1000}\right]$ are the relative frequencies of 2 Heads, Heads and Tails, and 2 Tails, respectively. Over a great number of trials, these relative frequencies will approach the theoretical probabilities of these events. When the relative frequencies are replaced by probabilities, we call the average winnings, the expected value of the game.*

Expected Value = E = payoff × probability + payoff × probability + ...

Now we will look at the games played and compute theoretical expected values. For coins in a pocket, the probability of 2 pennies is $\frac{10}{15}$ the probability of a dime and penny is $\frac{5}{15}$. Thus the expected value is

$$2 \times \frac{10}{15} + 1 \times \frac{5}{15} = \frac{75}{15} = 5 \text{ cents}.$$

This confirms the student predictions based on the simulations.

For Over and Under, the probability of a 7 is 6/36 = 1/6; the probability of over is 15/36; the probability of under is also 15/36. For the games, assuming a $1 bet,

E (bet on 7) = 4 × 1/6 + (−1) × 5/6 = −1/6
= −.1666
E (bet on over) = 1 × (15/36) + (−1) × (21/36)
= −1/6 = −.1666

The student predictions based on the simulations cited earlier might be −.26 for a bet on 7 and −.16 for a bet on under. After the theoretical discussion, the students might decide to simulate the experiment again using the computer for a larger number of trials.

For Gambler's Choice I and II, the possible outcomes can be listed and their relative frequencies (i.e. theoretical probabilities) can be determined. For example in Game 1, we get the following theoretical frequencies

| Sum | 3 | 4 | 5 | 6 | 7 | 8 | 9 | 10 | 11 | 12 | 13 |
|---|---|---|---|---|---|---|---|---|---|---|---|
| Frequency | 1 | 3 | 6 | 10 | 15 | 21 | 25 | 27 | 27 | 25 | 21 |

| 14 | 15 | 16 | 17 | 18 |
|---|---|---|---|---|
| 15 | 10 | 6 | 3 | 1 |

$$E \text{ (game 1)} = (-7) \times \left(\frac{1}{216}\right) + (-6) \times \left(\frac{3}{216}\right) + (-5) \times \left(\frac{6}{216}\right)$$
$$+ (-4) \times \left(\frac{10}{216}\right) + (-3) \times \left(\frac{15}{216}\right) \ldots \text{etc.}$$

E (game 1) = .50

These calculations are tedious and serve as a motivation for the theoretical analysis of expected value of independent events. If X,Y and Z are the values showing on a die, then E(X + Y + Z) = E(Y) + E(Z) and E(XY) = E(X)E(Y). The expected value on the face of a die is $\frac{1+2+3+4+5+6}{6} = 3.5$. For game 1, subtracting the $10 to play, the expected value is 3.5 + 3.5 + 3.5 − 10 = .5 or $.50. For game 2 it is (3.5) × (3.5) − 12 = 12.25 − 12 = .25. From the simulations, the former value was predicted by many students; the .25 did not become apparent even after 10,000 trials.

## TRIALS UNTIL FIRST SUCCESS

One of the first applications of expected value is determining the mean number of trials until first success. The students first use the computer to simulate several examples. The program used is:

```
LIST
100 PRINT "NUMBER OF TRIALS UNTIL FIRST
        SUCCESS"
110 PRINT "ENTER P, THE PROBABILITY OF
        SUCCESS, AS A DECIMAL"
120 INPUT P
130 PRINT "ENTER N, THE NUMBER OF
        EXPERIMENTS"
140 INPUT N
150 T=O
160 FOR J=1 TO N
170 K=O
180 X=RND(-1)
190 K=K+1
200 IF X < P THEN 220
210 GO TO 180
220 T=T+K
230 NEXT J
240 PRINT "AVERAGE NUMBER OF TRIALS UNTIL
        SUCCESS IS ";T/N
250 END
```

Then if p is the probability of success and q = 1 − p the probability of failure, the expected number of trials is

$$E = 1 \cdot p + 2q \cdot p + 3q^2 \cdot p + 4q^3 \cdot p + \ldots,$$

where $1 \cdot p$ stands for 1 trial × probability of success, $2 \cdot q \cdot p$ stands for 2 trials × probability of 1 failure then success, $3 \cdot q^2 \cdot p$ stands for 3 trials × probability of 2 failures then success. A standard manipulation, computing $E - qE$, summing the geometric series and solving for E, leads to

$$E = \frac{1}{1-q} = \frac{1}{p}$$

If the probability of success is 1/6, e.g., as in rolling a 4, then it will take on the average $1/p = 6$ tries until the first success, i.e., until the first 4.

**BASEBALL CARDS:** An application of this result is the number of baseball cards needed before acquiring a complete set. We assume there is an unlimited supply of N different baseball cards. The expected number of cards necessary to have a complete set is

$$\frac{N}{N} + \frac{N}{N-1} + \frac{N}{N-2} + \cdots + \frac{N}{1} = N(1 + \frac{1}{2} + \frac{1}{3} + \cdots + \frac{1}{N}).$$

For N = 10, E = 29.29. For N = 50, E = 224.96. The program listing is

```
100 DIM C(100)
110 PRINT "BASEBALL CARDS. THERE ARE N CARDS
        IN THE SERIES. ENTER N"
120 INPUT N
130 PRINT "FROM HOW MANY TRIALS DO YOU WANT
        TO COMPUTE THE AVERAGE?"
140 INPUT K
150 S=O
160 FOR I=1 TO K
170 D=O
180 FOR J=1 TO N
190 C(J)=O
200 NEXT J
210 X=INT(N*RND(-1))+1
220 C(X)=C(X)+1
230 IF C(X)=1 THEN 250
240 GO TO 210
250 D=D+1
260 IF D=N THEN 280
270 GO TO 210
280 T=O
290 FOR J=1 TO N
300 T=T+C(J)
310 NEXT J
320 PRINT T;" CARDS"
330 S=S+T
340 NEXT I
350 PRINT "AVERAGE NUMBER OF CARDS IS ";S/K
360 END
```

Sample output from the BASEBALL CARDS program is:
```
BASEBALL CARDS. THERE ARE N CARDS IN THE
SERIES. ENTER N
10
FROM HOW MANY TRIALS DO YOU WANT TO COM-
PUTE THE AVERAGE?
15
 29 CARDS
 31 CARDS
 34 CARDS
 17 CARDS
 64 CARDS
 44 CARDS
 20 CARDS
 30 CARDS
 23 CARDS
 38 CARDS
 30 CARDS
 16 CARDS
 31 CARDS
 23 CARDS
 28 CARDS
AVERAGE NUMBER OF CARDS IS 30.5333333333
```

Let us return to the Chevalier de Mere's problem and compute the expected winnings of a bet of a louis d'or in each situation.

The probability of at least one six in four rolls of a fair die is $1-(5/6)^4 = .5177$, so that the expected winnings are:

$$E = 1 \times .5177 + (-1) \times .4823 = .0354.$$

But, the probability of a double six in twenty four rolls of a pair of fair dice is

$$1 - \left(\frac{35}{36}\right)^{24} = .4914$$

so that the expected winnings are

$$E = 1 \times .4194 + (-1) \times .5086 = -.0172$$

Computer simulations make one wonder how much time the Chevalier spent gambling since the difference in the two bets is not perceivable in 10,000 trials!

*For this example,

$$E = 5 \times \left(\frac{1}{4}\right) + 2 \times \left(\frac{1}{2}\right) + (-20) \times \left(\frac{1}{4}\right)$$

$$= -\frac{11}{4} = -3.75$$

In the (very) long run you will average a loss of $2.75 per game.

# Two Million Frantic Frenchmen:
# A study in probability

### N.B. Winkless, Jr.

Young Stanislaus heard his printer grinding, and went bounding up the stairs to his room. His father was sitting at the machine, happily watching the action. "Ah ha! Caught ya!" said Stan. "I hope you're going to buy me some paper and a new ribbon."

"C'mon, kid," said his Pop." It's only a two-page run, program and all."

"Whatsit?"

The machine ground out a final line and stopped. "Take a look," said his father.

And he explained: this is an exercise in probabilities. Suppose you're flipping a coin. Since HEADS and TAILS are equally probably, you'd expect each to come up the same number of times — maybe not immediately, but **after a while**, as you keep on flipping.

In the 17th century, a mathematician named Edmund Borel did a calculation to determine what "after a while" really means. He said that if the two million residents of Paris all started flipping coins, and flipped them at the rate of once per second, and if each stopped only when he had flipped into "equipartition," an equal number of HEADS and TAILS, there would still be a thousand games going after ten years.

"I thought I might check that out," said Pop.

"Hmm," said Stan." Let me see your program.

Pop typed LIST and there it was.

At 91, Pop explained, are the data to start with. Consider: if you flip a coin twice, there are four possible combinations of results, two HEADS, two TAILS, one HEAD and one TAIL, or one TAIL and one HEAD. (If he'd been writing it, he'd have said HH, TT, HT, or TH.) Now, HT and TH represent "equipartition," and the flippers who get those results quit. But HH and TT are misses, **half** of the four possible combinations, and the players who get those continue to flip. So in our data at line 91, we make that **half** a probability factor by letting X equal 1 and Y equal 2, for the X over Y, as you see it at line 100. Clear? Still at 91, P is of course the number of Parisians at play; O is to remember the Original number.

"Pop," said Stan, "I think this is simpler than you make it sound."

"At 92, 95 I print my headings. At 100, I'm counting **pairs** of flips, then reducing P to a new quantity of survivors by multiplying P by the probability factor X/Y; then I figure what percentage those survivors are in relation to the Original number, and I call that N for Net.

N.B. Winkless, Jr., 11745 Landale St.,
No. Hollywood, CA 91607.

"At 102-108 I'm rounding out the value of P, so that the moire between hexadecimal and decimal doesn't give me fractional Frenchmen.

"Whew!" said Stan.

"And dropping the digits beyond the second decimal place in the N values. Very handy, this formatting in Micropolis Basic."

"Yes, yes," said Stan. "But what are you doing at 120?"

"Giving new values to X and Y, the probability factors. Now, when we do the printout…"

"Hold it," said Stan. "How do you know X is supposed to become Y plus one, and Y become X plus one?"

"Take my word for it," said Pop. "Now, when we do the printout…"

"HOLD IT," said Stan. "I understand everything else you've got there. But how did you light on this X and Y stuff?"

Pop smiled. "Proud of you, son. That is indeed the hard part. I'm sure there must be a standard mathematical routine for it, but —"

"I know, Pop, I know. You were mathematically deprived as a kid. Meanwhile, how'd you do it?"

"I got a pattern by Brute Force." Pop showed his notes:

| Pairs Of Flips | Outcomes | Hits | Ratio, Non-Hits To Outcomes |
|---|---|---|---|
| 1 | HH TT HT TH | * * | 2/4 |
| 2 | HHHH TTTT | -- | |
|   | HHHT TTTH | -- | |
|   | HHTH TTHT | -- | |
|   | HHTT TTHH | * * | 6/8 |
| 3 | HHHHHH TTTTTT | -- | |
|   | HHHHHT TTTTTH | -- | |
|   | HHHHTH TTTTHT | -- | |
|   | HHHHTT TTTTHH | -- | |
|   | HHTHHH TTHTTT | -- | |
|   | HHTHTH TTHTHT | -- | |
|   | HHTHHT TTHTTH | -- | |
|   | HHTHTT TTHTHH | * * | |
|   | HHHTHH TTTHTT | -- | |
|   | HHHTTH TTTHHT | -- | |
|   | HHHTHT TTTHTH | -- | |
|   | HHHTTT TTTHHH | * * | 20/24 |

"Well?" said Pop, as Stan stood studying. "Do you see the pattern?"

"Wot?"

"Whew!" said Pop. "I was afraid it would be obvious to you. It wasn't to me."

"Wot?" said Stan.

"You see, I've made the X/Y ratio the probability factor. Well, that means that those fractions — the series 2/4, 6/8, 20/24 — are X/Y. Reduce the fractions and what do you get? **You get 1/2, 3/4, 5/6…**"

"Hmm," said Stan. "And it goes on — 7/8, 9/10…?"

"Brute Force says yes. So that's what's happening at 120."

"But why?"

"Beats me. Maybe they'll tell you at school."

Stan looked at the printout. "When does Borel say we'll run out of flipping Frenchmen?"

"He says that after a thousand years, there'll still be ten games going."

"Hmmm. We'd run out of paper before then."

```
1 PRINT CHAR$(6)
5 PRINT: PRINT
9 PRINT CHAR$(16)
10 PRINT "BOREL'S BUSY FRENCHMEN"
20 PRINT: PRINT "TWO MILLION FRENCHMEN"
30 PRINT "START FLIPPING SOUS, ONE FLIP"
40 PRINT "PER SECOND.  WHEN ONE HAS GAINED"
50 PRINT "'EQUIPARTITION' -- WHEN ONE'S HEADS"
60 PRINT "EQUAL ONE'S TAILS — HE QUITS."
70 PRINT "QUESTION: DO THEY DROP AWAY QUICKLY?"
90 PRINT
91 X=1: Y=2: P=2*10^6: O=P
92 PRINT "PAIRS          NUMBER OF        % OF"
95 PRINT "OF FLIPS       SURVIVORS        ORIGINALS"
100 T=T+1: P=P*X/Y: N=P*100/O
102 P=INT(P+.5)
105 P$=FMT(P,"ZZZZZZZV")
108 N$=FMT(N,"ZZV.99")
110 IF M=100 THEN PRINT CHAR$(16)
115 IF N<1 THEN 1000
120 X=Y+1: Y=X+1
130 IF T<30 THEN PRINT T;TAB(15);P$;TAB(29);N$
140 IF T>=30 THEN GOSUB 200
150 IF T/30=INT(T/30) THEN M=T/30: PRINT M;TAB(15);P$;
155 IF T/30=INT(T/30) THEN PRINT TAB(29);N$
157 IF M=10 THEN PRINT CHAR$(15)
160 GOTO 100
200 IF J=0 THEN PRINT "MINUTES     SURVIVORS     % OF ORIGINALS":J=1
210 RETURN
1000 PRINT "AFTER ";M;" MINUTES, "
1010 PRINT P$;" FRENCHMEN ARE STILL SHOOTING."
1020 END
READY    BOREL'S BUSY FRENCHMEN

TWO MILLION FRENCHMEN
START FLIPPING SOUS, ONE FLIP
PER SECOND.  WHEN ONE HAS GAINED
'EQUIPARTITION' -- WHEN ONE'S HEADS
EQUAL ONE'S TAILS — HE QUITS.
QUESTION: DO THEY DROP AWAY QUICKLY?
```

| PAIRS OF FLIPS | NUMBER OF SURVIVORS | % OF ORIGINALS |
|---|---|---|
| 1 | 1000000 | 50.00 |
| 2 | 750000 | 37.50 |
| 3 | 625000 | 31.24 |
| 4 | 546875 | 27.34 |
| 5 | 492188 | 24.60 |
| 6 | 451172 | 22.55 |
| 7 | 418945 | 20.94 |
| 8 | 392761 | 19.63 |
| 9 | 370941 | 18.54 |
| 10 | 352394 | 17.61 |
| 11 | 336376 | 16.81 |
| 12 | 322360 | 16.11 |
| 13 | 309962 | 15.49 |
| 14 | 298892 | 14.94 |
| 15 | 288929 | 14.44 |
| 16 | 279900 | 13.99 |
| 17 | 271668 | 13.58 |
| 18 | 264122 | 13.20 |
| 19 | 257171 | 12.85 |
| 20 | 250742 | 12.53 |
| 21 | 244772 | 12.23 |
| 22 | 239209 | 11.96 |
| 23 | 234009 | 11.70 |
| 24 | 229134 | 11.45 |
| 25 | 224551 | 11.22 |
| 26 | 220233 | 11.01 |
| 27 | 216155 | 10.80 |
| 28 | 212295 | 10.61 |
| 29 | 208635 | 10.43 |

| MINUTES | SURVIVORS | % OF ORIGINALS |
|---|---|---|
| 1 | 205158 | 10.25 |
| 2 | 145366 | 7.26 |
| 3 | 118772 | 5.93 |
| 4 | 102892 | 5.14 |
| 5 | 92048 | 4.60 |
| 6 | 84041 | 4.20 |
| 7 | 77816 | 3.89 |
| 8 | 72795 | 3.63 |
| 9 | 68637 | 3.43 |
| ⋯ | ⋯ | ⋯ |
| 102 | 20399 | 1.01 |
| 103 | 20309 | 1.01 |
| 104 | 20219 | 1.01 |
| 105 | 20129 | 1.00 |
| 106 | 20039 | 1.00 |

```
AFTER  106  MINUTES,
    20000 FRENCHMEN ARE STILL SHOOTING.
```

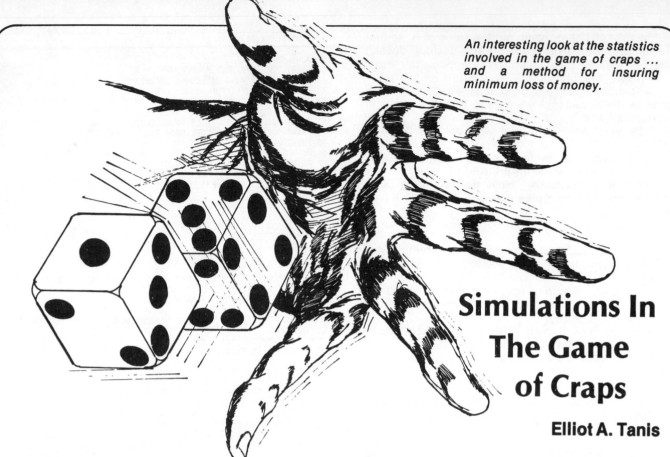

*An interesting look at the statistics involved in the game of craps ... and a method for insuring minimum loss of money.*

# Simulations In The Game of Craps

### Elliot A. Tanis

## Introduction

Almost every state in the United States permits some form of gambling - bingo, horse racing, jai alai, casinos, or lotteries. In some games of chance, the expected loss per play for a gambler is large, in other games the expected loss per play is small.

In this paper we shall use simulation techniques to examine the probability of losing and the expected number of plays before losing $5 in a casino game that is almost fair.

## Craps

In the game of craps a pair of dice is rolled and the sum of the spots is observed. The player wins with a 7 or 11, loses with a 2, 3, or 12 on the first roll. If any other number comes up - that's the players "point" and the player continues to roll the dice. If the player's "point" comes up again before a 7 is thrown, the player wins, otherwise the player loses.

## Probability of Winning in Craps

The probability that the player wins when rolling the dice in craps is 0.49293. Thus the probability of losing is 0.50707. Since the probability of winning is almost equal to 0.5, this game is almost fair.

---

Elliot A. Tanis, Math Dept., Hope College, Holland, MI 49423.

The game of craps can be simulated on the computer to illustrate empirically that the probability of winning is 0.49293. A program that can be used for this simulation follows along with a typical run (Listing 1). This program was run on a Tektronix 4051 graphics system. However it can be run on any computer that uses BASIC.

If you run this program on your computer, the proportion of wins should be close to 0.493. However the probability that the proportion of wins is exactly equal to 0.493 is not very large. Probability theory does permit us to say that 95% of the time the proportion of wins in 1000 plays will lie in the interval

$$0.493 - 1.96\sqrt{(0.493)(0.507)/1000}$$

to $0.493 + 1.96\sqrt{(0.493)(0.507)/1000}$

or 0.462 to 0.524.

If the number of plays is increased to 2000, 95% of the time the proportion of wins will lie in the interval 0.471 to 0.515.

## Expected Number of Plays and Rolls

Now suppose that a player has $5 and plans to place $1 bets in the game of craps until this $5 is lost. There are two questions for which we might like to obtain answers using simulation techniques. (1) How many times, on the average, can a player expect to place $1 bet before losing $5? (2) How many times, on the average, can a player expect to roll the dice before losing the $5?

Theoretically it is possible to find the answers. To answer Question 1, we note that for each $1 bet, a dollar is won with probability 0.49293 and a dollar is lost with probability 0.50707. Thus for each bet, a player can expect to lose, on the average,

$1(0.49293) - $1(0.50707) = $-0.01414.

That is, on the average 1.414 cents is lost on each bet. This means that a player can expect to place on the average,

$$n = \frac{5}{0.01414} = 353.61$$

one dollar bets before losing $5.

To answer Question 2, it can be shown that on the average it requires 3.38 rolls of the dice to determine whether the bettor has won or lost [1]. Thus on the average, it requires (3.38) (353.61) = 1,195.20 rolls of the dice to lose $5.

Both of the above answers can be illustrated empirically using simulation techniques on the computer. In addition to illustrating the answers to these two questions in our simulation

## Simulations, con't....

program, we shall also find the maximum number of dollars that the player's capital had attained when beginning with $5.

This program (Listing 2) was again written for the Tektronix 4051 graphics system. In line 110, Z9 is for the output device, that is, the screen or plotter.

## Summary

Two simulations (Figure 1 and 2) are given which illustrate an interesting phenomenon about the game of craps. Although on the average a player can expect to place 353.61 one dollar bets before losing $5, the simulations show that most of the time, the number of bets placed is less than 353. It requires a "lucky streak" to get a number of plays greater than 353.

Instead of losing $5, it would be nice to know when we have reached our maximum capital. We could then quit and be a winner. Since that information is unknown, we can only guarantee that we are never losers by never gambling. □

### Reference

1. Armand V. Smith, Jr., "Some probability problems in the game of 'craps' ", The American Statistician, Vol. 22, No. 3 (1968), pp. 29, 30.

### Listing 1

```
100 INIT
110 REM N WILL KEEP TRACK OF THE NUMBER OF WINS.
120 N=0
130 REM SIMULATE 1000 PLAYS IN THE GAME OF CRAPS.
140 FOR K=1 TO 1000
150 REM SIMULATE ROLLING A PAIR OF DICE.
160 REM EACH OF D1 AND D2 CAN EQUAL 1,2,3,4,5, OR 6.
170 REM THEIR OUTCOMES ARE INDEPENDENT.
180 D1=INT(6*RND(-1)+1)
190 D2=INT(6*RND(-1)+1)
200 S1=D1+D2
210 REM THE PLAYER WINS IF S1 = 7 OR 11.
220 IF S1=7 OR S1=11 THEN 350
230 REM THE PLAYER LOSES IF S1 = 2 OR 3 OR 11
240 IF S1=2 OR S1=3 OR S1=12 THEN 360
250 REM IF S1 = 4,5,6,8,9, OR 10, ROLL DICE AGAIN.
260 D1=INT(6*RND(-1)+1)
270 D2=INT(6*RND(-1)+1)
280 S2=D1+D2
290 REM IF S2 = 7, THE PLAYER LOSES
300 IF S2=7 THEN 360
310 REM IF S2 = S1, THE PLAYER WINS
320 IF S2=S1 THEN 350
330 REM OTHERWISE ROLL THE DICE AGAIN.
340 GO TO 260
350 N=N+1
360 NEXT K
370 PRINT "THE NUMBER OF WINS OUT OF 1000 PLAYS IS ";N
380 P=N/1000
390 PRINT "THE PROPORTION OF WINS IS ";P
400 END

RUN

THE NUMBER OF WINS OUT OF 1000 PLAYS IS 488
THE PROPORTION OF WINS IS 0.488
```

### Listing 2

```
100 INIT
110 INPUT Z9
120 PAGE
130 REM THIS PROGRAM WILL SIMULATE THE NUMBER OF PLAYS TO
140 REM LOSE $5. THE SIMULATION IS REPEATED 25 TIMES
150 REM N WILL HOLD THE NUMBER OF PLAYS.
160 REM R WILL HOLD THE NUMBER OF ROLLS.
170 REM M WILL HOLD THE MAXIMUM CAPITAL.
180 DIM N(25),R(25),M(25)
190 PRINT #Z9:" "
200 PRINT #Z9:"  TRIAL      NUMBER     NUMBER    MAXIMUM"
210 PRINT #Z9:"  NUMBER    OF PLAYS   OF ROLLS   CAPITAL"
215 PRINT #Z9:
220 FOR K=1 TO 25
230 N(K)=0
240 R(K)=0
250 M(K)=5
260 REM MAXIMUM CAPITAL AT BEGINNING IS $5.
270 REM D WILL EQUAL NUMBER OF DOLLARS HELD BY PLAYER WHO
280 REM BEGINS WITH $5.
290 D=5
300 N(K)=N(K)+1
310 R(K)=R(K)+1
320 REM SIMULATE THE ROLL OF A PAIR OF DICE.
330 D1=INT(6*RND(-1)+1)
340 D2=INT(6*RND(-1)+1)
350 S1=D1+D2
360 IF S1=7 OR S1=11 THEN 490
370 IF S1=2 OR S1=3 OR S1=12 THEN 560
380 REM ROLL DICE AGAIN SINCE S1 = 4,5,6,8,9, OR 10
390 R(K)=R(K)+1
400 D1=INT(6*RND(-1)+1)
410 D2=INT(6*RND(-1)+1)
420 S2=D1+D2
430 REM IF S2 = 7, THE PLAYER LOSES.
440 IF S2=7 THEN 560
450 REM IF S2 = S1, THE PLAYER WINS.
460 IF S2=S1 THEN 490
470 REM OTHERWISE ROLL THE DICE AGAIN.
480 GO TO 390
490 REM PLAYER WINS
500 D=D+1
510 IF D>M(K) THEN 530
520 GO TO 300
530 REM THERE IS A NEW MAXIMUM.
540 M(K)=D
550 GO TO 300
560 REM PLAYER LOSES
570 D=D-1
580 IF D=0 THEN 600
590 GO TO 300
600 REM PLAYER HAS LOST $5.
620 PRINT #Z9: USING 630:" ";K;" ";N(K);" ";R(K);" ";M(K)
630 IMAGE 1A,4D,6A,5D,6A,5D,5A,5D
640 NEXT K
650 REM FIND THE AVERAGE OF THE NUMBER OF PLAYS.
660 REM FIND THE AVERAGE OF THE NUMBER OF ROLLS.
670 REM FIND THE AVERAGE OF THE MAXIMUM CAPITOL.
680 A1=0
690 A2=0
700 A3=0
710 FOR K=1 TO 25
720 A1=A1+N(K)
730 A2=A2+R(K)
740 A3=A3+M(K)
750 NEXT K
760 A1=A1/25
770 A2=A2/25
780 A3=A3/25
790 PRINT #Z9:" "
800 PRINT #Z9: USING 810:" AVERAGES   ";A1;"   ";A2;"   ";A3
810 IMAGE 11A,5D.2D,3A,5D.2D,2A,5D.2D
820 END
```

| TRIAL NUMBER | NUMBER OF PLAYS | NUMBER OF ROLLS | MAXIMUM CAPITAL |
|---|---|---|---|
| 1 | 69 | 224 | 14 |
| 2 | 17 | 52 | 6 |
| 3 | 15 | 63 | 7 |
| 4 | 13 | 42 | 6 |
| 5 | 9 | 33 | 5 |
| 6 | 19 | 72 | 6 |
| 7 | 31 | 120 | 7 |
| 8 | 95 | 336 | 18 |
| 9 | 129 | 469 | 13 |
| 10 | 4393 | 14623 | 65 |
| 11 | 431 | 1472 | 24 |
| 12 | 73 | 236 | 12 |
| 13 | 15 | 54 | 6 |
| 14 | 11 | 50 | 5 |
| 15 | 29 | 109 | 6 |
| 16 | 5 | 18 | 5 |
| 17 | 129 | 440 | 16 |
| 18 | 17 | 49 | 6 |
| 19 | 107 | 329 | 12 |
| 20 | 1461 | 4986 | 39 |
| 21 | 639 | 2077 | 35 |
| 22 | 9 | 47 | 5 |
| 23 | 7 | 11 | 5 |
| 24 | 49 | 131 | 9 |
| 25 | 109 | 402 | 11 |
| AVERAGES | 315.24 | 1057.80 | 13.72 |

Figure 1

| TRIAL NUMBER | NUMBER OF PLAYS | NUMBER OF ROLLS | MAXIMUM CAPITAL |
|---|---|---|---|
| 1 | 19 | 39 | 6 |
| 2 | 141 | 525 | 12 |
| 3 | 1345 | 4546 | 60 |
| 4 | 209 | 694 | 16 |
| 5 | 487 | 1703 | 21 |
| 6 | 43 | 166 | 9 |
| 7 | 89 | 272 | 14 |
| 8 | 91 | 301 | 11 |
| 9 | 9 | 32 | 5 |
| 10 | 4289 | 14629 | 94 |
| 11 | 31 | 118 | 7 |
| 12 | 1229 | 4079 | 38 |
| 13 | 11 | 25 | 6 |
| 14 | 17 | 67 | 6 |
| 15 | 297 | 1033 | 26 |
| 16 | 11 | 33 | 5 |
| 17 | 9 | 20 | 6 |
| 18 | 39 | 159 | 7 |
| 19 | 9 | 19 | 6 |
| 20 | 1085 | 3793 | 30 |
| 21 | 15 | 51 | 6 |
| 22 | 35 | 108 | 8 |
| 23 | 343 | 1164 | 22 |
| 24 | 145 | 455 | 17 |
| 25 | 7 | 25 | 6 |
| AVERAGES | 400.20 | 1362.24 | 17.76 |

Figure 2

# HOW LATE CAN YOU SLEEP IN THE MORNING?

### David H. Ahl

Probabilities and expected values are a vital part to writing almost any game or simulation. Here are two real-life problem situations (I face them practically daily) which can be solved with simple statistics. Be warned: the second is considerably more difficult than the first.

## GETTING TO WORK

Driving to work, you can take one of two routes. Route 1 is 5 miles long and has 4 traffic lights. Each light is on a 1-minute cycle but with different intervals. Light 1 is green in your direction for 30 sec., red for 30 sec. Light 2 is 20 sec. green, 40 sec. red. Light 3 is 25 sec. green, 35 sec. red. Light 4 is 40 sec. green, 20 sec. red.

Route 2 is 5.2 miles long with only 1 traffic light which is green your way 20 sec., red 40 sec.

Speed limit in the town is 35 mph.

1. Which is the best route and what is the expected time difference between the two?

2. Route 2 also takes you by a factory loading dock. If a truck is just arriving (occurs 1 day out of 30) you will be held up an average of 3 minutes. Does this change your answer?

## AND ONCE THERE

Parking your car, you rush into the 10-story Morristown AT&T Building for your 8:30 am appointment with the vice president whose office is on the 9th floor. As you reach the elevators you glance at your watch and see that it is 8:29. From past experience you know that there is a 40% chance of an elevator stopping at any given floor; a stop takes an average of 10 seconds. The elevator passes from one floor to another in 6 seconds. There are 3 elevators all of which have an indicator on the 1st floor (where you are) that shows the location and direction of that elevator. At the point of your arrival, each elevator is equally likely to be on any of the 10 floors going on either direction.

1. Assuming you take the elevator, what is the probability that you will make your appointment on time? What is the probability that you will be less than 1 minute late? Less than 2 minutes late?

2. Under the conditions stated, exactly when would you expect to arrive on the 9th floor?

3. You also know from past experience that you can run up the stairs to the 2nd floor in 10 seconds. The 2nd to 3rd takes you 10% more time (11 sec.) and 3rd to 4th 10% more time (12.1 sec.) and so on. What position of elevators upon your arrival would cause you to run up all 9 floors? (Many answers are possible — select one "break even" combination).

4. Assuming your office is on the 3rd floor and you are faced with the same situation as above (time 8:29 with 8:30 appointment on the 9th floor), but with no elevator indicators, what is your best strategy to be on time for the meeting or as little late as possible? That is, do you run up or wait for the elevator?

## PARTIAL ANSWERS

"Getting to Work." 1. Route 2 is approximately 0.47 sec. faster. 2. The arriving truck has an expected value of –6 sec. per day on Route 2, hence Route 1 now has the edge by 5.53 sec.

"And Once There." Once you get on an elevator, it is a fairly simple matter to determine how long it will take to get to the ninth floor (8 floors x 6 sec. = 48) plus (40% chance of stopping x 10 sec. per stop x 8 floors = 32) equals 80 sec. But figuring when the elevator will arrive on the first floor is something else again. Any elevator can be at any floor going in either direction. Hence, Elevator 1 has a 0.056 probability of being at, say, Floor 3 going up*. How long until it returns to Floor 1? Well, if we know it goes to the 10th floor, that's easy, but it has only a 0.4 chance of going to the 10th floor, 0.4 chance of going to 9 and so on. Multiplied by 18 different possible starting positions and by 3 elevators, this is a nasty problem. In a situation like this you have to ask yourself whether a heuristic, or rule of thumb, or best guess wouldn't provide an adaquate answer. For example, you might want to make the assumption that at least one elevator is at the 4th floor (or below) and heading down.

It is sometimes easier to come up with a solution if you think of the problem in entirely different terms. For example, think of the elevator as a one-way trolley on a circular track — station 1 is Floor 1, station 2 is Floor 2 going up, and so on. Station 18 is Floor 2 coming down and then back to station 1 again. Using this approach *may* make it easier to work the problem.

By the way, it should be apparent that a computer isn't much help in solving this particular problem. However, if this were part of a much larger simulation in which the output of one part provided the input to the next (a very typical situation), a computer would be almost vital to the solution.

If you're still with me and want to read a fun little book on the subject, get "Flaws and Fallacies in Statistical Thinking" by Stephen Campbell published by Prentice-Hall.

---

*If there are 10 floors and the elevator has an equal chance of being at any floor going in either direction, why isn't the probability of being at Floor 3 going up 0.1 x 0.5 = 0.05? Simply because Floor 1 and 10 do not have a direction associated with them, hence there are really only 18 locations. Actually, that's over-simplified because the elevator may not even reach Floor 10, or 9 or 8 etc.

# Mathematical Miscellany

Including circular functions, regression analysis, flowcharts, magic squares, Pascal's triangle, differential equations, series, double precision and much more!

# The World of Series — Playoff That Is

James Reagan
Stevenson High School
Sterling Heights, Mich.

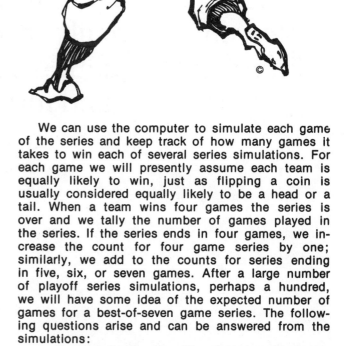

## What Are Playoffs?

Traditionally the professional sports of baseball, basketball, and hockey have held what are called championship playoffs to determine the particular sport league champion. Comparable playoff systems are used after the regular season ends by such professional sport organizations as Major League Baseball, the National Football League, the National Hockey League, and the National and American Basketball Associations. An even number of teams qualify for the championship playoffs determined by their regular season record and/or by their comparative standing within a particular division of the league.

The playoff setup may be illustrated by Major League Baseball. Each of the division winners qualifies for the playoffs. The winners of the two American League divisions, East and West, play each other in a best-of-five games series while the National League divisional winners play their best-of-five series. (In a best-of-five series the first team to win three games is the winner of the series.) The winner of the World Series is considered the best team in baseball.

There has been a trend in the major professional sports to expand—add teams to the respective leagues. This has changed many previous playoff systems to include more teams qualifying for the championship playoffs. For example, the National Basketball Association now has a first round series in which there are two best-of-three series before the semi-final and final playoffs.

Some of the criticism of professional sport expansion is that the entire championship structure is designed to add revenue to the pockets of the team owners and the best team is not necessarily the one that wins.

Let us investigate some of the questions and results of employing a championship playoff system.

## First Investigation

If we consider that playoff series are a money generating activity, how much money can the competing teams, not necessarily the players, expect to earn? The earnings are reflected in the number of games that can be expected to be played. In a best-of-seven series, there could be four, five, six, or seven games played. Which number of games should we expect?

Certainly games cannot or should not be fixed so that there can be more games in the series; that is both illegal and unethical. Just let things go naturally and see what happens.

We can use the computer to simulate each game of the series and keep track of how many games it takes to win each of several series simulations. For each game we will presently assume each team is equally likely to win, just as flipping a coin is usually considered equally likely to be a head or a tail. When a team wins four games the series is over and we tally the number of games played in the series. If the series ends in four games, we increase the count for four game series by one; similarly, we add to the counts for series ending in five, six, or seven games. After a large number of playoff series simulations, perhaps a hundred, we will have some idea of the expected number of games for a best-of-seven game series. The following questions arise and can be answered from the simulations:

1. Is a prediction of a "four game sweep" reasonable?
2. Is a prediction that the series will go six or seven games really going out on a limb?
3. How well do the results of the simulation agree with actual outcomes of Major League World Series or other best-of-seven championship playoffs?

Many preliminary playoffs are not best-of-seven. Some are best-of-five and some are best-of-three. How should the series results be expected to be distributed with these kinds of playoffs?

We originally assumed that the probability that a team would win any game was 0.5; but, seldom are the two teams equally likely of winning a given game. There are many factors affecting the *a priori* probability of a team winning a game. Some of the factors are:

1. The place where the game is played. Does the home team have the advantage?

2. The season records of the teams. The team with the better season record may have a better than 0.5 probability of winning a game.
3. The winner of the previous game. A psychological advantage is usually associated with the team that has won the previous game.

An investigation stating the probability influencing factors and the simulation will lead to some interesting results.

*Second Investigation*

An equally important investigation is one which will help answer the question: "Just because Team A wins a series, does that mean that Team A is the better of the two teams?" Certainly if we agree to use the criterion that the winner is the better team the answer to the question is "Yes". But, most sports fans have a preconceived idea of which team is the better of the two. Now I rephrase the question: "If Team A is better than Team B, what is the probability that Team A will win a best-of-seven series from Team B?"

By Team A being better than Team B I mean that for any game that they play, the probability that Team A will win, P(A), is greater than the probability that Team B will win, P(B). If we first consider P(A) = 0.55, what is the probability that Team A will win the series? The question can be answered using various probabilities for Team A.

Further, if Team A is better than Team B, what is the probability that Team A will win a best-of-five series? I recently heard a sportscaster say, "Anything can happen in a five game series." Finally, what happens in a best-of-three series as in the NBA?

# How Many Ways Can You Make Change For A Dollar?

by Brian Hess

Even with all the nickels and dimes and pennies running around this country, somebody still always needs change for a dollar for this or that. Assuming that you carried around enough change, how many ways could you help out someone who needed the right change for some infernal vending machine? You can use half-dollars, quarters, dimes, nickels, or pennies to make the right change. For example, you could give him 1 half, 1 quarter, 1 dime, 2 nickels, and 5 pennies. Get it?

There are a few different ways to solve the problem. One is to break it down into smaller problems, easily solved (e.g., how many ways can you make change for a quarter?) and then combine the answers to get the "big" answer. Another mathematical method would be to write out a series of equations relating each piece of change to each other and the dollar and then solve them. Finally, you could do the problem by exhaustion.

Solving a problem by exhaustion means writing down all the answers until all the possibilities of solution are exhausted (or until you are exhausted, whichever comes first). Fortunately, you *Creative Computing* readers can exhaust a computer rather than yourselves. Write a program to figure out how many ways you can make change for a dollar. Print the ways as well as a final total. (WARNING: Printing takes time on a TTY—if you are in a hurry [or being charged] don't bother printing all the ways, just the final total.)

*Hints:* 1) If you use loops, counting one by one, it will probably take close to 20 minutes to compute all of it (even without printing all the combinations). Do you have to index the "pennies-counter" by one? Once the half-dollar counter reaches 2, what happens to all the other nested coin-counters? What about 4 quarters? Dimes?

2) If your program doesn't come out with well over 100 ways to make change for a dollar, it has something wrong with it. (I'm not going to tell you the exact answer—work it out for yourself!)

3) Once you have gotten the answer, ask some friends to guess at what they think it is. You'll hear some very interesting numbers. Use the computer to tabulate them, etc.

4) Write some sort of applications changes for this program. Look at how the number of combinations changes. For example, nobody uses half-dollars in vending machines, so restrict the number of halves to 1. Also, who wants more than 25 pennies? Only parking meters and gumball machines use them. Finally, include at least 1 dime in the change so that your changeless friend can make a phone call!

*Brian is a high school student in Western Springs, Illinois.*

# KEEPING THE LOAN ARRANGER HONEST

James A. Warden*

Most of us have borrowed money at some time, whether to buy a house, to pay for a car or a large appliance, or to extend payment on a revolving charge account. The truth-in-lending laws may have made us aware that loans do cost money and that we can shop around for a place to borrow, but the details of computing the monthly payment schedule may still seem a bit mysterious. Yet these calculations are in fact quite easy and anyone with access to a computer (or even a calculator) can crank out loan schedules at will.

Let's get a few basic facts straight first. Usually we borrow an amount of money known as the principal amount, which is to be paid back in monthly installments. We must pay interest on the loan each month, which is based on an annual percentage of the principal, or the annual interest rate. We might wish to know answers to these questions about the loan: If I want to pay it back in N months, what is the monthly payment? If I want to pay back so much per month, how long will it take to repay it? What is the total amount I will pay?

There are standard loan formulas which can provide the answers, and these formulas really are not difficult to use. but you can't generate a payment schedule with them, and you may not wish to bother with the mathematics necessary to derive them. Instead, we can consider a simple recursive procedure which will provide all the answers to allow us to see what is going on at the same time. If we borrow an amount P and pay it back monthly at an annual interest rate of R (per cent), our monthly payment M is used first to pay the interest which has accrued on the principal during the month. Expressed mathematically, the interest due on the principal remaining is I = P*R/1200. The 1200 comes in because we must divide the annual rate by 12 to get the monthly rate, and we must divide the percentage by 100 to obtain a fraction. The monthly payment reduces the principal by M-I, leaving P+I-M to be repaid. Expressing this in the form of a "recursive relation", we have that P(new) = P+I-M, which tells us how much we have left to pay (the new principal amount) after making a payment on the old P. To find out how much we will owe next month, we take P(new), insert it in place of P, and repeat the calculation. Eventually, P(new) will be reduced to nothing, and we can "burn the note". Of course, P(new) may turn up negative on a particular payment, meaning that the payment M will overpay the loan. In this case, the final payment will be whatever principal is left plus the interest owed on it, or P+I. This algorithm is illustrated in flowchart form below.

To find out how many months it will take to repay the loan, we simply have to choose a monthly payment, start with the principal amount owed, compute the new principal owed after a month, and repeat the operation until the principal drops to zero, printing the result each cycle. If we also put a counter in the loop and display payment numbers, the number of payments will be obvious. Finding a monthly payment which pays back the loan in exactly N payments requires a bit of guesswork. Here we pick a reasonable payment, run the calculation until the principal drops to zero, note the number of months required to pay, and then adjust our guess of the monthly payment to come out closer to the desired N on the next calculation. In either case, simply summing the monthly payments made will give the total cost of the loan. That's all there is to the technique! If it still sounds a bit vague, study the flowchart. This algorithm can be written up in BASIC, FORTRAN or a calculator procedure.

If you would like some practice or you wish to introduce this technique to someone else, the BASIC program LOANER may be of help. This program is one of a series of routines used at Wabash College to generate random exercises for the elementary computer science classes. LOANER will make up loan problems at random and use standard loan formulas to estimate values for number of payments for a given monthly payment and the total amount paid.

I would like to thank Prof. T. Mielke of the Wabash Mathematics Department for some of the ideas developed in this article.

---
* Cragwall Computer Center Wabash College Crawfordsville, Indiana 47933

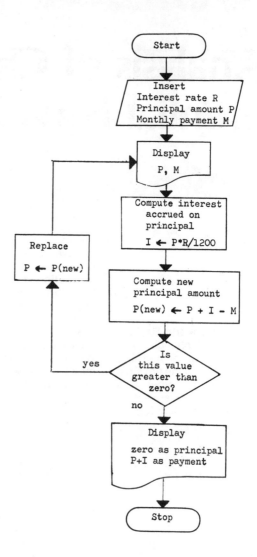

Fig. 1 -- Flowchart for an algorithm which generates an amortization schedule

```
100 REM LOANER -- AN EXERCISE GENERATOR
120 REM J. WARDEN         WABASH COLLEGE
140     RANDOMIZE
160 REM GENERATE VALUES OF PRINCIPAL, RATE, AND PAYMENT
180     P=INT(RND*500)*10+1000.
200     R=INT(RND*20+20)*.5
220     M=INT(P*(.07+RND*.05))
240 PRINT
260 PRINT"WRITE A BASIC PROGRAM WHICH WILL GENERATE AN AMORTIZATION"
280 PRINT"SCHEDULE OF MONTHLY PAYMENTS FOR A LOAN OF  $";P
300 PRINT"TO BE REPAID AT AN ANNUAL INTEREST RATE OF";R;"PER CENT."
320 PRINT"ASSUME THAT THE MONTHLY PAYMENT IS   $";M" ."
340 PRINT
360 PRINT"INCLUDE IN THE SCHEDULE THE PAYMENT NUMBER, THE PRINCIPAL"
380 PRINT"REMAINING, INTEREST, AMOUNT PAID TO PRINCIPAL, AND THE"
400 PRINT"PAYMENT.  AT THE END, DISPLAY THE TOTAL AMOUNT PAID."
420 REM GENERATE COMPARISON VALUES FOR NUMBER OF PAYMENTS
440 REM AND TOTAL AMOUNT PAID FOR THE LOAN
460     A = 1 + R/1200
480     N=INT(  (LOG(M)-LOG(M-P*(A-1.)))/LOG(A) )
500     A2=(1.-A^N)/(1-A)
520     T=P+(R/1200)*(P*A2 - M*(N-A2)/(1-A))
540     T=INT(T+.49)
560 PRINT
580 PRINT"(HINT: FOR THE FIGURES GIVEN, THE LOAN SHOULD REQUIRE";N
600 PRINT"PAYMENTS AND A TOTAL OF ABOUT   $";T;"WILL BE PAID.)"
620 END

READY
```

# An Analysis of Change: Differential Equations

Bruce D. Barnett

> We live in a changing world. Besides the obvious examples of change ... other less evident examples of change are evolution of animal and bacteriological populations, and money "growing" as it earns interest in a bank. These changing events are expressed by what is known as differential equations.

We live in a changing world. Besides the obvious examples of change, night into day, your hot cup of morning coffee turning cold and all the physical laws describing motion, other less evident examples of change are evolution of animal and bacteriological populations, and money "growing" as it earns interest in a bank. These changing events are expressed by what is known as differential equations. These equations specify the rate of change of a variable (quantity that is allowed to vary) in terms of a known function or expression. A few examples will help clarify this concept and perhaps will enable you to construct some differential equations of your own choosing. You can use the program that accompanies this article to solve the equations you derive.

In setting up a differential equation, first choose any letter or symbol you desire to represent the quantity that will be changing. A dot placed over that symbol will signify that this is a time rate of change of that quantity; that is, it represents how fast the quantity is changing. This in turn is to be equated to a given expression that actually specifies the rate of change. This expression may or may not include the quantity itself and/or time explicitly. Hence it is quite easy to write differential equations. It is quite another matter however, to solve the resultant equation, which requires knowledge of a branch of mathematics known as the calculus. In some cases the equation is easily solved, in others considerable ingenuity must be exercised, while in yet other cases, one must resort to a solution that consists of adding an infinite number of terms or use some approximate numerical method to arrive at an answer. Now for some examples.

Bruce D. Barnett, RD 2, Box 213, Blairstown, NJ 07825.

### Example I — Population Growth and Related Matters

Consider the population of rabbits (or even rats). A rather safe assumption one could make about their population is that "the more there are, the more there will be," so that for example if there are 100 rabbits in existence today, there may be 150 rabbits a month from now — an increase of 50 rabbits, while if there are 200,000 rabbits now, the population may have grown by another 100,000 in one month. To write this as a differential equation, let N represent the number of rabbits at any given time. Naturally N will change in time. Let the rate of change of N be denoted by $\dot{N}$ and let this rate of change be proportional to the number of rabbits in existence at any given time, (recall "the more there are, the more there will be"). The differential equation governing this is written as $\dot{N} = KN$. Here K is the factor of proportionality. If K is positive the population will increase; in fact the larger the value substituted for K, the more rapidly the population will expand. For K negative, which denotes a negative rate of change, the population will decrease. One final bit of information is required to provide a single solution to $\dot{N} = KN$ that is characteristic of all differential equations. At any selected time you choose, you need to know a value for the quantity that is changing. In this example you need know N, the population size at any one given time. For example at $t = 0$ which could represent the present, let's assume there are 100,000 rabbits (N = 100,000). This assignment of values is called an initial condition and is essential to obtain a unique solution to a differential equation. Once having this solution however, you can determine the population of rabbits at any time hence. Figure 1 presents a graph of the number of rabbits vs. time for several values of K, assuming there are 100,000 rabbits at $t = 0$. These curves and many others can be generated using the accompanying program which is explained in a later section of this article.

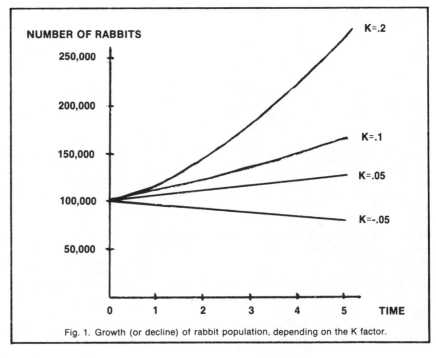

Fig. 1. Growth (or decline) of rabbit population, depending on the K factor.

Believe it or not, but many other seemingly unrelated problems are governed by this same differential equation! This illustrates one of the beauties of mathematics whereby one equation or even an entire theory may apply to many different fields. Some examples that are described by $\dot{N} = KN$ are radioactive decay, mixing of solutions and continuous compounding of interest. Let's look at the latter example more carefully. Assume that you have invested $100 in a savings account that earns 6 percent interest. How much money would you have at the end of 10 years if interest is compounded annually, semi-annually, daily, ... continuously? Table I gives answers to these questions. The differential equation that applies to the continuous compounding is $\dot{P} = .06P$. Here P represents the principal at any time t. Note again that the larger P becomes, the larger $\dot{P}$, the rate of change of P, becomes and the faster your money will grow. As the initial condition, P = $100 at t = 0 was used; t is expressed in years, and .06 represents the annual interest rate. Again, curves similar to that of the population expansion can be generated using the appended program in which you can substitute different interest rates and determine how much money you can save for different principals at any future time.

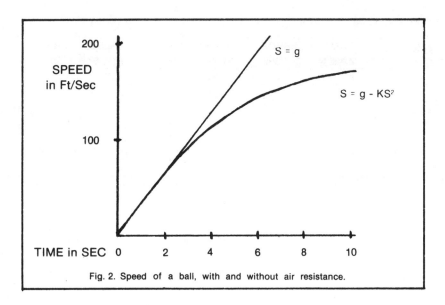

Fig. 2. Speed of a ball, with and without air resistance.

| Compounding Interval | Amount At End Of 10-Year Period |
|---|---|
| Annually | $179.08 |
| Semi-Annually | $180.61 |
| Daily | $182.20 |
| Continuously | $182.21 |

Table I. Ten years of interest on $100, compounded at various intervals.

### Example II — Various Motions of a Ball

As a second example, let's consider the motion of a ball that is moving at a constant speed as might be the case if it is rolling on a level plane under ideal conditions. Constant speed means that the rate of change of speed is zero. The differential equation would then simply be $\dot{S} = 0$, where S denotes the ball's speed. If the ball is dropped from a window, however, the picture will be quite different and the ball continually gains speed as it falls. In the absence of air resistance, the rate of change of the ball's speed is proportional to the gravitational attraction, thus the governing equation is $\dot{S} = g$ (where g is a given constant). Solving this equation you will find that the speed of the ball will increase without limit, that is, will continue to accelerate. The most interesting case arises when air resistance is introduced. Air resistance tends to slow the ball somewhat; in fact the faster the ball travels, the more effect air resistance has on its motion. What will the motion of the ball be like now? Will it still increase its speed without limit? One can account for air resistance by introducing a term that depends on the ball's speed. Physical experiments suggest adding a term that is proportional to the square of the speed, namely $KS^2$. Here K is a mixture of physical constants that involve the shape, size and weight of the ball as well as the density and viscosity of the air. This term, when added to the gravitational influence, produces the differential equation $\dot{S} = g - KS^2$. The term $KS^2$ is introduced with a negative sign since it acts to decrease S. Figure 2 results when using a value of 32.2 ft/sec² for g and a value of .001 for K. For an initial condition, the ball is assumed to start from rest, at t = 0, S = 0. The solution for the equation $\dot{S} = g$ is shown also for the same initial condition as above. The reader may want to experiment with other laws such as assuming the air resistance is proportional to the cube of the speed, which applies under certain circumstances, to see what motion results for that case also.

What was presented is only a small view of how and where differential equations arise. Other examples described by such equations are mass-spring systems, electric circuits, deflection of beams and orbital mechanics. One can truly go on and on. Differential equations are necessary and do indeed help describe the "changing" world we live in.

### Using the Program

The accompanying program solves differential equations numerically, hence any solution so obtained will only approximate the true solution. For differential equations that cannot be solved directly — and a numerical solution must be resorted to — the mathematician will not only be interested in obtaining the approximate solution but will ask how well the numerical solution indeed represents the true solution. The accompanying program was written to allow you to experiment and exercise your own judgement regarding this question of accuracy. Typically the numerical solution of a differential equation is more accurate the nearer one is to the initial condition; thus as one progresses further from this given condition the less accurate the result. This phenomenon is not unlike the situation where a story or rumor is passed on in turn from person to person, where the more people involved (the further away the story can travel from the source), the more distorted the story becomes. What can be done about this? Well, besides performing a tedious error analysis to limit the error, you can usually still judge how good the solution is simply by rerunning the program using a smaller step size. In fact, continue this process until the solution, at the point you are interested in, changes very little from run to run. Your final solution should then be a good estimate of the true solution. The smaller step size is somewhat analogous to using more capable people who can transmit the "story" more accurately. Table II illustrates these concepts for the equation $\dot{N} = 3N$ for the given initial condition t = 0, N = 1.

| STEP SIZE | t = .5 | | t = 1.5 | | t = 3. | |
|---|---|---|---|---|---|---|
| (H) | VALUE | % ERROR | VALUE | % ERROR | VALUE | % ERROR |
| .5 | 4.39844 | 1.86 | 85.0933 | 5.47 | 7240.87 | 10.64 |
| .1 | 4.48134 | .01 | 89.9958 | .02 | 8099.25 | .05 |
| .05 | 4.48166 | 0 | 90.0156 | 0 | 8102.81 | 0 |
| TRUE SOLUTION (N=e3t) | 4.48169 | | 90.01713 | | 8103.08 | |

Table II. Showing how the error percentage is minimized when using a smaller step size and smaller value for t.

```
10 PRINT "INPUT A VALUE FOR STEP SIZE";
20 INPUT H
30 PRINT "INPUT INITIAL VALUE FOR INDEPENDENT VARIABLE";
40 INPUT T0
50 PRINT "INPUT INITIAL VALUE FOR DEPENDENT VARIABLE";
60 INPUT Y
70 PRINT "T","Y"
75 PRINT T0,Y
80 LET N1 = 0
90 LET N1 = N1 + 1
100 LET T1 = T0 + H + (N1 - 1)*H*10
110 FOR T = T1 TO T1 + 9*H STEP H
120 LET I = T
130 LET D = Y
140 GOSUB 500
150 LET K1 = H*F
160 LET I = T + .5*H
170 LET D = Y + K1/2
180 GOSUB 500
190 LET K2 = H*F
200 LET D = Y + K2/2
210 GOSUB 500
220 LET K3 = H*F
230 LET I = T + H
240 LET D = Y + K3
250 GOSUB 500
260 LET K4 = H*F
270 LET Y = Y + (K1 + 2*K2 + 2*K3 + K4)/6
280 PRINT T,Y
290 NEXT T
300 PRINT "DO YOU WANT TO CONTINUE THIS PROBLEM?"
310 PRINT "TYPE 1 IF YES, 0 IF NO"
320 INPUT N
330 IF N = 1 THEN 90
340 PRINT "WANT TO SELECT ANOTHER INITIAL CONDITION OR"
350 PRINT "SELECT ANOTHER STEP SIZE? (1 = YES,  0 = NO)";
370 INPUT N
380 IF N = 1 THEN 10
390 PRINT "HOPE YOU HAD FUN   ...BYE"
400 GOTO 999
500 F = 3*D
510 RETURN
999 END
```

```
RUN
INPUT A VALUE FOR STEP SIZE? .5
INPUT INITIAL VALUE FOR INDEPENDENT VARIABLE? 0
INPUT INITIAL VALUE FOR DEPENDENT VARIABLE? 1
T           Y
0           1
.5          4.39844
1           19.3463
1.5         85.0933
2           374.278
2.5         1646.24
3           7240.87
3.5         31848.5
4           140084
4.5         616149
5           2.7101E+06
DO YOU WANT TO CONTINUE THIS PROBLEM?
TYPE 1 IF YES, 0 IF NO
? 1
5.5         1.19202E+07
6           5.24302E+07
6.5         2.30611E+08
7           1.01433E+09
7.5         4.46146E+09
8           1.96234E+10
8.5         8.63125E+10
9           3.7964E+11
9.5         1.66982E+12
10          7.34462E+12
DO YOU WANT TO CONTINUE THIS PROBLEM?
TYPE 1 IF YES, 0 IF NO
? 0
WANT TO SELECT ANOTHER INITIAL CONDITION OR
SELECT ANOTHER STEP SIZE? (1 = YES,  0 = NO)? 0
HOPE YOU HAD FUN   ...BYE
Ok
```

Note that for a step size of .5 and at t = .5, a 1.86% error is made which grows to 5.47% at t = 1.5. Notice also that the same approximate answer is obtained at t = 1.5 for both H = .1 and .05. This suggests that 90.0156 is a good estimate of the true solution.

To run the program, follow the instructions as they are printed out for you using the sample run as a guide that helped produce Table II.

The independent variable called for in the program is the variable that you usually have no control over; this very often is time. The dependent variable is the quantity that changes as the independent variable changes. In the examples that were presented, the population size and speed would be considered as dependent variables that depend upon time in order to have a specific value. To change the differential equation in the program simply change line 500 in accordance with the following conventions. Use D to denote the dependent variable, I to denote the dependent variable and the symbol F to denote the rate of change of the dependent variable. Thus the equation $\dot{N} = 3*N$ would be written as $F = 3*D$ which is in the current listing of the program at line 500. As a second example, the equation $\dot{X} = (X*T - 6)/2*T^2$, where T is considered as the independent variable would be entered as $F = (D*I - 6)/(2*I*I)$.

Happy solving!!!

Reference:
Ralston, A., (1965) A First Course in Numerical Analysis, McGraw-Hill Inc., (page 200).

A page of a newspaper is about 0.003 inches thick. If you piled up $2^{50}$ sheets of newspaper, you would have 1,125,-899,906,842,624 sheets. This pile would be over 53,000,000 miles (85,-000,000 kilometers) high.

The distance from the earth to the moon is about 240,000 miles, or 386,-000 kilometers.

# Progression Problems

### Charles A. Reeves

Last week we grew paramecium in a hay infusion, as described in the experiment from our science book (*Today's Basic Science*). We put some hay in a bucket of tap water, and left it sitting by itself for 7 days. At the end of that time, we had a bucket full of the things.

The book also mentioned that paramecium reproduce by cell division about every 5 hours. Assume that there was only 1 paramecium in the bucket when we started — how many would there be at the end of the 7th day?

For those who want more: Have the computer print the number of paramecium at the end of the 4th, 5th, 6th, and 7th days, all in one run!

On page 194 of *Today's Basic Science*, you will find: "The female grasshopper is especially adapted for egg-laying. The female lays from 20 to 100 eggs. It lays the eggs in the ground or perhaps in a rotted log. A structure at the tip of the abdomen enables the female to dig a hole in the ground or in rotted wood. This structure is called the 'ovipositer'."

Assume for a moment that you are a scientist, doing an experiment with grasshoppers over a ten-year period. You are applying for a grant from the U. S. government, and so you have to plan how much money you will spend on food, tags, etc. for these animals.

You have to first find out how many grasshoppers you will have in a ten year period (you are starting the experiment with only 1 pair, a male and a female). Have the computer calculate and report to you approximately how many grasshoppers will be born from that one pair. Grasshoppers live only one year, so the females will lay eggs only once in their lives. Assume also that half of those born will be males.

Write a program that you can use to find the average of a given set of numbers. We will use this program to find the class average on tests, and to find the average height and weight of the class. You will want to tell the computer to save this program for future usage.

It takes nature about 500 years to produce 1 inch of topsoil. Many years ago our country had an average depth of almost 9 inches of this good dirt, but now we are down to 6 inches. This type of dirt is necessary, of course, for growing food.

Careless management of our soil causes about 1% per year to erode away, and then it's lost forever. Once we get down to less than 3 inches, it will be impossible to grow crops on a major scale. Have the computer calculate and report to you the year that our country will have less than 3 inches of topsoil, assuming that it continues to erode away at 1% per year.

Will you be alive then? Will your children be alive?

Jack got a pair of bunnies as a New Year's Day present in 1972. This pair became a pair of young rabbits in February, and a pair of adult rabbits in March. A pair of adult rabbits produces a pair of bunnies each month from then on, and this growth cycle continues. The number of pairs of rabbits of each type is provided below, for the first six months:

|                  | J | F | M | A | M | J |    |
|------------------|---|---|---|---|---|---|----|
| Pairs of bunnies | 1 | 0 | 1 | 1 | 2 | 3 | ... |
| Pairs of young   | 0 | 1 | 0 | 1 | 1 | 2 | ... |
| Pairs of adults  | 0 | 0 | 1 | 1 | 2 | 3 | ... |
| Total pairs      | 1 | 1 | 2 | 3 | 5 | 8 | ... |

Have the computer tell you the maximum number of rabbits that Jack could have in three years. [Saving the "maximum number" means we are assuming that, of each pair born, one is a male and one a female, and that none of the rabbits die over this period of time.]

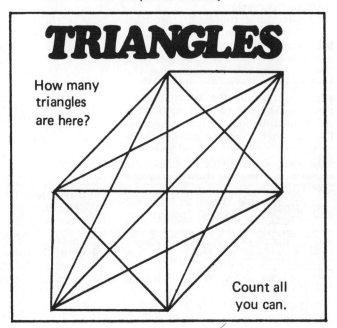

**TRIANGLES**

How many triangles are here?

Count all you can.

Analyzing statistical data is possible with TRS-80 Level I Basic or other computers with single-dimensional arrays.

# Multiple Regression Analysis — Simplified

### Dr. David M. Chereb

While the LEVEL I BASIC from Radio Shack is limited, it does have enough capability to handle some rather advanced programs. The example included here (and listed in Figure 4) is a multiple regression routine. The key to the program is the single allowable vector A(n). The length of this vector is limited only by memory size (memory/4).

Because multi-dimensional arrays are not supported by LEVEL I, special techniques are needed to simulate the N by K matrix which is the basic starting point for the multiple regression routine. Also since the program uses a modified Gauss-Jordan elimination technique, there are many sections which use nested FOR loops. Doing all this with a single A(n) vector brings one closer to understanding the nature of insanity. The basic technique is to manipulate the vector as if it *were* an N by K matrix, letting special counters do all the hard work of locating the correct numbers (see Figure 1).

**Multiple Regression**

The multiple regression program finds the statistical relationship between the K independent variables and a single dependent variable. What does that mean? It means that the statistical technique of multiple regression will give us the *best* equation possible. There are certain restrictions which must be met for this to be true. For our needs, we can assume these conditions are met in most practical situations. An example of a multiple regression problem is shown in Figures 2 and 3.

In the example shown in Figures 2 and 3, United States imports are analyzed. The variable to be explained is U.S. imports in constant 1972 dollars (the dependent variable). The explanatory variables are
1) the change in Gross National Product,
2) the size of the labor force, and
3) a variable representing time.

### FIGURE 1
### MATRIX MANIPULATION WITH THE A(n) VECTOR

Let's suppose we wanted to add two matrices B and C and call the resultant matrix D. The matrices must be of the same size to do this, so let's say they are both 4 by 3 (4 rows by 3 columns).

$$B = \begin{vmatrix} 1 & 0 & 1 \\ 1 & 1 & 0 \\ 1 & 2 & 3 \\ 1 & 3 & 1 \end{vmatrix} \qquad C = \begin{vmatrix} 1 & 3 & 4 \\ 0 & 1 & 0 \\ 2 & 0 & 1 \\ 1 & 2 & 0 \end{vmatrix}$$

If D = B + C then

$$D = \begin{vmatrix} 2 & 3 & 5 \\ 1 & 2 & 0 \\ 3 & 2 & 4 \\ 2 & 5 & 1 \end{vmatrix}$$

For LEVEL I BASIC this is done by:

Assume A(1) ........ A(12) = B matrix
A(13) ........ A(24) = C matrix
A(25) ........ A(36) = D matrix

$$\text{Then } B = \begin{vmatrix} A(1) & A(5) & A(9) \\ A(2) & A(6) & A(10) \\ A(3) & A(7) & A(11) \\ A(4) & A(8) & A(12) \end{vmatrix} = \begin{vmatrix} 1 & 0 & 1 \\ 1 & 1 & 0 \\ 1 & 2 & 3 \\ 1 & 3 & 1 \end{vmatrix}$$

The routine is:

```
FOR J=1 TO 3
    FOR I=1 TO 4
        K=I+(J-1)*4
        L=12+I+(J-1)*4
        M=24+I+(J-1)*4
        A(M)=A(K)+A(L)
    NEXT I
NEXT J
```

Notice that when J=1 then K=I, L=12+I and M=24+I so that the data is allocated to the correct element of the A(n) vector. While there are other ways to do this problem the one presented here closely simulates a two dimensional array program and can be easily modified to handle all forms of matrix operations with any number of columns and rows.

Dr. David M. Chereb, 4005 Locust Ave., Long Beach, CA 90807.

## FIGURE 2
### MULTIPLE REGRESSION INPUTS

Time Period: 1965 to 1974 by quarter—40 observations

Dependent Variable: U.S. imports in billions of 1972 dollars

Independent Variables (Chosen by author, based upon experience):
1) Change in Gross National Product in 1972 $ from the previous quarter * 10
2) U.S. labor force in millions
3) Time variable: 1000/time, where time=1 in 1957 first quarter and advances by one in each quarter.

These variables were chosen because they are the important variables which affect imports. The first variable is included because rapid growth in U.S. production usually causes imports to accelerate. The labor force variable accounts for the gross level of imports. If the labor force is only 50 million people instead of 100 million, then imports would be much less. The time variable represents a non-linear response over time for U.S. imports. A *negative* coefficient for this variable would mean that imports increase over time even if the GNP and the labor force did not increase. This reflects the generally increasing interdependence of national economies over time (the import share in GNP has doubled in the last 9 years).

The results reveal that all of the independent variables are important (t values over + 2) and that an increase in the labor force by one million people will cause imports to increase by $1.43286 billion. To get an estimate, values for all the independent variables are entered into the equation. If the values for the independent variables are

1) GNP-1 * 10 = $101 billion,
2) Labor Force = 84.2 million, and
3) 1000/Time = 20 (1969:2)

then next quarter the level of imports is expected to be $62.18 billion (at an annualized rate).

We must remember that this is only an estimate. The higher the t values and the coefficient of determination (the R squared adjusted), the more confidence we usually have in the validity of the equation. In this case the estimates of the next period's imports are accurate to within + $3.8 billion (90% confidence interval).

Since the LEVEL I BASIC has only six significant digits, there are roundoff problems. For most uses the estimated coefficients from the program are accurate to 3 or 4 places. Given our current knowledge about economic relationships, we rarely need more than three places of accuracy.

## FIGURE 3
### MULTIPLE REGRESSION RESULTS

Equation to be estimated:

**Imports** = B0 + B1 * GNP - 1 + B2 * Labor Force + B3 * 1000/Time
$\quad\quad\quad\quad\quad\quad\quad\quad\quad\quad$ (LF) $\quad\quad\quad\quad\quad$ (T)

$\quad\quad$ = -32.6641 + .0118321 * GNP - 1 + 1.43286 * LF - 1.34998 * 1000/T

Std. Errs $\quad$ (33.4041) $\quad$ (.0038404) $\quad$ (.314984) $\quad$ (.333134)
t Values $\quad$ (-.977845) $\quad$ (3.0809) $\quad\quad$ (4.549) $\quad\quad$ (-4/05235)

$R^2$ = .974
$R^2$(adj.) = .972
S.E. = 2.2594

Multiple regression uses the data to solve for the unknowns: B0, B1, B2, B3.

The actual technique is to minimize the sum of squared errors of the actual data points of the dependent variable and the estimated data points from the regression equation (i.e. $y - \hat{y}$). There are many textbooks which explain the details of regression analysis.

---

The Basic program uses 6K of memory with each data point adding 4 more bytes. For the example shown, less than 9K of memory is required. To execute this problem takes about 50 seconds. While this seems long in comparison to a large mainframe computer, it is generally acceptable since this is not a problem that is run many times with the same data. Once the equation has been estimated, the regression routine has done its job. From thereon U.S. imports can be easily estimated on any calculator using the equation in Figure 3. The only time the regression routine need be used again on *this* problem is when new information leads you to believe that forecasting accuracy can be improved by re-estimating the equation using the new information (usually more data or more variables). In a business forecasting environment, even higher accuracy would be the goal.

### The Future

Radio Shack's LEVEL II BASIC allows multi-dimensional arrays, transcendental routines and double precision variables. This will make statistical analysis much easier. It will be interesting to compare the execution speed of a multiple regression routine for LEVEL II versus the LEVEL I program shown here. In any event now that you've seen the power of the A(n) vector, perhaps you'll shun all multi-dimensional arrays in the future and stick with A(n). If you do, don't count me in with you. ∎

```
10 REM***********************************************
20 REM***    MULTIPLE REGRESSION ANALYSIS         ***
30 REM***      ( LEAST SQUARES ANALYSIS )          ***
40 REM***      BY    DAVID M. CHEREB  5/5/78       ***
50 REM***********************************************
52 REM
53 CLS : PRINT:PRINT TAB(10);"MULTIPLE REGRESSION ANALYSIS"
55 REM *** DATA FOR CONSUMPTION FUNCTION ***
56 REM *** VAR 1=CONSTANT  2=DIS. INCOME  3=CONSUMPTION(-1) ***
58 DATA "U. S. CONSUMP. "
60 DATA 1,1,1,1,1,1,1,1,1,1
62 DATA 1,1,1,1,1,1,1,1,1
64 DATA 2.862,2.877,2.910, 2.911,2.946,2.961,2.933
66 DATA 2.913,2.926,2.999,3.021, 3.059,3.125,3.113,3.132
68 DATA 3.154,3.203,3.210,3.201
70 DATA 2.632,2.637,2.634,2.669, 2.689,2.704,2.734,2.721
72 DATA 2.689,2.709,2.744,2.787, 2.838,2.897,2.908,2.928
74 DATA 2.954,2.995,2.986
76 REM *** DEP. VAR (CONSUMPTION) IS NEXT ***
78 DATA 2.637,2.634,2.669, 2.689,2.704,2.734,2.721
80 DATA 2.689,2.709,2.744,2.787, 2.838,2.897,2.908,2.928
82 DATA 2.954,2.995,2.986,2.996
90 PRINT : PRINT"THIS PROGRAM ESTIMATES A LEAST SQUARES EQUATION "
92 PRINT"OF THE FORM   Y = XB      WHERE   Y = (N * 1) VECTOR"
94 PRINT"                                  X = (N * K) MATRIX "
96 PRINT"                                  B = (K * 1) VECTOR"
98 PRINT" Y = DEPENDENT VAR.    X = INDEPENDENT VAR. "
101 REM
102 PRINT" N= NUMBER OF OBSERVATIONS    K= NUMBER OF INDEPENDENT VAR. "
103 PRINT:PRINT"A SAMPLE PROBLEM IS INCLUDED   TO SEE HOW THE PROBLEM WORKS"
104 PRINT"THE EXAMPLE HAS 19 OBS. & 2 INDEP. VAR. "
105 INPUT"DATA SOURCE : 1 = KEYBOARD     2 = EXAMPLE PROBLEM ";B
107 IF B=2 THEN N=19 : K=2 :GOTO 110
108 INPUT"INPUT  N  AND  K  ";N,K
109 INPUT"INPUT NAME OF DEPENDENT VAR. (UP TO 16 CHAR. )";A$
110 FOR I=1 TO 2*N*(K+2)+4*K*K: A(I)=0:NEXTI
111 IF B=2 THEN READ A$
112 IF B=2 THEN 120
```

## MULTIPLE LEAST SQUARES ANALYSIS PROGRAM

This program computes the least squares coefficients for the following equation:

Given Y = XB

where  Y=Nx1 vector of N observations of the dependent variable
X=NxK matrix of N observations of K independent variables
B=Kx1 vector of coefficients

Then the least squares solution is:

$$\hat{B} = (X'X)^{-1}(X'Y)$$

The specifics of the program are:

1) Will take K independent variables (where K is limited by system memory not program)
2) Will take N observations (where N is limited by system memory not program)
3) Allows the user to input the name of the dependent variable
4) A constant is assumed to be included in the equation and is automatically read into the A() vector (when program asks for K do not count the constant as one of the indep. var)
5) The program displays the $(X'X)^{-1}$ matrix. This helps in analyzing multicolinearity.
6) A complete example of a regression analysis is included in the program for examining the outputs. The data is United States quarterly figures from 1956:I to 1960:IV. The data is in dollars($\times 10^{11}$). The estimated equation is a consumption function which estimates consumer consumption using income(disposable) and consumption from the previous period as independent variables. The resulting estimated equation should be

$$C_t = -.34 + .76Y_t + .30C_{t-1}$$

$\qquad\qquad$ (.082) (.079) — standard errors)
$\qquad\qquad$ (9.27) (3.80) — t-value

This equation was reported by Zvi Griliches, et al., in the July 1962 *Econometrica* journal.

The TRS-80 results from the current program are acceptable given the difficulty of the problem (19 obs. & 3 coeff. to estimate and a 3x3 matrix inversion)

7) The outputs from the route are:
   a) name of dependent variable
   b) least squares coefficients (indep.var.)
   c) standard errors of coefficients (est.of $\sigma B$)
   d) t-value of coefficients
   e) R-squared (coefficient of determination)
   f) R-squared (adj. for degrees of freedom)
   g) standard error of equation) est.of $\sigma$)

```
114    FOR J=2 TO K+1
116    PRINT"VAR # ";J-1 :GOTO 140
120    FOR J=1 TO K+1
130    PRINT"VAR # ";J
140    FOR I=1 TO N
142    IF B=2 THEN READ A(I+(J-1)*N+N):GOTO 180
160      INPUT"INPUT OBS.   ";A(I+(J-1)*N+N)
170    A(I+N)=1
180    NEXTI
200    NEXTJ
220    PRINT" NOW INPUT Y "
230    FOR I=1 TO N
232    IF B=2 THEN READ A(I):GOTO 250
240    INPUT A(I)
250    NEXT I
400    CLS
500    PRINT:PRINT" COMPUTING  X'X MATRIX ELEMENT # "
600    REM  ***
690    G=0
695    K=K+1
700    FOR H=1 TO K
800      FOR J=1 TO K
810        G=G+1
820        PRINT@ 95,G
900        FOR I=1 TO N
1000         L=H*N+I
1100         P=J*N+I
1200         Q=K*N+G+N
1300         A(Q)=A(L)*A(P)+A(Q)
1400       NEXTI
1500     NEXTJ
1600   NEXTH
1700   REM
1800   CLS
3020     CLS
3030   PRINT@ 0," INVERTING MATRIX            J ="
3100   GOSUB 25000
3200   GOSUB 26000
3210   REM
3220   REM  ****************************************
3230   REM  *  B-HAT FROM  (X'X)INV. & (X'Y)       *
3240   REM  ****************************************
3245   REM
3247   CLS
3250   PRINT:PRINT"THE DEPENDENT VAR. = ";A$
3260   G=0
3270   FOR J=1 TO R
3300     G=G+1
3310     FOR I=1 TO R
3320       L=N*(R+1)+R*R*I+(J-1)*R
3330       P=N*(R+1)+R*R*2+I
3340       Q=N*(R+1)+R*R*2+1+(R-1)+G
3400       A(Q)=A(L)*A(P)+A(Q)
3410     NEXT I
3600   NEXT J
4000   REM
4010   REM  **** GOTO SUB. FOR SUMMARY STATISTICS ****
4020   GOSUB 27000
4030   REM
4940   IF B<>2 THEN 9990
9990   PRINT : INPUT"AGAIN (TYPE 1 ELSE 0)";B
9992   IF B=1 THEN 101
9999   END
25000  REM  ****************************************
25002  REM  *  INVERSE BY GAUSS-JORDAN METHOD  *
25004  REM  ****************************************
25006  REM
25010  REM  **  INPUT IS A  (K,K) MATRIX   OF  X'X  **
25020  REM  **  STARTING POINT OF MATRIX IS N*(K+1)  **
25110    R=K
25120  FOR J=1+N*(R+1) TO R+N*(R+1)
25150    M=R*R+J+(J-(N*(R+1))-1)*R
25160      A(M)=1
25170  NEXT J
25175  REM
25180  REM  ***  INVERT MATRIX  ***
25190  REM
25210  REM
25220  REM  ****  BIG LOOP STARTS ****
25230  REM
25235   G=0
25240  FOR J=1+N*(R+1) TO R+N*(R+1)
25242     G=G+1 : PRINT@ 29,G
25250     FOR I=J TO R+N*(R+1)
25260       H=I+(J-(N*(R+1))-1)*R
25270       IF A(H) <> 0 THEN 25310
25280     NEXT I
25290     PRINT:PRINT"SINGULAR MATRIX  -  CANNOT BE INVERTED "
25300     GOTO 9990
25310     FOR K=1+N*(R+1) TO R+N*(R+1)
25320       M=K+(J-(N*(R+1))-1)*R
25330       S=A(M)
25340       V=K+(I-(N*(R+1))-1)*R
25350       A(M)=A(V)
25360       A(V)=S
25370       S=A(M+R*R)
25380       A(M+R*R)=A(V+R*R)
25390       A(V+R*R)=S
25400     NEXT K
25410     P=J+(J-(N*(R+1))-1)*R
25420     T=1/A(P)
```

```
25430     FOR K=1+N*(R+1) TO R+N*(R+1)
25440       M=K+(J-(N*(R+1))-1)*R
25450       A(M)=T*A(M)
25455       A(M+R*R)=T*A(M+R*R)
25460     NEXT K
25470     FOR L=1+N*(R+1) TO R+N*(R+1)
25480       IF L=J THEN 25570
25490       M=J+(L-(N*(R+1))-1)*R
25500       T=-A(M)
25510       FOR K=1+N*(R+1) TO R+N*(R+1)
25520         V=K+(L-(N*(R+1))-1)*R
25530         Q=K+(J-(N*(R+1))-1)*R
25540         A(V)=A(V)+T*A(Q)
25550         A(V+R*R)=A(V+R*R)+T*A(Q+R*R)
25560       NEXT K
25570     NEXT L
25580 NEXT J
25590 REM
25600 REM ***  PRINT RESULTANT MATRIX  ***
25610 REM
25615 PRINT:PRINT"  (X'X) INVERSE IS : "
25620 FOR I=1+N*(R+1) TO R+N*(R+1)
25630   FOR J=1+N*(R+1) TO R+N*(R+1)
25640     L=J+(I-(N*(R+1))-1)*R+R*R
25650     PRINT A(L);" ";
25660   NEXT J
25670   PRINT
25680 NEXT I
25700 INPUT"PRESS ENTER TO CONTINUE";Q
25990     RETURN
26000 REM
26100 REM ********************************************
26120 REM *  SUBROUTINE FOR COMPUTING X'Y          *
26130 REM ********************************************
26140   REM
26200   G=0
26210   FOR J=1 TO R
26220     G=G+1
26225     FOR I=1 TO N
26230       L=J*N+I
26240       P=I
26250       Q=(R+1)*(R+N)+(R-1)*R+G
26260       A(Q)=A(L)*A(P)+A(Q)
26270     NEXT I
26280   NEXT J
26990 RETURN
27000 REM
27010 REM ********************************************
27020 REM *  SUBROUTINE FOR COMPUTING LS SUMMARY STATS *
27030 REM
27040 REM ********************************************
27100 REM ***   R SQR.  ***
27105 K=R : S=0
27110 FOR I=1 TO N
27120   S=S+A(I)
27130 NEXT I
27140 M=S/N
27154 V=0:S=0
27160 FOR I=1 TO N
27178   Y=0
27180   FOR J=1 TO K
27184     D=N+I+(J-1)*N
27186     C=N*(K+1)+K*K*2+K+J
27200     Y=Y + A(C)*A(D)
27210   NEXT J
27220   E=A(I) - Y
27240   E=E*E
27250   S=S+E
27260   T=A(I)-M
27270   T=T*T
27280   V=V+T
27290 NEXT I
27300 R=1 - (S/V)
27302 S=S/(N-K)
27340 REM
27342 PRINT
27345 PRINT"   LS COEFF.     STD. ERROR     T-VALUE"
27347 PRINT"   ----------     ----------    ----------"
27350 FOR I=1 TO K
27360   Q=N*(K+1)+K*K+(I-1)*K + I
27370   L=N*(K+1)+K*K*2+K + I
27400   PRINT"B(";I-1;") = ";A(L),
27402   X=S*A(Q) : GOSUB 28100
27404   PRINTY,
27406   PRINTA(L)/Y
27410 NEXT I
27430 PRINT : PRINT"R-SQR =";INT(R*1000+.5)/1000
27432 R=1-(1-R)*(N-1)/(N-K)
27440 PRINT"R-SQR (ADJ) = ";INT(R*1000+.5)/1000
27450 X=S : GOSUB 28100
27460 PRINT"STD. ERROR = ";Y
27990   RETURN
28000 REM
28010 REM ********************************************
28020 REM * SUB FOR SQUARE ROOT  -  X IN  Y OUT *
28030 REM ********************************************
28040 REM
28100 Y=X/2 : Z=0
28110 W=(X/Y-Y)/2
28120 IF (W=0) + (W=Z) THEN RETURN
28130 Y=Y+W : Z=W :GOTO 28110
```

8) The program proceeds as follows:

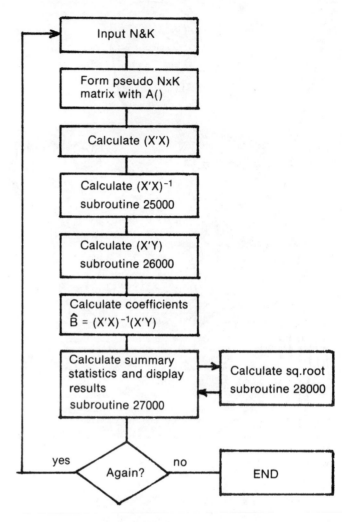

"Where did you learn to debug a program, Haverstraw?"

# A ROMAN'S ASSIGNMENT PROBLEM

## John M. Anderson

One of the classical problems encountered by students of computer applications is that of how to assign or match one group of objects to another group of objects in the best possible way. Dubbed the "assignment problem," a procedure for solving the problem was developed almost fifty years ago by the Hungarian mathematician E. Egervary. Our purpose here is to tantalize the dabbler in algorithms with a challenge to discover further applications of this amazing "Hungarian Algorithm."

Consider the problem of Dilirius the Roman promoter—namely that of matching mean and nasty gladiators against wild and vicious beasts in the Coliseum of ancient Rome. A good contest was appreciated by the affluent Romans who demonstrated their pleasure with cheers and a shower of coins on the winner, who later split the loot with the promoter; unless of course the winner was a dumb animal, in which case the promoter was left to handle all of the winnings. A poor contest was met with booing and whistling and a shower of less valuable missiles such as half-eaten pizzas. On a given day, Dilirius tried his best to provide a group of contests which pleased the crowd which he hoped would increase the total loot thrown and with it his share

It was clear to Dilirius that the weight of a spectator's purse at the end of the day was directly related to the total

John M. Anderson, Department of Business and Economics, University of North Carolina at Wilmington, Wilmington, North Carolina 28406

level of his booing for all of the contests, so he sought to match the gladiators and animals in such a way that the overall level of booing was kept to a minimum. As a first step toward solving the problem, Dilirius used his judgement and an arbitrary numerical scale of one to one-thousand to estimate the level of booing that would result from a particular match. For example, if he matched Clodius and a frothing opossum, his estimate might have been recorded as "Booing CCCMX LVII." (The reader should not confuse the arbitrary heuristic described here with the more refined concepts of Boolean algebra.) Estimates for all possible matches were scratched on a clay tablet, an example of which appears below:

Figure I
SAMPLE BOOING TABLET*

TABLET
OF
BOOS

|  | OPOSSUM | LION | WILDEBEEST | TIGER |
|---|---|---|---|---|
| CLODIUS | 6 | 2 | 8 | 4 |
| GAUDIUS | 5 | 9 | 4 | 3 |
| PLODIUS | 5 | 12 | 8 | 9 |
| SNODIUS | 15 | 30 | 20 | 18 |

*Ratings are translated to Arabic sympols for this example. They are further expressed in hundreds.

If he chose the matches in such a way that the total of all booing for the day was a minimum, then it was likely that the spectators' purses were as light as possible at the end of the day and he was quite well off indeed. On the other hand, a group of matches which increased the total level of booing reduced the promoter's returns. Dilirius studied the tablet shown in Figure I and began in row 1 matching Clodius with the lion, thereby minimizing the boos for that match. Moving to row 2, he matched Gaudius with the tiger which again kept the boos to a minimum for a match. Plodius was then matched with the opossum and Snodius was left with the wildebeest. This rather simple method of solving the problem resulted in a total booing level of 30. It seemed natural to Dilirius to start in the upper left-hand corner of the tablet (northwest corner) and work his way down, row by row. Such an approach to finding quick and easy solutions is still widely used in many such problems. In fact, over the years this algorithm has come to be known as the "Greedy Algorithm." (I have often wondered if it was named for Horace Greedy...I seem to recall a famous admonishment: "Go northwest young man!" Alas, the name really comes from the manner in which you move systematically through the tablet, grabbing greedily the best matches you can as you go.)

As it turns out, Dilirius could have scheduled the matches in Figure I with a total booing level of 28, thus bringing him a few more mites. Using trial and error on a small problem such as this, even the wildebeest could have found the best set of matches. Try your hand at it...

Clodius vs. _____ = ____ boos
Poldius vs. _____ = ____ boos
Gaudius vs. _____ = ____ boos
Snodius vs. _____ = ____ boos
Total Boos = ____ boos

On more festive occasions such as the Caesar's birthday, Dilirius was faced with scheduling as many as twenty matches. Such a list of matches could be scheduled in over 2,430,000,000,000,000,000 ways! Even the penurious Dilirius could not evaluate all possible schedules for this event. If he could evaluate one different schedule each second (which is unlikely), it would take him over 77,000,000,000 years to finish the job—and that's without taking a break for wine! Unfortunately for Dilirius, more than two thousand years passed before an easy way of selecting the best tournament schedule emerged. Let's look at how modern promoters might solve such a problem.

To begin, assume that no wild beast was compelled to combat more than one wild gladiator and vice versa. Therefore, a schedule always consisted of a single entry in each row and each column of the booing table. Since there must be one entry chosen in each row (or column), we can conclude that it is the relationship between those entries that must be important to the problem and not the absolute size of any particular entry. Therefore, we can reduce all of the entries in a row (or column) by the same amount without disturbing the relationships between those entries—the smallest before the reduction is still the smallest after the reduction; the largest before the reduction is still the largest after the reduction; and so forth. Also, notice that the relationship between rows and columns is not disturbed. If we systematically reduce each row and column by an amount equal to the smallest element in a row or column, then we will produce a new table of what could be called "relative boos." In fact, there will be at least one element in each row and column which represents zero relative boos.

If we could find a set of those zeroes which made up a complete schedule of matches, then we would have a total of zero relative boos for all of the matches. That would certainly be the best we could do. The Hungarian Algorrithm systematically reduces the entries in the table until the table contains the minimum number of zero entries needed for the construction of a zero-relative-boo schedule, and then identifies such a schedule.

This writer's program, which is based upon a modified version of the Hungarian Algorithm developed by J. Munkres, is listed below. The program, called ASSIGN, is written in Hewlett-Packard BASIC. A sample run solving the problem in Figure I follows the program listing. You can enter any N by N table, where N is no larger than 20. If larger tables are needed, then the DIM statements must be changed to set up additional storage.

As a footnote to the above example, you may be interested to know that poor scheduling was the eventual undoing of Dilirius who was attacked by a frothing gladiator after the gladiator was subdued by a frustrated opossum. One could say that it certainly wasn't the boos that made him Diliruis, but it was the boos that led to his downfall. ∎

Program Next Page

```
LIST
ASSIGN

10    DIM A[20,20],P[20,2],R[20],C[20],Z[40,2],X[20,20]
11    DIM W[20,20]
20    PRINT "FOR AN NXN MATRIX, INPUT THE VALUE OF N: ";
30    INPUT N
40    PRINT "ENTER THE MATRIX ONE ROW AT A TIME, WITH COMMAS"
50    PRINT "SEPARATING EACH ELEMENT:"
60    MAT  INPUT X[N,N]
70    FOR I=1 TO N
80    FOR J=1 TO N
90    W[I,J]=X[I,J]
100   NEXT J
110   NEXT I
120   PRINT " "
130   PRINT "THE COST MATRIX IS:"
140   PRINT " "
150   MAT  PRINT X;
160   PRINT " "
170   FOR I=1 TO N
180   M1=9999
190   FOR J=1 TO N
200   IF X[I,J]<M1 THEN 220
210   GOTO 230
220   M1=X[I,J]
230   NEXT J
240   FOR J=1 TO N
250   X[I,J]=X[I,J]-M1
260   NEXT J
270   NEXT I
280   FOR J=1 TO N
290   M1=9999
300   FOR I=1 TO N
310   IF X[I,J]<M1 THEN 330
320   GOTO 340
330   M1=X[I,J]
340   NEXT I
350   FOR I=1 TO N
360   X[I,J]=X[I,J]-M1
370   NEXT I
380   NEXT J
390   MAT A=ZER[N,2]
400   MAT C=ZER[N]
410   MAT Z=ZER
420   MAT P=ZER
430   MAT R=ZER[N]
440   L=1
450   FOR I=1 TO N
460   J=1
470   IF J>N THEN 640
480   IF X[I,J]=0 THEN 510
490   J=J+1
500   GOTO 470
510   IF I=1 THEN 600
520   L1=L-1
530   FOR K=1 TO L1
540   IF (A[K,2])#J THEN 560
550   GOTO 580
560   NEXT K
570   GOTO 600
580   J=J+1
590   GOTO 470
600   A[L,1]=I
610   A[L,2]=J
620   C[J]=1
630   L=L+1
640   NEXT I
650   L=L-1
660   IF L=N THEN 1600
670   M=1
680   FOR I=1 TO N
690   J=1
700   IF J>N THEN 860
710   IF X[I,J]#0 THEN 750
720   IF C[J]#0 THEN 750
730   IF R[I]#0 THEN 750
740   GOTO 770
750   J=J+1
760   GOTO 700
770   P[M,1]=I
780   P[M,2]=J
790   M=M+1
800   FOR K=1 TO L
810   IF (A[K,1])=I THEN 840
820   NEXT K
830   GOTO 880
840   R[I]=1
850   C[A[K,2]]=0
860   NEXT I
870   GOTO 1380
880   K2=M-1
890   R1=P[K2,1]
900   C1=P[K2,2]
910   K3=L
920   K=1
930   S=1
940   IF K=1 THEN 1110
950   GOTO S OF 960,1040
960   FOR J=1 TO K3
970   IF A[J,2]=C1 THEN 1010
980   NEXT J
990   K=K-1
1000  GOTO 1150
1010  R1=A[J,1]
1020  S=2
1030  GOTO 1110
1040  FOR J=1 TO K2
1050  IF P[J,1]=R1 THEN 1090
1060  NEXT J
1070  K=K-1
1080  GOTO 1150
1090  C1=P[J,2]
1100  S=1
1110  Z[K,1]=R1
1120  Z[K,2]=C1
1130  K=K+1
1140  GOTO 940
1150  K5=1
1160  IF K5=K THEN 1260
1170  FOR I=1 TO L
1180  IF A[I,1]#Z[K5+1,1] THEN 1210
1190  IF A[I,2]#Z[K5+1,2] THEN 1210
1200  GOTO 1220
1210  NEXT I
1220  A[I,1]=Z[K5,1]
1230  A[I,2]=Z[K5,2]
1240  K5=K5+2
1250  GOTO 1160
1260  L=L+1
1270  A[L,1]=Z[K,1]
1280  A[L,2]=Z[K,2]
1290  IF L=N THEN 1600
1300  MAT P=ZER
1310  MAT R=ZER
1320  MAT C=ZER
1330  FOR I=1 TO L
1340  C[A[I,2]]=1
1350  NEXT I
1360  M=1
1370  GOTO 680
1380  M1=9999
1390  FOR I=1 TO N
1400  IF R[I]#0 THEN 1470
1410  FOR J=1 TO N
1420  IF C[J]#0 THEN 1460
1430  IF X[I,J]<M1 THEN 1450
1440  GOTO 1460
1450  M1=X[I,J]
1460  NEXT J
1470  NEXT I
1480  FOR I=1 TO N
1490  FOR J=1 TO N
1500  IF R[I]#0 THEN 1540
1510  IF C[J]#0 THEN 1540
1520  X[I,J]=X[I,J]-M1
1530  GOTO 1570
1540  IF R[I]#1 THEN 1570
1550  IF C[J]#1 THEN 1570
1560  X[I,J]=X[I,J]+M1
1570  NEXT J
1580  NEXT I
1585  PRINT "INTERMEDIATE MATRIX"
1586  MAT  PRINT X;
1590  GOTO 680
1600  Q=0
1610  FOR I=1 TO N
1620  Q=Q+W[A[I,1],A[I,2]]
1630  NEXT I
1640  PRINT " "
1650  PRINT "MINIMUM TOTAL COST IS:";Q
1660  PRINT " "
1670  PRINT "THE SOLUTION PAIRS FOLLOW, WITH ROW
                                         INDEX FIRST:"
1680  PRINT " "
1690  MAT PRINT A;
1700  PRINT " "
1710  PRINT "THE FINAL MATRIX FORM FOLLOWS:"
1720  PRINT " "
1730  MAT  PRINT X;
1740  STOP
1750  END
```

```
RUN
ASSIGN

FØR AN NXN MATRIX, INPUT THE VALUE ØF N: ?4
ENTER THE MATRIX ØNE RØW AT A TIME, WITH CØMMAS
SEPARATING EACH ELEMENT:
?6,2,8,4
??5,9,4,3
??5,12,8,9
??15,30,20,18

THE CØST MATRIX IS:

    6     2     8     4
    5     9     4     3
    5    12     8     9
   15    30    20    18


INTERMEDIATE MATRIX
    4     0     3     0
    4     8     0     0
    0     7     0     2
    0    15     2     1
```

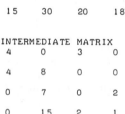

```
MINIMUM TØTAL CØST IS: 28

THE SØLUTIØN PAIRS FØLLØW, WITH RØW INDEX FIRST:

    1     2
    3     3
    4     1
    2     4


THE FINAL MATRIX FØRM FØLLØWS:

    4     0     3     0
    4     8     0     0
    0     7     0     2
    0    15     2     1

DØNE
```

# DAYS AND DATES

**James Reagan**
**Mathematics Teacher**
**Stevenson High School**
**Sterling Heights, Michigan**

Dates become important and remembered because of their importance. You remember your birthday, that perfect date, a confirmation or bar mitzvah, a marriage, divorce, death, birth, or graduation date of yourself or your love. These are personal. Remember the dates? Sure. Remember the day of the week? No? I didn't think so. But now, to take you back in your memory lane to that fond or dreaded day there is a find-the-day-of-the-week formula known as Zeller's Congruence.

If you don't care about your personal past, how knowledgeable are you about your historical past? Try the quiz to see.

**An Illustrative Quiz**

Provide the date and the day of the week for each of the following events.

1. The stock market crashes beginning the Great Depression.
2. The Second Continental Congress adopts the Declaration of Independence.
3. Japan attacks Pearl Harbor.
4. President Lincoln is assssinated at Ford's Theater.
5. The bombardment of Fort Sumter begins the Civil War.
6. General Custer makes his "last stand" at Little Big Horn.
7. Russia launches Sputnik I, the first artificial satellite.
8. The United States of America drops an atomic bomb on Hiroshima, Japan.
9. President Kennedy is slain by an assassin's bullet in Dallas, Texas.
10. The oceanliner Lusitania is sunk by German U-boat torpedoes killing 1198 persons including 124 Americans.
11. German armies invade Poland starting World War Two.
12. The United States Supreme Court rules in the case of *Brown v Board of Education* that separate schools based upon skin color are inherently unequal.
13. South Korea is invaded by North Korean troops.
14. Richard M. Nixon resigns as President of the United States of America.
15. D-day. Allied troops land in Normandy, France.

Scoring: Count 1 point each for month, day of the month, and year; count 5 points for correct day of the week. There are a possible 8 points for each event with 120 possible points for the quiz. If you scored 0-10 points you are about average; 11-20 points above average; 21-40 points superior; 41-80 points unbelievable; 81-120 points an historical nut — congratulations!!!

For those of you who need help 1) reread history books for the date and 2) utilize Zeller's Congruence to determine the day of the week for any particular date. The formula is:

$$F = (\lfloor 2.6m - 0.2 \rfloor + k + d + \frac{d}{4} + \frac{c}{4} - 2c) \bmod 7.$$

In this formula, F will have a value 0, 1, 2, 3, 4, 5, or 6; the corresponding day of the week is Sunday, Monday, Tuesday, Wednesday, ..., or Saturday. The modulus 7 can be thought of as the remainder when the value of the parenthetical expression is divided by 7.

The righthand side of the congruence contains the variables described as follows:
- k is the day of the month,
- c is the number of hundreds in the year,
- d is the year in the century, and
- m is the month number, but not the layman's month number.
  January and February are month numbers 11 and 12 of the preceding year (affecting d and possibly c described above), March is month number 1, April is month 2, May is 3, ..., and December is month number 10.

The square brackets, $\lfloor \ \rfloor$, indicate that the "greatest integer value" is to be applied to the included expression. A specific example follows.

### Example

The date is October 12, 1956. In layman's terms the date is expressed as 10, 12, 1956. For Zeller's Congruence we use m = 8, k = 12, c = 19, and d = 56.

Substituting these values into the right side of the congruence we have

$$F = (\lfloor 2.6*8 - 0.2 \rfloor + 12 + 56 + \frac{56}{4} + \frac{19}{4} - 2*19) \bmod 7$$

$$= (\lfloor 20.8 - 0.2 \rfloor + 12 + 56 + \lfloor 14 \rfloor + \lfloor 4.75 \rfloor - 38) \bmod 7$$

$$= (20 + 12 + 56 + 14 + 4 - 38) \bmod 7$$

$$= (68) \bmod 7$$

$$= 5 \bmod 7.$$

Thus, we conclude that the day of the week is Friday.
In the application of the formula the following mapping may be a helpful study guide.

| LAYMAN'S NOTATION | FORMULA REQUIRES | F VALUE COMPUTED | DAY OF THE WEEK |
|---|---|---|---|
| 10-12-1956 | 8,12,19,56 | 5 | Friday |
| 9-18-1963 | 7,18,19,63 | 3 | Wednesday |
| 12-25-1972 | 10,25,19,72 | 1 | Monday |
| 3- 9-1929 | 1, 9,19,29 | 6 | Saturday |
| 2- 6-1976 | 12, 6,19,75 | 5 | Friday |
| 1-13-1970 | 11,13,19,69 | 2 | Tuesday |
| 1- 1-2000 | 11, 1,19,99 | 6 | Saturday |

It might be helpful to understand that the month numbers for the application of the congruence begin with March = 1 and continue to the following February = 12; in this way any leap year day is placed at the end of the formula year.

### The First Problem

Write a program that will accept any date in layman's terms and print the corresponding day of the week. The program must provide the translation for the application of the variables used in the congruence. For example, if one types 1,13,1974 the program must translate these values to 11,13,19,73 for the corresponding values of m, k, c and d, respectively. Using this program you may verify the days of the week for the dates of the Illustrative Quiz.

### The Second Problem

Superstitions have developed over the history of man. Many people are superstitious of certain events; those who are not superstitious have some knowledge of the superstitions. Some of the events associated with "bad luck" are: walking under a ladder, having a black cat cross one's path, and breaking a mirror. Perhaps the most well known of all superstitions involves "Black Friday," the description of Friday the Thirteenth.

This year, 1976, has two Friday the Thirteenths; one occurred in February and the other in August. This may be verified by a search of the calendar or by observation of a perpetual calendar.

The second problem becomes one of modifying the program produced to solve the first problem: produce a list of Friday the Thirteenths over a given interval of years. For example, produce a list of Friday the Thirteenths for the years from 1977 to 1980.

### The Third Problem

This third problem might be investigated using the computer program produced for the second problem. However, there is also a rigorous mathematical proof of the conjectures motivated by the computer investigation.

The problem is stated in the form of two questions:
1. What is the most number of Friday the Thirteenths in any given year?
2. Is there any year that does not have at least one Friday the Thirteenth?

### The Fourth Problem

Some workers are paid bi-weekly, that is they are paid every-other week. The traditional payday is Friday. In a given year there are some months that have 5 Fridays; two of these months occur so that there are 3 paydays in that month, one on each of the first, third and fifth Fridays. The month of February has 4 of each day of the week except in years that are leap years; then one day occurs five times. If that day that occurs five times is Friday, there is a possibility that three paydays may occur in that month.

In what years will February have five Fridays? How often does this occur? If one has bi-weekly pay-periods and one of them does occur on the first Friday of a leap year February beginning on Friday, will the same situation occur again in the worker's lifetime?

### Answers To Illustrative Quiz

1. October 24, 1929; Thursday
2. July 4, 1776; Sunday
3. December 7, 1941; Sunday
4. April 14, 1865; Tuesday
5. April 12, 1861; Friday
6. June 25, 1876; Sunday
7. October 4, 1957; Friday
8. August 6, 1945; Monday
9. November 22, 1963; Friday
10. May 7, 1915; Friday
11. September 1, 1939; Friday
12. May 17, 1954; Monday
13. June 25, 1950; Monday
14. August 9, 1974; Friday
15. June 6, 1944; Tuesday

### ANSWERS TO COMPUTER LITERACY QUIZ

| | | | | |
|---|---|---|---|---|
| 1. F | 7. F | 13. 4 | 19. F | 25. F |
| 2. F | 8. F | 14. T | 20. T | 26. T |
| 3. 1 | 9. T | 15. T | 21. F | 27. T |
| 4. T | 10. 1 | 16. 2 | 22. F | 28. 4 |
| 5. 5 | 11. T | 17. 3 | 23. F | 29. 4 |
| 6. 2 | 12. 3 | 18. T | 24. 4 | 30. F |

# Non-Usual Mathematics for Computer Solution
James Reagan

## Introduction

Mathematics instruction generally proceeds sequentially and deductively. This instructional procedure creates some misconception of the mathematics. Mathematics is not totally deductive logic; the deductive proof of any hypothesis is developed after one has become quite certain that the conjecture is true. One investigates enough specific cases to become somewhat sure that the observed cases generalize or that the proper limits on the conjecture have been found. Thus, there is a contradiction between the mathematics in its instruction and mathematics in its historical development. In formal instruction in mathematics the discovery of the theorems, rules, and properties are taught as though they were bestowed upon man as were the two tablets containing the Ten Commandments; the time and effort expended are seldom discussed. Most mathematics courses offer the student the deductive process in developing the material when, historically, the deductive process was employed late in the development of the topic.

Because of mathematics instruction's dependence upon deductive development, certain topics fall before or after certain other topics; and mathematics instruction has become characterized by its sequential approach. It is true that there are certain foundations upon which some topics rest; these pre-requisites are necessary for the development of the vocabulary and the organization of latter theorems. With the use of computers in many schools, some of the latter topics can be studied out of sequence.

Agreed, there is what might be called mathematics sophistication before one can *master* certain topics, but how much mathematics sophistication is required to *understand* and *appreciate* the material? It is this writer's experience that students in the secondary school can investigate topics and solve problems prior to the traditional time location of the topic or problem in the instructional sequence. Many topics commonly deferred to the college curriculum are suitable and interesting for the secondary school student.

What follows in this series are examples of such problems that have been studied and solved by high school students in computer programming classes at Stevenson High School and many other high schools having computer access.

## Infinitely Many Primes

### Background of the Problem

The great mathematician of the third century B. C., Euclid, proved that there are infinitely many primes. Euclid's proof leads to interesting problems some 2000 years later.

First the proof and then the problems.

The proof is by *reductio ad absurdum,* an indirect proof.

Suppose there are finitely many prime numbers. Then, these $n$ primes can be listed in order.

$$2, 3, 5, \ldots, P_n.$$

Form a number N by adding 1 to the product of the $n$ primes:

$$N = 2*3*5*\ldots*P_n + 1.$$

Either N is prime or N is composite. Each of these results for N leads to a contradiction that $P_n$ is the largest prime.

First, if N is prime, then it is clearly greater than $P_n$ and $P_n$ is not the greatest prime.

Second, suppose N is composite. It has a prime factor $p$. This prime factor $p$ cannot be one of the primes $2, 3, 5, \ldots, P_n$, since dividing each of the primes in the list into N leaves a remainder of 1. Thus, $p$ must be a prime greater than $P_n$.

Therefore, there are infinitely many prime numbers. QED.

### Statement of the Problem

The creation of the number N in the proof by Euclid leads to many interesting questions.

Create a set of numbers by the recursive definition:

$P_1 = 2$
$P_2 = 3$
$P_3 = P_1*P_2 + 1 = 2*3 + 1 = 7$
$P_4 = P_1*P_2*P_3 + 1 = 2*3*7 + 1 = 43$
.
.
.
$P_{n+1} = P_1*P_2*P_3*\ldots*P_n + 1.$

Are each of the numbers in the set prime?

If some number in the list is not prime, is a prime factor of it greater than the preceding number in the list? For example, if $P_6$ is composite, is a prime factor of $P_6$ greater than $P_5$?

These same questions apply to a second set that can be created by subtracting 1 instead of adding 1 to the previous list product.

Create a second set as follows:

$P_1 = 2$
$P_2 = 3$
$P_3 = P_1*P_2 - 1 = 2*3 - 1 = 5$
.
.
.
$P_{n+1} = P_1*P_2*\ldots*P_n - 1.$

Answer the same questions as with the first set. Finally, create a third set and investigate.

$P_1 = 2$
$P_2 = 3$
$P_3 = 5$
$P_4 = P_1*P_2*P_3 + 1 = 2*3*5 + 1 = 31$
.
.
.
$P_{n+1} = P_1*P_2*\ldots*P_n + 1.$

### Hints

1. One of the constraints on the computations is the number of significant digits of the computing machine. The numbers in each of the sets become large rapidly and can soon overflow the significant digit capacity of the machine. An extended precision routine may be needed to investigate far into the sets. [See "Computing Factorials — Accurately" by Walter Koetke, *Creative Computing* Vol 1, No 3, pp 9–11.]

2. To save on variable storage during execution of the program, instead of using high dimensioned vectors, it might be helpful to create a file in which to store new numbers in the set and from which to read out previous numbers.

# Computerized Sports Predictions

## Don Smith

An interesting application for home computing is forecasting the outcomes of sporting events. Since large sums of money are wagered on these outcomes, a great deal of effort goes into obtaining information and working out prediction strategies. Professional oddsmakers have entire staffs working to obtain the latest information on the teams. Computers have sometimes been used to keep track of the data and make predictions. The prediction algorithms can be very complex, taking into account much detailed information.

This article presents a sports prediction algorithm developed with the following goals in mind:
1. It should be simple enough for implementation on a small home computer.
2. It should require a minimum of data entry from week to week.
3. It should develop numerical ratings for each team so that the predicted point spread for a game is the difference in the ratings of the two teams involved.
4. It should "learn," adjusting team ratings as the season progresses.
5. It should consider the quality of opponents in making and adjusting its ratings.
6. It should consider home-team advantage.

Since the author is interested in professional football, and since there are a reasonable number of teams (28) all playing exclusively among themselves, NFL football seemed a likely test bed for the development of the prediction algorithm. The program listing shows the algorithm as applied to NFL football, implemented on the Tektronix 4051. The sample run shows the three output tables generated by the program for the fourth week of the 1977 NFL season. Table 1 gives the results for week 4, Table 2 lists the updated ratings of the 28 teams sorted in ratings order, and Table 3 lists the prediction for week 5.

The 1977 matchups are stored in a matrix M, where M(I,J) is the number of the opponent of team I in week J. Team numbers are as follows:

Don Smith, PhD., G519 Plaza Parkway #123, Ft. Worth, Texas 76116.

| AFC | | | NFC | |
|---|---|---|---|---|
| 1. BALTIMORE | | | 15. NEW YORK GIANTS | |
| 2. NEW ENGLAND | | | 16. DALLAS | |
| 3. MIAMI | | EAST | 17. PHILADELPHIA | |
| 4. BUFFALO | | | 18. ST. LOUIS | |
| 5. NEW YORK JETS | | | 19. WASHINGTON | |
| 6. PITTSBURGH | | | 20. CHICAGO | |
| 7. HOUSTON | | | 21. GREEN BAY | |
| 8. CLEVELAND | | CENTRAL | 22. DETROIT | |
| 9. CINCINNATI | | | 23. MINNESOTA | |
| | | | 24. TAMPA BAY | |
| 10. OAKLAND | | | 25. ATLANTA | |
| 11. DENVER | | | 26. NEW ORLEANS | |
| 12. KANSAS CITY | | WEST | 27. LOS ANGELES | |
| 13. SEATTLE | | | 28. SAN FRANCISCO | |
| 14. SAN DIEGO | | | | |

NFL Team Numbers

If an element of M is negative, that number represents the home team. For example, M(1,3) = 4 means that Baltimore (1) plays Buffalo (4) the third week of the season at Baltimore. M(5,1) = -7 means that the New York Jets (5) play at Houston (-7) the first week. Typing the entire season of matchups into DATA statements is a little work, but it avoids having to enter the matchups each week. Note that each DATA statement (lines 140-410) gives the 14 opponents of a team for the 1977 season.

After the user enters the week number (I), the program reads the current team ratings into array E from tape, and adds the home advantage (F1) where applicable, to obtain array G. The "actual deltas," or point spreads, for week I are read into array D. For example, if Houston beat Buffalo by 10 points, D(7) would be 10, and D(4) would be -10. If Miami lost to New England by 16, then D(3) = -16 and D(2) = 16. The program computes the predicted point spread A1 in line 810 and compares it to the actual point spread, incrementing counter C when the prediction was correct. Note that the program predicts no ties — if a team is favored by only 0.01 points it is still the predicted winner. If the favored team wins, the prediction is counted as correct, even if the predicted point spread was wrong (as it usually will be). For example, in Table 1, even though the predicted point spread for Baltimore over Miami was 1.53, and the actual point spread was 17, the prediction is counted as correct.

The team ratings are adjusted in lines 880 and 890. Since this adjustment is the heart of the algorithm, it deserves some explanation. The "ratings adjustment equation" is

880 E(K1) = E(K1) + H * (M1 - A1)

where

K1 = team index
E  = ratings array
H  = adjustment factor
M1 = actual point spread of game, limited to ± N2
A1 = predicted point spread

N2 is a maximum point spread for adjusting team ratings. M1 is calculated in line 790. This maximum spread is used because in a runaway game, beyond a certain number of points, the point spread is no longer significant.

The essence of line 880 is that if a team does better than expected, its rating is increased, and if it does worse than expected, its rating is decreased. The adjustment factor H is chosen large enough to make the algorithm sensitive to actual results, but not so large as to make the algorithm erratic.

## PROGRAM LISTING

This adjustment algorithm satisfies the six goals in the following ways:
1. It is simple.
2. It requires only actual point spreads from the games of the previous week as input.
3. The ratings automatically converge to numbers which yield predicted point spreads, since the diference between the predicted and the actual point spread drives the adjustment.
4. It learns. The program can be started with all teams equally rated at the beginning of a season, and it will learn which teams are good and which are bad.
5. It automatically considers the quality of the opposition. For example, if Tampa Bay lost to Pittsburgh by only one point, the algorithm would increase Tampa Bay's rating, and decrease Pittsburgh's rating.
6. It considers home-team advantage.

Three parameters in the program, F1 (home-advantage), H (adjustment factor), and N2 (maximum point spread) are given numerical values in lines 440 and 450. How were these values obtained? The 1976 NFL season schedule and results were put into a separate test program. The prediction algorithm was then cycled through all 14 games of the 1976 season for various values of F1, H and N2. At the beginning of each run, the ratings array E was set to 10, so all teams were initially considered equal. The values of F1, H and N2 were optimized to produce the highest percentage of correct predictions. The table below shows some typical parameter values and the resulting percentage of correct predictions.

```
100 REM    - - -    NFL PREDICTION PROGRAM    - - -
110 REM                BY DON SMITH
120 REM --- PUT 1977 MATCHUPS INTO ARRAY M ---
130 DIM D(28),M(28,14),E(28),Y(28),G(28)
140 DATA -13,-5,4,3,-12,-2,6,19,-4,5,-11,-3,22,2
150 DATA 12,-8,-5,13,-14,1,5,4,-3,-4,17,-25,3,-1
160 DATA -4,-28,7,-1,5,13,14,-5,2,-9,-18,1,-2,4
170 DATA 3,-11,-1,5,25,8,-13,-2,1,2,-10,19,-5,-3
180 DATA -7,1,2,-4,-3,10,-2,3,13,-1,6,-26,4,-17
190 DATA 28,10,-8,-7,9,7,-1,-11,8,16,-5,13,-9,-14
200 DATA 5,-21,-3,6,8,-6,-9,20,-10,-13,12,11,-8,9
210 DATA -9,2,6,10,-7,-4,12,9,-6,-15,27,-14,7,-13
220 DATA 8,13,-14,-21,-6,11,7,-8,-23,3,15,-12,6,-7
230 DATA 14,-6,-12,-8,11,-5,-11,13,7,-14,4,-27,23,12
240 DATA 18,4,-13,12,-10,-9,10,6,-14,-12,1,-7,14,-16
250 DATA -2,14,10,-11,1,-14,-8,21,-20,11,-7,9,13,-10
260 DATA 1,-11,11,-2,24,-3,4,-10,-5,7,14,-6,-12,8
270 DATA -10,-12,9,-26,2,12,-3,-22,11,10,-13,8,-11,6
280 DATA 19,-16,-25,17,28,-19,-18,16,-24,8,-9,18,-17,20
290 DATA -23,15,24,-18,19,-17,22,-15,18,-6,-19,17,-28,11
300 DATA 24,-27,-22,-15,18,16,-19,26,19,-18,-2,-16,15,5
310 DATA -11,20,-19,16,-17,26,15,-23,-16,17,3,-15,19,-24
320 DATA -15,25,18,-24,-16,15,17,-1,-17,21,16,-4,-18,27
330 DATA 22,-18,26,27,-23,25,-21,-7,12,23,-22,-24,21,-15
340 DATA -26,7,-23,9,-22,-24,20,-12,27,-19,23,22,-20,28
350 DATA -20,26,17,-23,21,-28,-16,14,-25,24,20,-21,-1,23
360 DATA 16,-24,21,22,20,-27,-25,18,9,-20,-21,28,-10,-22
370 DATA -17,23,-16,19,-13,21,-28,-27,15,-22,25,20,-26,18
380 DATA 27,-19,15,-28,-4,-20,23,28,22,-26,-24,2,-27,21
390 DATA 21,-22,-20,14,-27,-18,27,-17,28,25,-28,5,24,-25
400 DATA -25,17,28,-20,26,23,-26,24,-21,-28,-8,10,25,-19
410 DATA -6,3,-27,25,-15,22,24,-25,-26,27,26,-23,16,-21
420 READ M
430 REM --- SET HOME ADVANTAGE,ADJUSTMENT
          FACTOR & MAX POINT SPREAD ---
440 DATA 1.5,0.05,20
450 READ F1,H,N2
460 PRINT "LWEEK NUMBER?   ";
470 INPUT I
480 IF I>1 THEN 530
490 PRINT "INITIALIZE TAPE DATA?   ";
500 INPUT Z$
510 IF Z$<>"Y" THEN 530
520 GOSUB 1620
530 A2=1977
540 REM --- READ CURRENT TEAM RATINGS FROM TAPE FILE ---
550 FIND 5
560 C=0
570 READ @33:E
580 PRINT "UPDATE BACK-UP DATA FILE?   ";
590 INPUT Z$
600 IF Z$<>"Y" THEN 630
610 FIND 6
620 WRITE @33:E
630 D=-9999
640 REM --- COMBINE HOME ADVANTAGE WITH RATING FOR EACH TEAM ---
650 FOR K1=1 TO 28
660 F=0
670 IF M(K1,I)<0 THEN 690
680 F=F1
690 G(K1)=E(K1)+F
700 NEXT K1
710 REM --- READ IN ACTUAL DELTAS FOR LATEST GAMES ---
720 FOR K1=1 TO 28
730 IF D(K1)<>-9999 THEN 900
740 PRINT "ENTER DELTA FOR ";
750 Z1=K1
```

The funny Ls and Js in PRINT statements are inserted for controlling the printer. Ls generate a top-of-page, and Js create line feeds.

This part of the program is designed for reading and writing the array E using a mass storage device. Since this feature varies greatly from one BASIC to another, look in your user's manual to find out how to load and save arrays on tape or disk.

It was found that F1 = 1.5 points, H = 0.05, and N2 = 20 points are optimal. Interestingly, the optimization was also run over the first 6 games of the 1977 season, with the same results. This would seem to indicate a consistency in the dynamics of the National Football League.

| F1  | H    | N2 | % CORRECT FOR 1976 |
|-----|------|----|--------------------|
| 1.5 | 0.05 | 20 | 68.62              |
| 1.4 | 0.05 | 20 | 68.11              |
| 1.5 | 0.04 | 20 | 67.60              |
| 1.5 | 0.05 | 15 | 67.09              |

```
760 GOSUB 1740
770 INPUT D(K1)
780 K=ABS(M(K1,I))
790 M1=D(K1) MIN N2 MAX -N2
800 D(K)=-D(K1)
810 A1=G(K1)-G(K)
820 Y(K1)=A1
830 Y(K)=-A1
840 REM --- COUNT THE NUMBER (C) OF CORRECT PREDICTIONS ---
850 IF SGN(D(K1))<>SGN(A1) THEN 880
860 C=C+1
870 REM --- ADJUST TEAM RATINGS ---
880 E(K1)=E(K1)+H*(M1-A1)
890 E(K)=E(K)+H*(A1-M1)
900 NEXT K1
910 REM --- PRINT 'RESULTS' TABLE ---
920 PRINT "L        RESULTS FOR WEEK ";I;" ";A2
930 PRINT USING "42X,FA":"JPRED    ACTJ"
940 FOR J=1 TO 28
950 IF Y(J)=1000 THEN 1090
960 M1=J
970 K=ABS(M(J,I))
980 IF M(J,I)<0 THEN 1010
990 M1=K
1000 K=J
1010 Z1=M1
1020 GOSUB 1740
1030 PRINT "AT ";
1040 Z1=K
1050 GOSUB 1740
1060 PRINT USING "3D.2D,3X,3D":Y(M1),D(M1)
1070 Y(M1)=1000
1080 Y(K)=1000
1090 NEXT J
1100 PRINT USING "/3D.2D,S":C*100/14
1110 PRINT " % CORRECT FOR WEEK ";I
1120 COPY
1130 I=I+1
1140 REM --- PRINT NEW RATINGS TABLE ---
1150 PRINT "LAVERAGES AFTER WEEK ";I-1;"  -  ";A2;"JJ"
1160 PRINT "       TEAM          RATINGJ"
1170 Y=E
1180 FOR J=1 TO 28
1190 PRINT USING "2D,2X,S":J
1200 M1=1
1210 FOR K=1 TO 28
1220 IF Y(K)<=Y(M1) THEN 1240
1230 M1=K
1240 NEXT K
1250 Z1=M1
1260 GOSUB 1740
1270 PRINT USING "3D.2D":E(M1)
1280 Y(M1)=-1000
1290 NEXT J
1300 COPY
1310 IF I>14 THEN 1600
1320 REM --- PRINT PREDICTION TABLE ---
1330 PRINT "L       PREDICTION FOR WEEK ";I;", ";A2;"JJ"
1340 Y=0
1350 FOR J=1 TO 28
1360 IF Y(J)=1 THEN 1510
1370 M1=J
1380 K=ABS(M(J,I))
1390 IF M(J,I)<0 THEN 1420
1400 M1=K
1410 K=J
1420 Z1=M1
1430 GOSUB 1740
1440 PRINT "AT ";
1450 Z1=K
1460 GOSUB 1740
1470 A1=E(M1)-E(K)-F1
1480 PRINT USING "3D.2D":A1
1490 Y(M1)=1
1500 Y(K)=1
1510 NEXT J
1520 COPY
1530 PRINT "UPDATE TAPE? ";
1540 INPUT Z$
1550 IF Z$="Y" THEN 1580
1560 END
```

> The MIN and MAX functions choose the greatest of least of two expressions. X MIN Y is X is X Y, and Y if X Y. For instance, 5 MAX 10 is 10, and 5 MIN 10 is 5.

> PRINT USING also varies quite a bit from one BASIC to another, and some don't even have it. In this BASIC, 3D.2D means to print a number as three digits, then the decimal point, then two digits. 3X means to print three spaces. If you can't figure out how this works, then just use PRINT statements with TABs to adjust the output as shown in the sample run, or to suit your own preference.

Starting the 1977 season with a reasonable estimate for the team ratings, the program has predicted 73.5% of the games correctly through the first 10 weeks. The worst week was week 7 with 42.8% and the best was week 10 with 92.8%. No claim is made that this is better than human predictors can do, especially with more detailed information, but it seems pretty good for such a simple algorithm.

Note that although this prediction program was written specifically for NFL football, the basic algorithm could be applied just as easily to college football, baseball, basketball, hockey, etc. It is probably true, however, that the optimal parameter values and the prediction accuracies will be different for different sports.

For those who want to achieve higher prediction accuracies at the expense of increased program complexity and data requirements, many possible improvements could be explored. If many factors are considered, the computation of G could take the form

$$G(I) = N1 * F1 + N2 * F2 + \ldots + NK * FK$$

where K is the number of factors, the F's are the factors and the N's are weights. The N's would be found by optimization to produce the greatest possible per cent of correct predictions. The additional factors might include team standings, injuries to key players, the biorhythms of the players,...?

∎

```
                RESULTS FOR WEEK 4 1977

                                              PRED     ACT
MIAMI            AT   BALTIMORE              -1.53     -17
SEATTLE          AT   NEW ENGLAND           -23.32     -31
NEW YORK JETS    AT   BUFFALO                 4.01       5
PITTSBURGH       AT   HOUSTON                 6.18     -17
OAKLAND          AT   CLEVELAND               1.90      16
CINCINNATI       AT   GREEN BAY               3.48      10
KANSAS CITY      AT   DENVER                -17.41     -16
SAN DIEGO        AT   NEW ORLEANS             6.43      14
PHILADELPHIA     AT   NEW YORK GIANTS         1.24      18
DALLAS           AT   ST. LOUIS               4.22       6
WASHINGTON       AT   TAMPA BAY               8.58      10
LOS ANGELES      AT   CHICAGO                13.86      -1
DETROIT          AT   MINNESOTA              -6.68      -7
ATLANTA          AT   SAN FRANCISCO           3.24       7

      85.71 % CORRECT FOR WEEK 4
```

TABLE 1.   RESULTS

```
1570 REM --- WRITE NEW TEAM RATINGS TO TAPE FILE ---
1580 FIND 5
1590 WRITE @33:E
1600 END
1610 REM --- SUBROUTINE FOR INITIALIZING TEAM RATINGS ---
1620 PRINT "LENTER TEAM RATINGSJ"
1630 FOR Z1=1 TO 28
1640 GOSUB 1740
1650 INPUT E(Z1)
1660 NEXT Z1
1670 PRINT "ALL RIGHT?  ";
1680 INPUT Z$
1690 IF Z$="Y" THEN 1710
1700 GO TO 1620
1710 FIND 6
1720 WRITE @33:E
1730 RETURN
1740 REM --- SUBROUTINE FOR PRINTING TEAM NAMES ---
1750 GO TO Z1 OF 1790,1810,1830,1850,1870,1890,1910,1930,1950,1970,1990
1760 GO TO Z1-11 OF 2010,2030,2050,2070,2090,2110,2130,2150,2170,2190
1770 GO TO Z1-21 OF 2210,2230,2250,2270,2290,2310,2330
1780 RETURN
1790 PRINT "BALTIMORE         ";
1800 RETURN
1810 PRINT "NEW ENGLAND       ";
1820 RETURN
1830 PRINT "MIAMI             ";
1840 RETURN
1850 PRINT "BUFFALO           ";
1860 RETURN
1870 PRINT "NEW YORK JETS     ";
1880 RETURN
1890 PRINT "PITTSBURGH        ";
1900 RETURN
1910 PRINT "HOUSTON           ";
1920 RETURN
1930 PRINT "CLEVELAND         ";
1940 RETURN
1950 PRINT "CINCINNATI        ";
1960 RETURN
1970 PRINT "OAKLAND           ";
1980 RETURN
1990 PRINT "DENVER            ";
2000 RETURN
2010 PRINT "KANSAS CITY       ";
2020 RETURN
2030 PRINT "SEATTLE           ";
2040 RETURN
2050 PRINT "SAN DIEGO         ";
2060 RETURN
2070 PRINT "NEW YORK GIANTS   ";
2080 RETURN
2090 PRINT "DALLAS            ";
2100 RETURN
2110 PRINT "PHILADELPHIA      ";
2120 RETURN
2130 PRINT "ST. LOUIS         ";
2140 RETURN
2150 PRINT "WASHINGTON        ";
2160 RETURN
2170 PRINT "CHICAGO           ";
2180 RETURN
2190 PRINT "GREEN BAY         ";
2200 RETURN
2210 PRINT "DETROIT           ";
2220 RETURN
2230 PRINT "MINNESOTA         ";
2240 RETURN
2250 PRINT "TAMPA BAY         ";
2260 RETURN
2270 PRINT "ATLANTA           ";
2280 RETURN
2290 PRINT "NEW ORLEANS       ";
2300 RETURN
2310 PRINT "LOS ANGELES       ";
2320 RETURN
2330 PRINT "SAN FRANCISCO     ";
2340 RETURN
```

AVERAGES AFTER WEEK 4 - 1977

|    | TEAM            | RATING |
|----|-----------------|--------|
| 1  | DENVER          | 30.15  |
| 2  | LOS ANGELES     | 24.88  |
| 3  | NEW ENGLAND     | 24.49  |
| 4  | DALLAS          | 23.95  |
| 5  | OAKLAND         | 23.13  |
| 6  | PITTSBURGH      | 22.61  |
| 7  | BALTIMORE       | 19.37  |
| 8  | SAN DIEGO       | 19.13  |
| 9  | CLEVELAND       | 18.32  |
| 10 | ST. LOUIS       | 18.05  |
| 11 | MIAMI           | 17.80  |
| 12 | ATLANTA         | 17.26  |
| 13 | HOUSTON         | 17.25  |
| 14 | NEW YORK JETS   | 15.86  |
| 15 | KANSAS CITY     | 14.38  |
| 16 | CINCINNATI      | 14.18  |
| 17 | PHILADELPHIA    | 13.45  |
| 18 | SAN FRANCISCO   | 12.14  |
| 19 | WASHINGTON      | 11.43  |
| 20 | CHICAGO         | 11.00  |
| 21 | NEW ORLEANS     | 10.44  |
| 22 | MINNESOTA       | 10.39  |
| 23 | BUFFALO         | 10.25  |
| 24 | NEW YORK GIANTS | 9.03   |
| 25 | GREEN BAY       | 8.54   |
| 26 | DETROIT         | 5.17   |
| 27 | SEATTLE         | 3.01   |
| 28 | TAMPA BAY       | 1.21   |

TABLE 2.   TEAM RATINGS


PREDICTION FOR WEEK 5, 1977

| | | | |
|---|---|---|---|
| BALTIMORE     | AT | KANSAS CITY     | 3.49   |
| NEW ENGLAND   | AT | SAN DIEGO       | 3.87   |
| NEW YORK JETS | AT | MIAMI           | -3.44  |
| ATLANTA       | AT | BUFFALO         | 5.51   |
| CINCINNATI    | AT | PITTSBURGH      | -9.94  |
| CLEVELAND     | AT | HOUSTON         | -0.43  |
| DENVER        | AT | OAKLAND         | 5.52   |
| TAMPA BAY     | AT | SEATTLE         | -3.30  |
| SAN FRANCISCO | AT | NEW YORK GIANTS | 1.61   |
| WASHINGTON    | AT | DALLAS          | -14.02 |
| ST. LOUIS     | AT | PHILADELPHIA    | 3.10   |
| CHICAGO       | AT | MINNESOTA       | -0.88  |
| GREEN BAY     | AT | DETROIT         | 1.87   |
| NEW ORLEANS   | AT | LOS ANGELES     | -15.94 |

TABLE 3.   PREDICTIONS

# Sports Judging Made Faster ... and Easier

N.B. Winkless, Jr.

**Now you can use your home computer for calculating scores from multiple judges in some sporting events. Better yet, let the judges use it!**

Young Stanislaus ran into the living room and grabbed his father by the wrist. Pop sighed. "Now what?"

"Come sit down at the machine and see what I've come up with. You know how they score in the Games, with half a dozen judges holding up signs—'8½, 7½, 6½, 7, 8,'—all that? Hard to figure, right? Especially when they drop out the highest and lowest ratings."

"Unh-hunh. They've got *people* for that."

"And there's the 'difficulty factor.' They've got to multiply by some funny numbers—2.8 or 3.5..."

"So?"

"So type 'RUN' and make up some numbers."

N.B. Winkless, Jr., 11745 Landale St., No. Hollywood, CA 91607.

Pop sighed and obeyed. Here's the way it went. (See Figure 1).

Pop looked at the sheet. "Fast. What's the program look like?" Stan showed him. (See Figure 2)

"Hmm," said Pop. "Seems clear enough down to line 114. What's that?"

"I'm taking the scores as they're INPUT for each contestant, and adding them up as I go along. But I have to pay attention to which is highest, and which is lowest, if I'm going to subtract those from the totals. And I do. We've picked Option B at 50, and so A$ is not A. X(X) is an arbitrarily high number—actually 10∧3, as you see at line 3—so I try that out against each of the scores as they're reported, there at 114, and spot the lowest in the group. Same way with Y(X), the highest scores. Then I take them both out at 1500."

"Hmmm."

"At 1500, they're subtracted, as they should be. Are you following this?"

"Hmm. And what's going on in the 5000 area?"

"That's a SORT routine I made up. Y'see ... Uh, shall I explain it or would you rather figure it out for yourself?"

Pop groaned. "Never mind. How come you gave *me* the by-line?"

Stan grinned. "When I sell it, I don't want anybody to know it came from an eleven-year-old kid." ∎

```
RUN
ATHLETES: HOW THEY SCORE AND RANK.    FIGURE 1
(WE'LL PAUSE FOR CALCULATIONS...)

THIS PROGRAM FACILITATES SCORING
OF DIVING, GYMNASTICS, ETC.,
IN WHICH  SEVERAL JUDGES AND
SEVERAL DEGREES OF DIFFICULTY
MAY HAVE TO BE CALCULATED.

BY N. B. WINKLESS, JR., 1978

HOW MANY CONTESTANTS (LIMIT 100)? 5
HOW MANY JUDGES? 5
WHICH SCORING METHOD --
      USE ALL SCORES (A)... OR
      DROP HIGHEST & LOWEST (B)? B
OKAY, WE'LL DROP THOSE SCORES.

ENTER DIFFICULTY FACTOR FOR CONTESTANT # 1
(IF NONE, ENTER 0)? 2.3
ENTER JUDGES' SCORINGS (ALL 5 OF THEM, PLS.)--
JUDGE # 1   RATES CONTESTANT # 1   AT ? 6.5
JUDGE # 2   RATES CONTESTANT # 1   AT ? 6.0
JUDGE # 3   RATES CONTESTANT # 1   AT ? 7.0
JUDGE # 4   RATES CONTESTANT # 1   AT ? 6.0
JUDGE # 5   RATES CONTESTANT # 1   AT ? 5.5
OKAY, NEXT CONTESTANT --
ENTER DIFFICULTY FACTOR FOR CONTESTANT # 2
(IF NONE, ENTER 0)? 3.1
ENTER JUDGES' SCORINGS (ALL 5 OF THEM, PLS.)--
JUDGE # 1   RATES CONTESTANT # 2   AT ? 6.0
JUDGE # 2   RATES CONTESTANT # 2   AT ? 5.5
JUDGE # 3   RATES CONTESTANT # 2   AT ? 7.0
JUDGE # 4   RATES CONTESTANT # 2   AT ? 5.5
JUDGE # 5   RATES CONTESTANT # 2   AT ? 5.0
OKAY, NEXT CONTESTANT --
ENTER DIFFICULTY FACTOR FOR CONTESTANT # 3
(IF NONE, ENTER 0)? 1.8
ENTER JUDGES' SCORINGS (ALL 5 OF THEM, PLS.)--
JUDGE # 1   RATES CONTESTANT # 3   AT ? 8.0
JUDGE # 2   RATES CONTESTANT # 3   AT ? 7.5
JUDGE # 3   RATES CONTESTANT # 3   AT ? 9.5
JUDGE # 4   RATES CONTESTANT # 3   AT ? 8.5
JUDGE # 5   RATES CONTESTANT # 3   AT ? 7.5
OKAY, NEXT CONTESTANT --
ENTER DIFFICULTY FACTOR FOR CONTESTANT # 4
(IF NONE, ENTER 0)? 4.2
ENTER JUDGES' SCORINGS (ALL 5 OF THEM, PLS.)--
JUDGE # 1   RATES CONTESTANT # 4   AT ? 5.3
JUDGE # 2   RATES CONTESTANT # 4   AT ? 3.5
JUDGE # 3   RATES CONTESTANT # 4   AT ? 5.0
JUDGE # 4   RATES CONTESTANT # 4   AT ? 6.0
JUDGE # 5   RATES CONTESTANT # 4   AT ? 4.0
OKAY, NEXT CONTESTANT --
ENTER DIFFICULTY FACTOR FOR CONTESTANT # 5
(IF NONE, ENTER 0)? 3.6
ENTER JUDGES' SCORINGS (ALL 5 OF THEM, PLS.)--
JUDGE # 1   RATES CONTESTANT # 5   AT ? 7.5
JUDGE # 2   RATES CONTESTANT # 5   AT ? 7.0
JUDGE # 3   RATES CONTESTANT # 5   AT ? 6.5
JUDGE # 4   RATES CONTESTANT # 5   AT ? 7.0
JUDGE # 5   RATES CONTESTANT # 5   AT ? 5.5

THAT'S IT.
JUST A SECOND.  I'M THINKING...
C( 1 )- 42.55
C( 2 )- 52.7
C( 3 )- 43.2
C( 4 )- 60.06
C( 5 )- 73.8
# 5   CONTESTANT RANKS 1  AT 73.8   POINTS.
# 4   CONTESTANT RANKS 2  AT 60.06  POINTS.
# 2   CONTESTANT RANKS 3  AT 52.7   POINTS.
# 3   CONTESTANT RANKS 4  AT 43.2   POINTS.
# 1   CONTESTANT RANKS 5  AT 42.55  POINTS.

THAT COVERS OUR  5  CONTESTANTS
```

FIGURE 2

```
1 PRINT "ATHLETES: HOW THEY SCORE AND RANK."
2 PRINT "(WE'LL PAUSE FOR CALCULATIONS...)"
3 DIM X(100):FOR O=1 TO 100:X(O)=10^3:NEXT O
4 DIM C(100),D(100),H(100),R(100),Y(100),Z(100)
6 PRINT
9 DIM T(100)
10 PRINT "THIS PROGRAM FACILITATES SCORING"
12 PRINT "OF DIVING, GYMNASTICS, ETC.,"
14 PRINT "IN WHICH   SEVERAL JUDGES AND"
16 PRINT "SEVERAL DEGREES OF DIFFICULTY"
18 PRINT "MAY HAVE TO BE CALCULATED."
20 PRINT: PRINT "BY N. B. WINKLESS, JR., 1978"
25 PRINT
26 INPUT "HOW MANY CONTESTANTS (LIMIT 100)";C
27 INPUT "HOW MANY JUDGES";J
30 PRINT "WHICH SCORING METHOD --"
40 PRINT "       USE ALL SCORES (A)... OR"
50 INPUT "       DROP HIGHEST & LOWEST (B)";A$
55 IF A$="B" THEN PRINT "OKAY, WE'LL DROP THOSE SCORES.":PRINT:GOTO 80
70 PRINT "OKAY.  WE'LL COUNT ALL SCORES."
75 PRINT
80 X=X+1:PRINT"ENTER DIFFICULTY FACTOR FOR CONTESTANT #";X
82 INPUT "(IF NONE, ENTER 0)";D(X)
83 IF D(X)=0 THEN D(X)=1
84 PRINT "ENTER JUDGES' SCORINGS (ALL";J;" OF THEM, PLS.)--"
100 U=U+1
104 PRINT "JUDGE #";U;" RATES CONTESTANT #";X;" AT ";
105 INPUT C(X)
112 T(X)=T(X)+C(X)
114 IF X(X)>=C(X) THEN X(X)=C(X)
116 IF Y(X)<=C(X) THEN Y(X)=C(X)
118 IF U=J THEN C1=C1+1: GOSUB 1500
119 IF C1=C THEN 200
120 IF U=0 THEN PRINT "OKAY, NEXT CONTESTANT --": GOTO 80
125 GOTO 100
200 !REM -- GOING TO A FINISH
220 FOR V1=1 TO C
230 C(V1)=T(V1)
240 NEXT V1
300 GOTO 4800
400 PRINT "THAT'S IT.": PRINT
1500 IF A$<>"A" THEN T(X)=T(X)-(X(X)+Y(X))
1510 T(X)=T(X)*D(X)
1515 U=0
1516 PRINT
1520 RETURN
4800 PRINT: PRINT "THAT'S IT.": PRINT
4810 FOR K=1 TO 50
4820 NEXT K
4828 PRINT "JUST A SECOND.  I'M THINKING..."
4830 FOR K=1 TO 100: NEXT K
4900 PRINT
4901 FOR M5=1 TO X
4902 PRINT "C(";M5;")-";C(M5)
4903 NEXT M5
5000 Q=Q+1
5100 FOR Z=Q TO X
5105 A=C(Q)
5110 IF A<=C(Z+1) THEN A=C(Z+1): Q=Z+1
5115 IF Z+1=X THEN 5130
5120 NEXT Z
5130 T=T+1
5140 PRINT "#";Q;" CONTESTANT RANKS";T;" AT";C(Q);" POINTS."
5150 IF T=X THEN 5300
5160 FOR J=1 TO X
5170 C(Q)=0
5180 NEXT J
5190 Q=0: GOTO 5000
5300 PRINT: PRINT "THAT COVERS OUR ";X;" CONTESTANTS"
5310 PRINT: PRINT: END
```

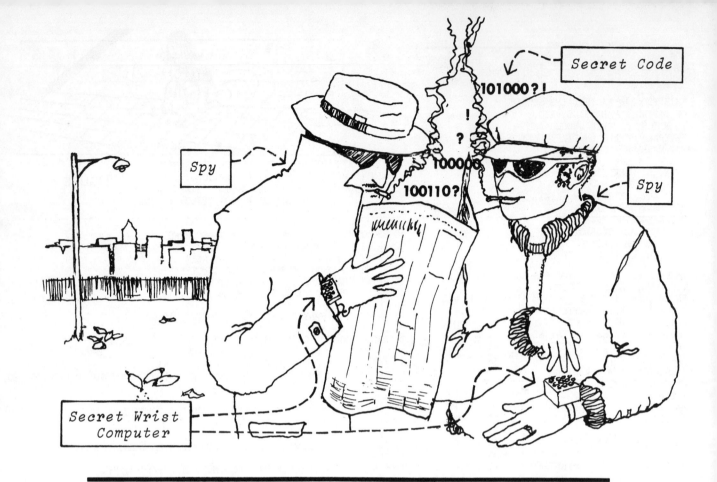

# Distance and Error - Correcting Codes

This is a reprint of one of the original Project Solo curriculum modules developed at the University of Pittsburgh. Project Solo was supported in part by the National Science Foundation, and it was directed by Tom Dwyer and Margot Critchfield. The modules were authored by various persons, including project staff, teachers, and students.

It should be kept in mind that Project Solo began in 1969 (which is probably before some of Creative's readers were born). Undoubtedly, many of the modules would be done differently today. There are also surely errors to be found, and neither Creative Computing, the authors, or NSF can warrant the accuracy of the reprints. But as a starting point for your own explorations, they should make a good (albeit slightly ancient) set of shoulders to stand upon. We hope you enjoy the view.

A binary code of length N is a string of N 0's or 1's. For example, if N = 3, all the possible binary codes are 000, 001, 010, 011, 100, 101, 110, and 111. We speak of these as 3 bit codes (1 and 0 are called bits).

These codes could be used to represent eight objects of any sort—the members of a musical octet, the digits 0,1,2,3,4,5,6,7 in a computer, or the letters A,B,C,D,E,F,G,H.

**Now for some intrigue**

Let's assume that we wish to assign binary code names to the agents of STICK (Society to Increase Contact for Keeps), an international ring of glue thieves.

Suppose we only have two agents but eight codes. Question: Can we assign codes so that:

   a. The computer will check code authenticity **without** knowing the correct codes?

   b. The computer can give the correct code even though the agent has deliberately changed one bit (to throw off eavesdroppers)?

To see how codes can be assigned to make this possible, let's place the codes at the vertices of a cube.

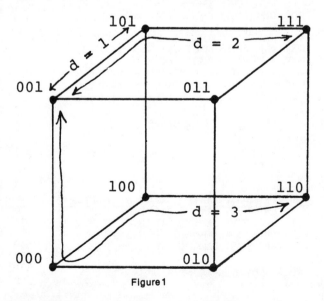

Figure 1

To be more precise, we should call Figure 1 a "3-dimensional cube." A picture of a "4-dimensional cube" (which has $2^4 = 16$ vertices) is shown in Figure 2. Thus we can associate a unique four-bit code with each vertex of a 4-D cube. Can you generalize this statement?

| NOTE: | $0 + 1 = 1$ and $1 + 0 = 1$ in binary arithmetic. |
|---|---|
| ALSO NOTE: | $0 + 0 = 0$, but $1 + 1 = 0$ (with carry of 1) |
| FURTHER: | $0*1 = 0, 1*0 = 0, 0*0 = 0$ and $1*1 = 1$ |

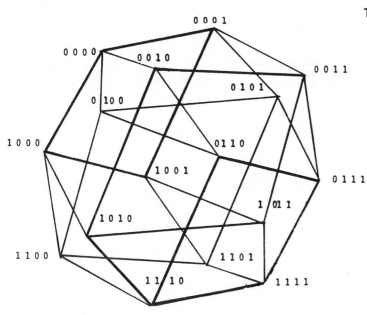

Figure 2

We will define the number of bits by which two codes differ as the "distance" ($=d$) between these codes. Thus, for example, the distance between 001 and 101 is $d = 1$, the distance between 001 and 111 is $d = 2$, and the distance between 001 and 110 is $d = 3$. Math students: Is this a legal use of the word distance? Notice that our picture has been drawn so that "distance" between codes corresponds to the number of edges of the cube you would have to walk along to get from one vertex to the other.

Let's assign our two authentic agents the codes 001 and 110 (which are a distance of three from each other). Now suppose one agent walks up to another and says, "My code is 101."

(a) How can we tell if it is an authentic code? One way would be to simply compare it to the list of authentic codes! However, if there were to be very many codes, such a search of the authentic list would be time consuming. Besides, we don't want this authentic list stored in too many places! There is another way to check authenticity.

In our example the two authentic codes have the property that if we add the first and third bits of the code we get 1, and this is also true if we add the second and third bits.

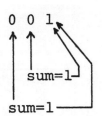  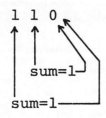

For all other codes this is false.

Thus the code 101 our agent gave is not authentic.

Since: $1 + 1 = 0$

(b) Suppose our agent deliberately changes one bit in his code when giving it verbally. Applying the above rule will detect the error, but can we figure out what the correct code should have been?
We can see the answer from our diagram. An authentic code which has only one bit changed is distance $d = 1$ from the original correct code, but distance $d = 2$ from the other correct code. Thus 101 has to be corrected back to 001, **not** to 110.
Try to develop an algorithm for making corrections in codes which have 1 bit in error: Here is how you might reason:

Let's call the 3 bits: B1, B2, and B3.

For 101   B1 + B3 = 0  WRONG
           B2 + B3 = 1  RIGHT

∴ Change B1, Correct code is: 001

For 111   B1 + B3 = 0  WRONG
           B2 + B3 = 0  WRONG

∴ Change B3, Correct code is: 110 etc.

---

Problems: Write programs for your wrist computer to handle the following:

1. The Two-Agent Problem
   Authentic Codes: 101 and 010
   
   INPUT: Any 3 bit code which is either an authentic code, or which contains an error in 1 bit.
   OUTPUT: The Message: "AUTHENTIC CODE"
             OR: "CODE IN ERROR
             CODE SHOULD BE _____"

2. Can use of a 4 bit code (see Figure 2) permit additional outputs for the above "Two-Agent" analysis program?

3. Four-Agent Problem
   Authentic Codes are: SMITH 0 0 0 0 0
                               BOND  1 1 1 0 0
                               SPIRO 0 0 1 1 1
                               JONES 1 1 0 1 1
   
   INPUT: Any code
   OUTPUT: The Message: "AUTHENTIC CODE"
             OR: "1 BIT ERROR —
             CORRECT CODE IS _____"
             OR: "ERROR >= 2 BITS —
             DOUBLE AGENT"

4. Here is a set of six bit codes to play with: 000000, 000111, 111000, 110110, 011011, 101101. (NOTE: $d >= 3$ for any two of these codes.)

```
10  S(1) = 0
20  S(2) = 1
30  PRINT "Two-Agent Problem"
40  PRINT "To end program, type the characters END when asked for a code."
50  PRINT
60  PRINT "Enter a 3-bit code."
70  INPUT B$
80  IF B$ = "END" THEN 280
90  GOSUB 300
100 S(3) = S1
110 S(4) = S2
120 IF (S(1) = S(3) AND S(2) = S(4)) THEN 260
130 IF (S(1) <> S(3) AND S(2) <> S(4)) THEN 180
140 X = 1-S(3)
150 IF (S(1) = X) THEN 200
160 X = 1-S(4)
170 IF (S(2) = X) THEN 220
180 B3$ = RIGHT$(STR$(1-VAL(B3$)),1)
190 GOTO 230
200 B1$ = RIGHT$(STR$(1-VAL(B1$)),1)
210 GOTO 230
220 B2$ = RIGHT$(STR$(1-VAL(B2$)),1)
230 PRINT "Code in error."
240 PRINT "Code should be: ";B1$+B2$+B3$
250 GOTO 50
260 PRINT "Authentic code."
270 GOTO 50
280 STOP
290 END
300 B1$ = LEFT$(B$,1)
310 X$ = RIGHT$(B$,2)
320 B2$ = LEFT$(X$,1)
330 B3$ = RIGHT$(B$,1)
340 S1 = VAL(B1$) + VAL(B3$)
350 S2 = VAL(B2$) + VAL(B3$)
360 IF S1 = 2  THEN LET S1 = 0
370 IF S2 = 2  THEN LET S2 = 0
380 RETURN
```

*Sample Solution — Problem 1*

```
LIST
10 PRINT "Four-Agent Problem"
20 PRINT "To end program, type the characters END when asked for a code."
30 PRINT
40 CLEAR :  REM Sets all variables to zero
50 PRINT "Enter a 5-bit binary code."
60 INPUT B$
70 IF B$="END" THEN 520
80 B1$ = LEFT$(B$,1)
90 X$ = RIGHT$(B$,4)
100 B2$ = LEFT$(X$,1)
110 X$ = RIGHT$(B$,3)
120 B3$ = LEFT$(X$,1)
130 X$ = RIGHT$(B$,2)
140 B4$ = LEFT$(X$,1)
150 B5$ = RIGHT$(X$,1)
160 S(1) = VAL(B1$) + VAL(B2$)
170 S(2) = VAL(B4$) + VAL(B5$)
180 S(3) = VAL(B1$) + VAL(B4$)
190 S(4) = VAL(B2$) + VAL(B5$)
200 FOR I=1 TO 4
210    IF S(I) = 2 THEN S(I) = 0
220 NEXT I
230 IF S(1) <> 0 THEN LET F1 = 1
240 IF S(2) = 0 THEN 270
250 IF F1 = 1 THEN 500
260 F2 = 1
270 IF VAL(B3$) <> S(3) THEN LET F3 = 1
280 IF VAL(B3$) <> S(4) THEN LET F4 = 1
290 K = F1 + F2 + F3 + F4
300 IF K = 0 THEN 480
310 IF K <> 2 THEN 500
320 IF (F1 + F3)<>2 THEN 350
330 B1$ = RIGHT$(STR$(1-VAL(B1$)),1)
340 GOTO 450
350 IF (F1 + F4)<>2 THEN 380
360 B2$ = RIGHT$(STR$(1-VAL(B2$)),1)
370 GOTO 450
380 IF (F3 + F4)<>2 THEN 410
390 B3$ = RIGHT$(STR$(1-VAL(B3$)),1)
400 GOTO 450
410 IF (F2 + F3)<>2 THEN 440
420 B4$ = RIGHT$(STR$(1-VAL(B4$)),1)
430 GOTO 450
440 B5$ = RIGHT$(STR$(1-VAL(B5$)),1)
450 PRINT
460 PRINT " 1 bit error--- Correct code is: ";B1$+B2$+B3$+B4$+B5$
470 GOTO 30
480 PRINT " Authentic code."
490 GOTO 30
500 PRINT " Error = 2 bits --- Double Agent!"
510 GOTO 30
520 STOP
530 END
```

*Sample Solution — Problem 3*

```
RUN
Two-Agent Problem
To end program, type the characters END when asked for a code.

Enter a 3-bit code.
? 000
Code in error.
Code should be: 010

Enter a 3-bit code.
? 001
Code in error.
Code should be: 101

Enter a 3-bit code.
? 010
Authentic code.

Enter a 3-bit code.
? 011
Code in error.
Code should be: 010

Enter a 3-bit code.
? 100
Code in error.
Code should be: 101

Enter a 3-bit code.
? 101
Authentic code.

Enter a 3-bit code.
? 110
Code in error.
Code should be: 010

Enter a 3-bit code.
? 111
Code in error.
Code should be: 101

Enter a 3-bit code.
? END


RUN
Four-Agent Problem
To end program, type the characters END when asked for a code.

Enter a 5-bit binary code.
? 00000
 Authentic code.

Enter a 5-bit binary code.
? 10000

 1 bit error--- Correct code is: 00000

Enter a 5-bit binary code.
? 00100

 1 bit error--- Correct code is: 00000

Enter a 5-bit binary code.
? 00010

 1 bit error--- Correct code is: 00000

Enter a 5-bit binary code.
? 11011
 Authentic code.

Enter a 5-bit binary code.
? 11010

 1 bit error--- Correct code is: 11011

Enter a 5-bit binary code.
? 10011

 1 bit error--- Correct code is: 11011

Enter a 5-bit binary code.
? 10101
 Error = 2 bits --- Double Agent!

Enter a 5-bit binary code.
? 10110
 Error = 2 bits --- Double Agent!

Enter a 5-bit binary code.
? END
```

# Accuracy Plus: Multiprecision Multiplication

**Bruce Barnett**

Did you know that

```
     15 241 578 774 577 047 238 073
  × 962 858 720 375 106 022 735 947

  = 146 754 870 353 856 322 393 154
    693 250 246 596 936 303 584 131
```

You can verify this result or accurately perform any of your favorite multiplication problems by using the BASIC program contained in this article. As possible applications, you might wish to calculate your monthly mortgage payment to within a millionth of a cent, precisely ascertain some gambling odds, arrive at an exact result in number theory or simply amaze your friends with the accuracy of your computer!

The accompanying program will allow you to multiply any two multi-digit numbers and give you an exact answer.[1] Once having this basic capability it then becomes an easy matter to continue the product, that is, to multiply the result of the previous multiplication by another number of your choice. If you choose to continue the product by multiplying each product by a constant, then exponentiation results, (i.e., 3x3 = $3^2$ = 9, 9x3 = $3^3$ = 27, 27x3 = $3^4$ = 81, ...). If, on the other hand, you choose to multiply each answer you obtain by consecutive integers starting at 1 then a factorial (!) results, (i.e., 1x1 = 1!, 1x2 = 2!, 1x2x3 = 3!, ...). This latter computation often arises in the art of counting when permutations and combinations are usually encountered. Though the program allows you to perform explicitly the above described computations, (e.g., calculate $2^{50}$ or 75!), the major benefit of the coding is that the actual multi-precision multiplication routine is written as a function, FNP$. It may be easily inserted into one of your own programs, for your own devilish purposes! It is written in Northstar BASIC but the program is sufficiently explained to enable you to modify it if your version of BASIC handles string variables somewhat differently.

The underlying algorithm is quite simple. It is essentially based upon repeated applications of the Distributive Laws of Multiplication:

a*(b + c) = a*b + a*c ; (d + e)*f = d*f + e*f

---

Bruce Barnett, RD#2, Box 213, Blairstown, NJ 07825.

To be specific, let's consider the following example,

```
        4 326 485
    ×   205 173

       12 979 455
      302 853 95
      432 648 5
   21 632 425
   00 000 00
  865 297 0

  887 677 906 905
```

The same answer can be obtained by grouping both the multiplier and multiplicand in three digit segments, following the usual multiplication procedure and arranging the computation as shown:

```
        4 326 485
    ×   205 173

           83 905    =  173*485
        56 398       =  173*326
       692           =  173*4
           99 425    =  205*485
     66 830          =  205*326
    820              =  205*4

    887 677 906 905
```

One can simplify the computation further by combining terms with the same starting place value as exhibited below:

```
           83 905    =  173*485
       155 823       =  173*326 + 205*485
    67 522           =  173*4   + 205*326
   820               =  205*4

   887 677 906 905
```

This is equivalent to:

$$(173*10^0) * (485*10^0)$$
$$(173*10^0) * (326*10^3) + (205*10^3) * (485*10^0)$$
$$(173*10^0) * (4*10^6)   + (205*10^3) * (326*10^3)$$
$$(205*10^3) * (4*10^6)$$

$$= 83,905*10^0 + 155,823*10^3 + 67,522*10^6 + 820*10^9$$

A little reflection shows that the multiplicand has been written as $4*10^6 + 326*10^3 + 485*10^0$ and the multiplier as $205*10^3 + 173*10^0$. The various subproducts result by using the Distributive Laws and having them earn their money!

The enclosed subroutine implements the above approach, whereby all multiplications are performed in three digit segments and are appropriately grouped.

## Multiplication, con't...

Since the number of digits in the multiplier and multiplicand can exceed the number of digits allowed in your BASIC, it is necessary to read these values as string variables. A$ and B$ serve this purpose and in fact form the inputs to the function FNP$(A$,B$,C$). The product is placed in the output string variable C$.

A rather unusual approach is used in the following example (to describe the subroutine) which enables you to see the various subproducts and subsums being formed and manipulated and stored. The same problem seen earlier is used.

```
L1 = 7          (Number of digits in multiplicand)
U1 = 3          (Multiplicand is divided into 3 segments
A(1) = 485      (1st segment of multiplicand)
A(2) = 326      (2nd    '         '       )
A(3) = 4        (3rd    '         '       )

L2 = 6
U2 = 2          (Similar information as above
B(1) = 173       for the multiplier)
B(2) = 205

C(2) = A(1)*B(1)                    = 485*173            =  83,905
C(3) = A(1)*B(2) + A(2)*B(1) = 485*205 + 326*173 = 155,823
C(4) = A(2)*B(2) + A(3)*B(1) = 326*205 +   4*173 =  67,522
C(5) = A(3)*B(2)                    =   4*205            =     820

FNR(C(2)) = 905         FNR(C(4)) = 522
FNM(C(2)) =  83         FNM(C(4)) =  67
FNL(C(2)) =   0         FNL(C(4)) =   0
FNR(C(3)) = 823         FNR(C(5)) = 820
FNM(C(3)) = 155         FNM(C(5)) =   0
FNL(C(3)) =   0         FNL(C(5)) =   0

D(1) = FNR(C(2)) + FNM(C(1)) + FNL(C(0))
     =    905    +     0     +     0     = 905

D(2) = FNR(C(3)) + FNM(C(2)) + FNL(C(1))
D(2) =    823    +    83     +     0     = 906
                        ...

D(3) =    522    +   155     +     0     = 677
D(4) =    820    +    67     +     0     = 877
D(5) =      0    +     0     +     0     =   0

D(1) = D(1) + FNM(D(0)) = 905 + 0 = 905
D(2) = D(2) + FNM(D(1)) = 906 + 0 = 906
                ...
D(5) = D(5) + FNM(D(4)) =   0 + 0 =   0
```

(These statements are necessary in case the appropriate sums exceed 3 digits. Thus they allow for any carryover

The remaining statements in the function, 1360-1560, check and force leading zeroes in each 3-digit segment of the answer, D( ) to be displayed, (since, for example, 053 ordinarily comes out as 53). Note that the above testing is performed in the reverse order from that which the product was calculated (see statement 1360). Therefore, testing starts from the most significant digits. S6 becomes non-zero in statement 1370 once the first non-zero partial answer is encountered and prevents a leading trio of zero digits from being displayed. You might elect to remove the REM from statement 1550 which would then place a blank between each three-digit segment of the answer for clarity in reading. This should be done only if you do not intend to perform any further operations on the answer.

Now what is ...

```
      398 133 358 946 552 076 341
  X   756 571 088 603 728 535 407
```

```
DO YOU WISH TO PRINT THE RESULTS ?   (Y OR N) ? N
DO YOU WISH TO ...
1-----MULTIPLY TWO NUMBERS
2-----EXPONENTIATE
3-----COMPUTE A FACTORIAL
4-----STOP

1, 2, 3 OR 4 ? 1

MULTIPLICAND ? 12345679
MULTIPLIER ? 27
PRODUCT IS       333333333

DO YOU WISH TO ...
1-----ENTER NEW MULTIPLICAND
2-----USE PRODUCT JUST OBTAINED AS THE MULTIPLICAND
3-----STOP
1, 2 OR 3 ? 2

MULTIPLIER ? 257689032257
PRODUCT IS       085896343999770322581

DO YOU WISH TO ...
1-----ENTER NEW MULTIPLICAND
2-----USE PRODUCT JUST OBTAINED AS THE MULTIPLICAND
3-----STOP
1, 2 OR 3 ? 3

DO YOU WISH TO ...
1-----MULTIPLY TWO NUMBERS
2-----EXPONENTIATE
3-----COMPUTE A FACTORIAL
4-----STOP

1, 2, 3 OR 4 ? 2

BASE ? 11
HIGHEST POWER ? 5

NUMBER      POWER       VALUE

11            1           11
11            2          121
11            3       001331
11            4       014641
11            5       161051

DO YOU WISH TO ...
1-----MULTIPLY TWO NUMBERS
2-----EXPONENTIATE
3-----COMPUTE A FACTORIAL
4-----STOP

1, 2, 3 OR 4 ? 3

HIGHEST FACTORIAL ? 8
NUMBER:  1
IT'S FACTORIAL: 001

NUMBER:  2
IT'S FACTORIAL: 002

NUMBER:  3
IT'S FACTORIAL: 006

NUMBER:  4
IT'S FACTORIAL: 024

NUMBER:  5
IT'S FACTORIAL: 120

NUMBER:  6
IT'S FACTORIAL: 720

NUMBER:  7
IT'S FACTORIAL: 005040

NUMBER:  8
IT'S FACTORIAL: 040320

DO YOU WISH TO ...
1-----MULTIPLY TWO NUMBERS
2-----EXPONENTIATE
3-----COMPUTE A FACTORIAL
4-----STOP

1, 2, 3 OR 4 ? 4

READY
```

## Multiplication, con't...

```
10  REM     THIS IS A MULTI-DIGIT PRECISION PROGRAM
20  REM     WHICH WILL ALLOW YOU TO MULTIPLY, EXPONENTIATE OR
30  REM     COMPUTE FACTORIALS ACCURATELY.
40  REM
50  REM     THIS PROGRAM IS WRITTEN IN NORTHSTAR BASIC,
60  REM     AND CAN OUTPUT RESULTS ON A SCREEN OR A PRINTER.
70  REM
80  REM     THE FOLLOWING TWO STATEMENTS EXPLAIN THE PRINT STATEMENTS
90  REM     IN THIS PROGRAM
100 REM     PRINT #0   CAUSES THE RESULTS TO APPEAR AT THE CRT
110 REM     PRINT #2   CAUSES THE RESULTS TO APPEAR ON THE PRINTER
120 REM
130 DIM A$(400),B$(400),C$(400),A(200),B(200),C(200),D(200)
140 INPUT"DO YOU WISH TO PRINT THE RESULTS ?   (Y OR N) ? ",P$
150 IF P$(1,1)="Y" THEN P=2 ELSE P=0
160 PRINT #P
170 PRINT #P,"DO YOU WISH TO ... "
180 PRINT #P,"1-----MULTIPLY TWO NUMBERS"
190 PRINT #P,"2-----EXPONENTIATE"
200 PRINT #P,"3-----COMPUTE A FACTORIAL"
210 PRINT #P,"4-----STOP"
220 PRINT #P
230 INPUT"1, 2, 3 OR 4 ? ",I9
240 PRINT #P
250 ON I9 GOTO 290,480,640,1570,160,160,160,160,160,160
260 REM
270 REM MULTIPLICATION
280 REM
290 INPUT"MULTIPLICAND ? ",A$
300 INPUT"MULTIPLIER ? ",B$
310 C$=FNP$(A$,B$,C$)
320 PRINT #P,"PRODUCT IS        ",C$
330 PRINT
340 PRINT"DO YOU WISH TO ... "
350 PRINT"1-----ENTER NEW MULTIPLICAND"
360 PRINT"2-----USE PRODUCT JUST OBTAINED AS THE MULTIPLICAND"
370 PRINT"3-----STOP"
380 INPUT"1, 2 OR 3 ? ",I9
390 PRINT
400 ON I9 GOTO 290,410,160
410 A$=C$
420 GOTO 300
430 REM
440 REM EXPONENTIATION
450 REM
460 PRINT #P
470 PRINT #P
480 INPUT"BASE ? ",L$
490 B$=L$
500 INPUT"HIGHEST POWER ? ",B9
510 PRINT #P
520 PRINT #P,"NUMBER",TAB(10),"POWER",TAB(20),"VALUE"
530 PRINT #P
540 PRINT #P,L$,TAB(11),"1",TAB(20),L$
550 FOR I9=2 TO B9
560 C$=FNP$(L$,B$,C$)
570 PRINT #P,L$,TAB(10),I9,TAB(20),C$
580 B$=C$
590 NEXT I9
600 GOTO 160
610 REM
620 REM FACTORIAL
630 REM
640 PRINT #P
650 B$="1"
660 INPUT"HIGHEST FACTORIAL ? ",N8
670 FOR I8=1 TO N8
680 PRINT #P,"NUMBER: ",I8
690 L$=STR$(I8)
700 C$=FNP$(L$,B$,C$)
710 PRINT #P,"IT'S FACTORIAL: ",C$
720 PRINT #P
730 B$=C$
740 NEXT I8
750 GOTO 160
760 DEF FNL(X) = INT(X/1000000)
770 DEF FNM(X) = INT((X-FNL(X)*1000000)/1000)
780 DEF FNR(X) = X-FNL(X)*1000000-FNM(X)*1000
790 DEF FNP$(A$,B$,C$)
800 Q=0
810 C(0)=0
820 D(0)=0
830 C(1)=0
840 FOR K=1 TO 200
850 C(K)=0
860 D(K)=0
870 NEXT K
880 L=0
890 S6=0
900 GOSUB 1260
910 L1=L2
920 U1=U2
930 FOR K = 1 TO U1
940 A(K) = B(K)
950 NEXT K
960 D$=" "
970 G$=" "
980 A$=B$
990 FOR K=1 TO 200
1000 C(K)=0
1010 D(K)=0
1020 NEXT K
1030 L=0
1040 S6=0
1050 GOSUB 1260
1060 FOR I = 1 TO U1
1070 FOR J = 1 TO U2
1080 C(I+J) = C(I+J) + A(I)*B(J)
1090 NEXT J
1100 NEXT I
1110 S = U1+U2
1120 FOR K = 2 TO S+1
1130 D(K-1) = FNR(C(K))+FNM(C(K-1))+FNL(C(K-2))
1140 NEXT K
1150 FOR K = 1 TO S
1160 D(K) = D(K) + FNM(D(K-1))
1170 NEXT K
1180 FOR K = 1 TO S
1190 D(K) = FNR(D(K))
1200 NEXT K
1210 FOR K = 1 TO S
1220 GOSUB 1360
1230 NEXT K
1240 RETURN C$
1250 FNEND
1260 L2 = LEN(A$)
1270 X = INT(L2/3)
1280 Y = 3*X - L2
1290 IF Y = 0 THEN U2 = X ELSE U2 = X+1
1300 FOR K = 1 TO U2
1310 W = L2 - 3*K + 1
1320 IF W <= 0 THEN W = 1
1330 B(K) = VAL(A$(W,L2-3*(K-1)))
1340 NEXT K
1350 RETURN
1360 X = D(S+1-K)
1370 S6 = X + S6
1380 IF S6 = 0 THEN 1560
1390 D$ = STR$(X)
1400 IF X > 99 THEN 1490
1410 IF X > 9 THEN 1470
1420 IF X > 0 THEN 1450
1430 G$ = "000"
1440 GOTO 1500
1450 G$ = "00" + D$(2,2)
1460 GOTO 1500
1470 G$ = "0" + D$(2,3)
1480 GOTO 1500
1490 G$ = D$(2,4)
1500 IF Q=0 THEN 1510 ELSE 1540
1510 C$=G$
1520 Q=1
1530 GOTO 1550
1540 C$=C$+G$
1550 REM C$=C$+" "
1560 RETURN
1570 END
```

**Footnotes**

1. One restriction is that for an 8 digit BASIC, the partial sums C(K), that are described later should not exceed 99,999,999. In an extreme case where both the multiplier and multiplicand each contain all 9's, both should not simultaneously exceed 300 digits.

# Double Precision: Does Your BASIC Have It? Now It Can...

### Delmar D. Hinrichs

A major shortcoming of TRS-80 Level II BASIC by Microsoft is that its math functions will not give double precision results. If you use any of the built-in math functions except ABS, INT, or FIX on your double precision variables, you get only a single precision result, with garbage filling out the less significant digits of your result. For many applications, six significant digits are not enough. Probably some of the other extended BASICs have this problem also. In addition, some BASICs do not have math functions, or have inaccurate functions.

What can you do about this? You can use BASIC language routines to calculate the math functions in double precision, or in whatever precision your BASIC provides. Program A contains short routines to calculate all of the common math functions: square root, sine, cosine, tangent, arctangent (in radians or degrees), logarithm, exponential (natural or base 10) and, power ($y^x$). While the program uses a menu selection for demonstration purposes, the routines are intended to be removed from the program and used as separate subroutines. All of the routines are written in a similar format to make conversion to subroutines easy. Program A requires about 7900 bytes of free memory to run.

Program A. Demonstration Program for Double Precision Mathematics Functions.

```
10 CLS
20 PRINT "<<<<<  DOUBLE PRECISION MATHEMATICS FUNCTIONS  >>>>>"
30  '  -   -   -   -     BY D. D. HINRICHS    -   -   -   -
40  '  -   -   -    FOR THE TRS-80 WITH LEVEL II BASIC   -   -
50 PRINT "    1.    SQUARE ROOT"
60 PRINT "    2.    SINE   (RADIANS)"
70 PRINT "    3.    SINE   (DEGREES)"
80 PRINT "    4.    COSINE (RADIANS)"
90 PRINT "    5.    COSINE (DEGREES)"
100 PRINT "    6.    TANGENT  (RADIANS)"
110 PRINT "    7.    TANGENT  (DEGREES)"
120 PRINT "    8.    ARCTANGENT  (RADIANS)"
130 PRINT "    9.    ARCTANGENT  (DEGREES)"
140 PRINT "   10.    LOGARITHM   (NATURAL)"
150 PRINT "   11.    LOGARITHM   (BASE 10)"
160 PRINT "   12.    EXPONENTIAL (NATURAL)"
170 PRINT "   13.    EXPONENTIAL (BASE 10)"
180 PRINT "   14.    POWER   (Y[X)"
190 INPUT "ENTER NO. FOR THE FUNCTION THAT YOU WANT TO USE" ; N
200 CLS : IF N>0 AND N<15 AND INT(N)=N GOTO 220
210 PRINT "YOUR ENTRY "; N ;" IS ILLEGAL.  TRY AGAIN" : GOTO 50
220 ON N GOTO 9790, 9800, 9810, 9820, 9830, 9840, 9860,
       9880, 9900, 9910, 9930, 9950, 9960, 9970
230 END
240  '
250  ' -  -  -  -  ALL ROUTINES AFTER THIS   -  -  -  -
260  '
9790  ' SQUARE ROOT ROUTINE
9791 DEFDBL P-Y : DEFINT I-N : J=2
9792 INPUT "ENTER X FOR SQUARE ROOT(X)"; X
9793 S=X : IF X=0 GOTO 9797
9794 IF X<0 PRINT "**** ERROR, SQ. ROOT NEGATIVE NO." :GOTO 9792
9795 S=SQR(X)                          ' PRELIMINARY VALUE OF SQUARE ROOT
9796 V=S :S=(X/S+S)/J :IF S<>V GOTO 9796        ' ADJUST VALUE
9797 PRINT "THE SQUARE ROOT OF  " ; X ; "  IS  " ; S
9798 PRINT : GOTO 9792                  ' RETURN FOR NEXT ENTRY
9799  '
9800  ' SINE ROUTINE, ANGLE IN RADIANS
9801 DEFDBL P-Y : DEFINT I-N : J=2 : PI=3.1415926535897932
9802 INPUT "ENTER ANGLE IN RADIANS FOR SINE(X) " ; X
9803 Y=X : R=0 : IF ABS(Y)>PI THEN R=FIX(Y/PI) : Y=Y-R*PI
9804 M=0 : N=1 : S=Y : T=Y : U=-Y*Y
9805 V=S :M=M+J :N=N+J :T=T*U/(M*N) :S=S+T :IF S<>V GOTO 9805
9806 IF INT(R/J) <> R/J THEN S=-S
9807 PRINT "SINE OF  " ; X ; "  RADIANS IS  " ; S
9808 PRINT : GOTO 9802                  ' RETURN FOR NEXT ENTRY
9809  '
9810  ' SINE ROUTINE, ANGLE IN DEGREES
9811 DEFDBL P-Y :DEFINT I-N :J=2 :P=180 : PI=3.1415926535897932
9812 INPUT "ENTER ANGLE IN DEGREES FOR SINE(X) " ; X
9813 Y=X : R=0 : IF ABS(Y)>P THEN R=FIX(Y/P) : Y=Y-R*P
9814 Y=Y*PI/P : M=0 : N=1 : S=Y : T=Y : U=-Y*Y
9815 V=S :M=M+J :N=N+J :T=T*U/(M*N) :S=S+T :IF S<>V GOTO 9815
9816 IF INT(R/J) <> R/J THEN S=-S
9817 PRINT "SINE OF  " ; X ; "  DEGREES IS  " ; S
9818 PRINT : GOTO 9812                  ' RETURN FOR NEXT ENTRY
9819  '
9820  ' COSINE ROUTINE, ANGLE IN RADIANS
9821 DEFDBL P-Y : DEFINT I-N : J=2 : PI=3.1415926535897932
9822 INPUT "ENTER ANGLE IN RADIANS FOR COSINE(X) " ; X
9823 Y=X : R=0 : IF ABS(Y)>PI THEN R=FIX(Y/PI) : Y=Y-R*PI
9824 M=-1 : N=0 : Q=1 : T=1 : U=-Y*Y
9825 V=Q :M=M+J :N=N+J :T=T*U/(M*N) :Q=Q+T :IF Q<>V GOTO 9825
9826 IF INT(R/J) <> R/J THEN Q=-Q
9827 PRINT "COSINE OF  " ; X ; "  RADIANS IS  " ; Q
9828 PRINT : GOTO 9822                  ' RETURN FOR NEXT ENTRY
9829  '
9830  ' COSINE ROUTINE, ANGLE IN DEGREES
9831 DEFDBL P-Y :DEFINT I-N :J=2 :P=180 : PI=3.1415926535897932
9832 INPUT "ENTER ANGLE IN DEGREES FOR COSINE(X) " ; X
9833 Y=X : R=0 : IF ABS(Y)>P THEN R=FIX(Y/P) : Y=Y-R*P
9834 Y=Y*PI/P : M=-1 : N=0 : Q=1 : T=1 : U=-Y*Y
9835 V=Q :M=M+J :N=N+J :T=T*U/(M*N) :Q=Q+T :IF Q<>V GOTO 9835
9836 IF INT(R/J) <> R/J THEN Q=-Q
9837 PRINT "COSINE OF  " ; X ; "  DEGREES IS  " ; Q
9838 PRINT : GOTO 9832                  ' RETURN FOR NEXT ENTRY
9839  '
9840  ' TANGENT ROUTINE, ANGLE IN RADIANS    (ALSO SIN & COS)
9841 DEFDBL P-Y : DEFINT I-N : I=1 : PI=3.1415926535897932
9842 INPUT "ENTER ANGLE IN RADIANS FOR TANGENT(X) " ; X
9843 Y=X : R=0 : IF ABS(Y)>PI THEN R=FIX(Y/PI) : Y=Y-R*PI
9844 M=I : Q=I : S=I : T=I : U=-Y*Y
9845 V=Q :W=S :M=M+I :T=T*U/M :Q=Q+T :M=M+I :T=T/M :S=S+T :
        IF Q<>V OR S<>W GOTO 9845
9846 IF INT(R/2) <> R/2 THEN Q=-Q : S=-S
9847 S=S*Y : TN=S/Q
9848 PRINT "TANGENT OF  " ; X ; "  RADIANS IS  " ; TN
9849 PRINT "ALSO SINE = " ; S ; "  COSINE = " ; Q
9850 PRINT : GOTO 9842                  ' RETURN FOR NEXT ENTRY
9859  '
9860  ' TANGENT ROUTINE, ANGLE IN DEGREES    (ALSO SIN & COS)
9861 DEFDBL P-Y :DEFINT I-N :I=1 :P=180 : PI=3.1415926535897932
9862 INPUT "ENTER ANGLE IN DEGREES FOR TANGENT(X) " ; X
9863 Y=X : R=0 : IF ABS(Y)>P THEN R=FIX(Y/P) : Y=Y-R*P
9864 M=I : Q=I : S=I : T=I : Y=Y*PI/P : U=-Y*Y
9865 V=Q :W=S :M=M+I :T=T*U/M :Q=Q+T :M=M+I :T=T/M :S=S+T :
        IF Q<>V OR S<>W GOTO 9865
9866 IF INT(R/2) <> R/2 THEN Q=-Q : S=-S
9867 S=S*Y : TN=S/Q
9868 PRINT "TANGENT OF  " ; X ; "  DEGREES IS  " ; TN
9869 PRINT "ALSO SINE = " ; S ; "  COSINE = " ; Q
9870 PRINT : GOTO 9862                  ' RETURN FOR NEXT ENTRY
```

Delmar Hinrichs, 2116 SE 377th Ave., Washougal, WA 98671.

```
9879 '
9880 '    ARCTANGENT ROUTINE, OUTPUT ANGLE IN RADIANS, DEGREES
9881 DEFDBL P-Y : DEFINT I-N : I=1 : J=2 : PI=3.1415926535897932
9882 INPUT "ENTER 'X' FOR ARCTANGENT(X) "; X
9883 U=X*X  Y=ABS(X) : N=I
9884 IF Y>.77 AND Y<1.18 THEN P=U+I : S=SQR(P) : GOTO 9887
9885 IF U<I THEN R=Y : T=Y : S=-U ELSE T=-I/Y :R=PI/J+T :S=-T*T
9886 V=R  :N=N+J : T=T*S  R=R+T/N : IF R<>V GOTO 9886 ELSE 9890
9887 V=S   S=(P/S+S)/J : IF S<>V GOTO 9887    ' ADJUST SQ. ROOT
9888 N=0 : R=Y/S : T=R : U=R*R                ' FOR ARCSINE
9889 V=R :N=N+J : T=T*U*(N-I)/N :R=R+T/(N+I) :IF R<>V GOTO 9889
9890 R=R*SGN(X) : IF DG=1 THEN R=R*180/PI
9891 PRINT "THE ARCTANGENT OF  "; X; "  IS  "; R;
9892 IF DG=1 PRINT " DEGREES" ELSE PRINT " RADIANS"
9893 PRINT : GOTO 9882                    ' RETURN FOR NEXT ENTRY
9899 '
9900 '    ARCTANGENT, OUTPUT IN DEGREES
9901 DG=1 : GOTO 9880
9909 '
9910 '    NATURAL LOGARITHM ROUTINE   (BASE E)
9911 DEFDBL P-Y : DEFINT I-N : I=1 : J=2 : Q=2.7182818284590452
9912 INPUT "ENTER 'X' FOR LN(X) "; X
9913 IF X<=0 PRINT "**** LN(X) ERROR.  X IS <= ZERO" :GOTO 9912
9914 Y=X : N=I : M=0 : S=I/Q
9915 IF Y>Q THEN Y=Y*S : M=M+I : GOTO 9915     ' SCALE INPUT
9916 IF Y<S THEN Y=Y*Q : M=M-I : GOTO 9916
9917 S=(Y-I)/(Y+I) : W=S  : U=S*S
9918 V=S  :N=N+J :W=W*U : S=S+W/N :IF S<>V GOTO 9918
9919 S=S*J+M
9920 PRINT "NATURAL LOG OF  "; X; "  IS  "; S
9921 PRINT : GOTO 9912                    ' RETURN FOR NEXT ENTRY
9929 '
9930 '    COMMON LOGARITHM ROUTINE   (BASE 10)
9931 DEFDBL P-Y : DEFINT I-N
9932 I=1 : J=2 : Q=2.7182818284590452 : R=.43429448190325183
9933 INPUT "ENTER 'X' FOR LOG(X) "; X
9934 IF X<=0 PRINT"**** LOG(X) ERROR.  X IS <= ZERO" :GOTO 9933
9935 Y=X  : N=I : M=0 : S=I/Q
9936 IF Y>Q THEN Y=Y*S : M=M+I : GOTO 9936     ' SCALE INPUT
9937 IF Y<S THEN Y=Y*Q : M=M-I : GOTO 9937
9938 S=(Y-I)/(Y+I) : W=S  : U=S*S
9939 V=S  :N=N+J :W=W*U :S=S+W/N :IF S<>V GOTO 9939
9940 S=S*J+M  : S=S*R
9941 PRINT "COMMON LOG OF  "; X; "  IS  "; S
9942 PRINT : GOTO 9933                    ' RETURN FOR NEXT ENTRY
9949 '
9950 '    E[X EXPONENTIAL ROUTINE
9951 DEFDBL P-Y : DEFINT I-N : I=1 : H=86.325
9952 INPUT "ENTER EXPONENT OF 'E' "; X
9953 Y=ABS(X) :IF Y>H PRINT "**** E[X ERROR.    X>86" :GOTO 9952
9954 S=I+Y : T=Y  : N=I : IF Y=0 GOTO 9957
9955 V=S  :N=N+I :T=T*Y/N :S=S+T :IF S<>V GOTO 9955
9956 IF X<0 THEN S=I/S
9957 PRINT "E TO THE  "; X; "  POWER IS  "; S
9958 PRINT : GOTO 9952                    ' RETURN FOR NEXT ENTRY
9959 '
9960 '    10[X EXPONENTIAL ROUTINE
9961 DEFDBL P-Y :DEFINT I-N :I=1 :H=37.49 :R=2.3025850929940457
9962 INPUT "ENTER EXPONENT OF 10 "; X
9963 Y=ABS(X) :IF Y>H PRINT"**** 10[X ERROR.    X>37" :GOTO 9962
9964 Y=Y*R  : S=I+Y : T=Y  : N=I : IF Y=0 GOTO 9967
9965 V=S  :N=N+I :T=T*Y/N :S=S+T :IF S<>V GOTO 9965
9966 IF X<0 THEN S=I/S
9967 PRINT "10 TO THE  "; X; "  POWER IS  "; S
9968 PRINT : GOTO 9962                    ' RETURN FOR NEXT ENTRY
9969 '
9970 '    Y[X POWER ROUTINE
9971 DEFDBL P-Y : DEFINT I-N : I=1 : J=2 : Q=2.7182818284590452
9972 INPUT "ENTER 'Y' AND 'X' FOR Y[X "; Y, X
9973 IF Y=0 AND X<=0 PRINT"**** Y[X ERROR.  Y=0, X<=0":GOTO 9972
9974 IF Y>0 OR INT(X)=X GOTO 9976
9975 PRINT "**** Y[X ERROR.   Y<0 AND X NOT INTEGER" : GOTO 9972
9976 IF Y=0 THEN R=0 : GOTO 9989
9977 IF X=0 THEN R=I : GOTO 9989
9978 P=ABS(Y) : N=I : M=0 : S=I/Q
9979 IF P>Q THEN P=P*S : M=M+I : GOTO 9979     ' SCALE Y ENTRY
9980 IF P<S THEN P=P*Q : M=M-I : GOTO 9980
9981 S=(P-I)/(P+I) : W=S  : U=S*S
9982 V=S  :N=N+J :W=W*U :S=S+W/N :IF S<>V GOTO 9982
9983 S=S*J+M  : N=I : U=ABS(S*X) : R=U+I : T=U   ' S = LN(Y)
9984 IF U>86.325 PRINT"**** Y[X ERROR.    X TOO LARGE" :GOTO 9972
9985 V=R  :N=N+I :T=T*U/N :R=R+T :IF R<>V GOTO 9985
9986 IF Y<0 AND INT(X/J) <> X/J THEN R=-R
9987 IF X<0 THEN R=I/R
9988 IF S<0 THEN R=I/R
9989 PRINT Y; " TO THE  "; X; "  POWER IS  "; R
9990 PRINT : GOTO 9972                    ' RETURN FOR NEXT ENTRY
9991 END
```

Most likely, only one or two of these routines would be used in any one of your programs. As an example of how these routines can be used as subroutines, I have written double precision polar-to-rectangular coordinate conversion and rectangular-to-polar coordinate conversion programs shown as Programs B and C.

These math function routines all iterate (repeatedly perform the same set of operations) until further iteration will not change the result. Therefore, they can be used without change for any number of significant digits that your BASIC supports. When used with double precision (16 digits), they normally give a result with 15 to 16 digits precision. The square root routine uses Newton's method of approximation and all of the other routines use Maclaurin's series expansions. Equations used for all routines are listed in Table 1. All of the routines use the input argument of X (except that X and Y are used for the power routine).

The execution time for these double precision routines written in BASIC is, of course, much longer than the built-in assembly language routines and varies with the argument. Maximum execution time is over 10 seconds for some arguments with the arctangent and exponential routines, as these routines may have to iterate 100 to 200 times. The other routines are faster. In any case, it is usually worth the wait to get a more accurate result.

### Square Root Routine

The square root routine starts with an initial estimate of the square root then adjusts that estimate to form the next estimate. It continues iterating until successive estimates are equal. In this routine, I have used the single precision square root function SQR to give the initial estimate. If your BASIC does not have a square root function, you may use an initial estimate of one (1). The routine will then have to iterate a little longer but it will still give the same result.

An error message is given if the argument is negative. In these demonstration programs, if an error occurs the program returns to ask for another input. When it is used as a subroutine the routine should probably be modified to halt on error.

### Sine Routines

The sine routines for radians or for degrees are the same, except that degrees are converted to radians before calculations. The input angle may have any value, but it is scaled into the range of $\pm \pi$ radians ($\pm 180°$) before calculations. As with the rest of the routines (except square root) the sine routine adds up a series of terms that get smaller and smaller until adding another term makes no difference in the result.

### Cosine Routines

The cosine routines are very similar to the sine routines, the main difference being in the starting point for calculation of the terms.

### Tangent Routines

The tangent is determined by first calculating both the sine and the cosine, then dividing the sine by the cosine. The routines for radians and for degrees are the same, except that degrees are converted to radians before calculations. Since sine and cosine calculations are so similar, they are done simultaneously with each contributing to the other. These tangent routines give both sine and cosine as byproducts so they may be used

## Precision, con't...

Table 1. Double Precision Mathematics Equations.

1. Square Root: To approximate $s = \sqrt{x}$

$$s = \frac{\frac{x}{s} + s}{2} \qquad x > 0$$

2. Sine:

$$\sin(x) = x - \frac{x^3}{3!} + \frac{x^5}{5!} - \frac{x^7}{7!} + \ldots$$

or:

$$\frac{\sin(x)}{x} = 1 - \frac{x^2}{3!} + \frac{x^4}{5!} - \frac{x^6}{7!} + \ldots$$

3. Cosine:

$$\cos(x) = 1 - \frac{x^2}{2!} + \frac{x^4}{4!} - \frac{x^6}{6!} + \ldots$$

4. Tangent:

$$\tan(x) = \frac{\sin(x)}{\cos(x)}$$

5. Arctangent:

$$\tan^{-1}(x) = x - \frac{x^3}{3} + \frac{x^5}{5} - \frac{x^7}{7} + \ldots \qquad x^2 < 1$$

$$= \frac{\pi}{2} - \frac{1}{x} + \frac{1}{3x^3} - \frac{1}{5x^5} + \ldots \qquad x^2 > 1$$

$$= \sin^{-1}\left(\frac{x}{\sqrt{1+x^2}}\right) = \sin^{-1}(z) \qquad x^2 \text{ near } 1$$

then:

$$\sin^{-1}(z) = z + \frac{1}{2} \cdot \frac{z^3}{3} + \frac{1 \cdot 3}{2 \cdot 4} \cdot \frac{z^5}{5} + \frac{1 \cdot 3 \cdot 5}{2 \cdot 4 \cdot 6} \cdot \frac{z^7}{7} + \ldots \qquad z^2 < 1$$

6. Logarithm:

$$\ln(x) = 2\left(\frac{x-1}{x+1} + \frac{1}{3}\left(\frac{x-1}{x+1}\right)^3 + \frac{1}{5}\left(\frac{x-1}{x+1}\right)^5 + \ldots\right) \qquad x > 0$$

$$\log(x) = \frac{\ln(x)}{\ln(10)}$$

7. Exponential:

$$e^x = 1 + x + \frac{x^2}{2!} + \frac{x^3}{3!} + \frac{x^4}{4!} + \ldots$$

$$10^x = e^{(x \ln(10))}$$

8. Power:

$$y^x = e^{(x \ln(y))}$$

Notes:

1) All trigonometric functions are in radian measure.

Degrees = Radians$(180/\pi)$ where $180° = \pi$ Radians

2) Factorial is symbolized by "!" that is, $3!$ means $1 \times 2 \times 3 = 6$

$5!$ means $1 \times 2 \times 3 \times 4 \times 5 = 120$

etc.

3) Natural logarithm of $x = \ln(x)$ logarithm base $e$, where $e = 2.71828$

Base 10 logarithm of $x = \log(x)$

where both sine and cosine of the same angle are needed as in polar-to-rectangular conversion (Program B).

Program B. Polar-to-Rectangular Coordinate Conversion.

```
10 CLS : PRINT "POLAR TO RECTANGULAR CONVERSION IN DEGREES"
20  ' BY D. D. HINRICHS IN TRS-80 LEVEL II BASIC
30 DEFDBL P-Y :DEFINT I-N :I=1 :P=180 : PI=3.1415926535897932
40 PRINT : INPUT "ENTER ANGLE IN DEGREES" ; X
50 INPUT "ENTER RADIUS FROM ORIGIN" ; RD
60 GOSUB 9860
70 PRINT "THE 'X' DISTANCE IS  " ; RD*Q
80 PRINT "THE 'Y' DISTANCE IS  " ; RD*S
90 GOTO 40
100 END
9860  ' TANGENT ROUTINE, ANGLE IN DEGREES   (ALSO SIN & COS)
9863 Y=X : R=0 : IF ABS(Y)>P THEN R=FIX(Y/P) : Y=Y-R*P
9864 M=I : Q=I : S=I : T=I : Y=Y*PI/P : U=-Y*Y
9865 V=Q :W=S :M=M+I : T=T*U/M :Q=Q+T :M=M+I :T=T/M :S=S+T :
     IF Q<>V OR S<>W GOTO 9865
9866 IF INT(R/2) <> R/2 THEN Q=-Q : S=-S
9867 S=S*Y : TN=S/Q
9868 RETURN
```

In these routines, **both** sine and cosine must be checked for no change with added terms. If, for example, the angle is near to 90° the sine is near to one while the cosine is near to zero. Thus a term that is not large enough to significantly change the sine may be large enough to change the cosine. If the angle is near to 0°, the situation is reversed.

### Arctangent Routines

The arctangent routines for radians or for degrees are actually the same routine, the only difference being in converting the output radians to degrees for the degree routine. This routine actually uses four different series to calculate the arctangent:

1. For arguments less than one the first arctangent series in Table 1.
2. For arguments greater than one the second arctangent series in Table 1.
3. Arguments near to one are calculated too slowly using arctangent routines (1) or (2), so the arctangent is then determined by first applying a trigonometric identity, then using an arcsine series calculation.
4. Using (3) requires use of the square root routine.

Routines for (1) and (2) are similar enough so that with suitable initialization, the same statements may be used. The cutoff points between using (1) and (3), or between using (2) and (3) were chosen to give the minimum calculation time.

Arcsine and arcosine are easily calculated from the arctangent (Table 2, equations 1 and 2). Instead, the arcsine routine may be removed from this arctangent routine and used by itself, and the arccosine calculated from the arcsine (Table 2, equation 3).

Table 2. Trigonometric Identities for Arcsine and Arccosine Calculations.

1. Arcsine:

$$\sin^{-1}(x) = \tan^{-1}\left(\frac{x}{\sqrt{1-x^2}}\right)$$

2. Arccosine:

$$\cos^{-1}(x) = \tan^{-1}\left(\frac{\sqrt{1-x^2}}{x}\right)$$

3. Arccosine:

$$\cos^{-1}(x) = \sin^{-1}\left(\sqrt{1-x^2}\right)$$

Note: The output is in radians. Multiply by $180/\pi$ to get degrees.

## Logarithm Routines

The natural logarithm (ln(X)) and the base 10 logarithm (log(X)) routines are the same, except that the natural logarithm is multiplied by a constant (l/ln(10)) to give the base 10 logarithm. The input argument is scaled, if necessary, so that it falls between l/e and e (e = 2.71828...). An error message is given if the input argument is zero or negative.

## Exponential Routines

The natural exponential ($e^X$) and the base 10 exponential ($10^X$) routines are the same, except that the input argument is first multiplied by a constant (ln(10)) if the output is to be for base 10. An error message is given if the absolute value of the input argument is so large that it would cause overflow during calculations (greater than about 86 for natural exponential, or about 37 for base 10 exponential).

## Power Routine

The power ($y^X$) routine makes use of logarithm and exponential routines, similar to the way that you could manually use a table of logarithms to calculate a power. It is most useful when the exponent is not an integer. If the exponent (X) is an integer, it is faster to just multiply the Y value by itself that many times.

Several checks for illegal entries are made and each gives its own error message. This routine requires two entries; if only one entry (Y) is made BASIC will ask for the second (X) with double question marks.

## Examples of Use as Subroutines

To show how these routines may be used as subroutines to a main program, I have written two short programs (Programs B and C) to convert double precision coordinates from polar to rectangular and from rectangular to polar. The equations used are given in Table 3. The easiest and most error-free way to write such programs is to load this double precision demonstration program into your computer from tape, then DELETE everything except the routine(s) that you will be using, then write your main program over it.

Program C. Rectangular-to-Polar Coordinate Conversion.

```
10 CLS : PRINT "RECTANGULAR TO POLAR CONVERSION IN DEGREES"
20 DEFDBL P-Y : DEFINT I-N : I=1 : J=2 : PI=3.1415926535897932
30 PRINT : INPUT "ENTER 'X' DISTANCE (+ OR -)" ; XD
40 INPUT "ENTER 'Y' DISTANCE (+ OR -)" ; YD
50 X=XD*XD+YD*YD : GOSUB 9790 : RD=S          ' FIND RADIUS
60 IF XD=0 THEN XD=1D-10                      ' XD CAN'T BE ZERO
70 X=YD/XD : GOSUB 9880 : W=R                 ' FIND ANGLE
80 IF XD<0 THEN W=W+180 : IF YD<0 THEN W=W-360 ' FOR QUADRANT
90 PRINT "RADIUS FROM ORIGIN IS  " ; RD
100 PRINT "ANGLE IS " ; W ; " DEGREES"
110 GOTO 30
120 END
9790  '   SQUARE ROOT ROUTINE
9793 S=X  : IF X=0 RETURN
9794 IF X<0 PRINT"**** ERROR, SQ. ROOT NEGATIVE NO. " : STOP
9795 S=1                         ' PRELIMINARY VALUE OF SQUARE ROOT
9796 V=S :S=(X/S+S)/J : IF S<>V GOTO 9796       ' ADJUST VALUE
9797 RETURN
9880  '  ARCTANGENT ROUTINE, OUTPUT ANGLE IN DEGREES
9883 U=X*X  : Y=ABS(X) : N=I
9884 IF Y>.77 AND Y<1.18 THEN P=U+I : S=1 : GOTO 9887
9885 IF U<I THEN R=Y :T=Y :S=-U ELSE T=-I/Y :R=PI/J+T :S=-T*T
9886 V=R  :N=N+J  :T=T*S  :R=R+T/N :IF R<>V GOTO 9886 ELSE 9890
9887 V=S  : S=(P/S+S)/J : IF S<>V GOTO 9887      ' ADJUST SQ. ROOT
9888 N=0  : R=Y/S  : T=R  : U=R*R                ' FOR ARCSINE
9889 V=R  :N=N+J  :T=T*U*(N-I)/N :R=R+T/(N+I) :IF R<>V GOTO 9889
9890 R=R*SGN(X)  : R=R*180/PI
9891 RETURN
```

Table 3. Polar-to-Rectangular and Rectangular-to-Polar Coordinate Conversion Equations.

1. Polar to Rectangular:   where: x, y = Rectangular coordinates

    $x = r \cos(a)$       r, a = Radius & angle of polar coordinates

    $y = r \sin(a)$

2. Rectangular to Polar:

    $r = \sqrt{x^2 + y^2}$

    $a = \tan^{-1}(y/x)$

For calculation of the proper quadrant for angle "a":

a) If $x = 0$, substitute a very small number for x, perhaps $10^{-10}$

b) If $x < 0$, add $180°$ to a     (or add $\pi$ radians)

c) If $x < 0$ and $y < 0$, subtract $180°$ from a  (or subtract $\pi$ radians)

To make it easy to avoid duplicate line numbers when using these routines as subroutines, I have compressed all line numbers for the routines into the 9790 to 9991 range. To convert these demonstration program routines into subroutines:

1. Delete the general initialization line(s) and put this initialization into the main program instead.
2. Delete the INPUT line and set the input argument to X (or to X and Y for the power routine) in the main program before calling the subroutine.
3. Delete the output PRINT statements and save the result as a variable in the main program, after calling the subroutine.
4. Delete the GOTO loop at the end of the routine and replace it with RETURN.
5. I suggest that you replace the GOTO after error messages with either STOP or END to halt program operation if an error occurs. If you do not want the program to halt, replace the GOTO with RETURN.

You must be sure that the subroutine variables do not conflict with variables used in the main program. One way to do this is to use only two-character variables in the main program. The only two-character variables used in the routines are PI and TN.

Compare the listings of the subroutines used in Programs B and C with the listings of the corresponding routines in Program A to see how this conversion to subroutines is done. Note that for the square root subroutine in Program B, and also for the square root routine embedded in the arctangent subroutine in Program C, it is assumed that BASIC does not have a SQR function. The initial guess of the square root is set to one (1).

The output angle for rectangular-to-polar conversion is given in the ±(0° to 180°) convention. The rectangular-to-polar coordinate conversion may fail with an overflow (?OV ERROR IN 9883) if either the X distance or the Y distance is very large or very small. This is most liable to occur if one is large and the other small. For example, if the X distance is zero (then set by the program to $10^{-10}$) and the Y distance is $10^{10}$, overflow will occur during the calculations.

## Precision, con't...

### Use with Other BASICs

This set of routines should be easy to translate to BASICs other than the TRS-80 Level II BASIC in which they were written. Following is an explanation of some of the special features used so that you can convert them to equivalent statements in your BASIC. CLS means "clear screen" and is not necessary, but gives a neater CRT display. The colon (:) is used to separate statements in multi-statement lines. The apostrophe (') is an abbreviation of REM. If your BASIC does not have them, the DEFDBL (define double) and the DEFINT (define integer) statements may be eliminated. Of course, if you cannot use double precision variables, you will not get double precision results. Extra program lines and GOTOs may substitute for ELSEs. If you do not have the FIX function, the statement: R = FIX(Y/PI) in line 9803 may be changed to: R = SGN(Y)*INT(ABS(Y)/PI) and similar changes made to all sine, cosine and tangent routines. Also note that in line 9803 and in similar lines in other routines, when the IF statement is false, **none** of the statements following it on the same line are performed. If you do not have a SQR function, set the initial estimate to one (1) in the square root routine and in the arctangent routine (i.e., S = 1).

### Conclusions

If you have avoided programming in those areas that require double precision math functions you no longer have an excuse. Accurate interest calculations, satellite orbit analysis and other such problems can now be done. It's up to you! □

---

# DOUBLING UP
## by FRANK TAPSON

TAKE a piece of paper—you may use any size you please—and fold it in half. Then fold it in half again, and yet again, and again... How many times do you think you can do this? If you have never met this problem before then try it before you read any further—you will very probably receive a surprise. Have a guess first before actually trying to do the folding and then see how far you get.

Many people on meeting this little problem for the first time are prepared to say that, provided the paper is large enough then it may be folded in half any number of times. Well, as you might have discovered by now, after 7 such foldings the task becomes extremely difficult, and if not impossible then it will almost certainly be after the next fold. It is interesting to look at what in fact happens.

After our first fold, the piece of paper we have to work on next is double the thickness of the original. Another fold of this piece doubles the thickness so that we now have 4x the thickness of the original. Folding again will once more double-up on the thickness so that we have (2 x 2 x 2) 8 thicknesses of paper. This is followed by 16 thicknesses after the next fold, then 32, then 64, and 128 after the 7th fold. Assuming that the paper we are using is one-thousandth of an inch thick (not the thinnest possible but still a flimsy paper) then after folding it 7 times we have a piece which is one-eighth of an inch thick, or about the thickness of a piece of stout card. Now such a card could certainly be folded in half generally, but there is an added difficulty. Just as the thickness has been doubled with each fold so the area has been halved, and after only 6 foldings we are usually trying to bend something which is not much bigger than an extra-large postage stamp, which is why that piece of 'stout card' is so difficult to fold.

It is interesting to wonder how far the process might be taken if a piece of super-large paper were used. Let us assume it is still one-thousandth of an inch thick, but that we can start with a piece the size of a football-pitch. Go on—have a guess, how many times would you manage to fold it in half?

Some might wish to argue about the precise stage at which the task becomes impossible, but if the 13th folding can be made, it produces something which is about four feet square and eight inches thick. Now think about bending that!

Once we start folding by speculation (and not by actually trying to do it) it becomes fascinating to go on with the process. For instance, just suppose we were able to get an extremely large piece of paper and fold in in half exactly 100 times and, having done that we wished to stand on top of it—how long a ladder would we need to get to the top? By now you have no doubt some idea of what to expect—or have you? After the 26th fold we have a "piece of paper" which is just over a mile thick so you might think we are going to need a fairly tall ladder for 100 folds. Keep going—the 53rd fold gets us just past the sun, and if you think that we are at least over half way then you have failed to see what doubling is all about. The 83rd fold gets us somewhere near the centre of our galaxy, from which it follows that the 84th fold puts us out on the other side and still going. And there we will let the matter rest, if anyone can work out 'precisely' where the top of our work will be after the 100th fold do let us know. We might be able to use it as a navigational aid for inter-stellar travel!

This simple concept of the growth of the doubling sequence has had a fascination for those concerned with the lighter side of mathematics for many years. Perhaps the most famous is the story told around the invention of the chess-board, how the king was so pleased that he offered the inventor any reward that the inventor cared to name. This was expressed as 'one grain of corn on the first square of the board, two grains on the second square, four grains on the third square and so on...' The king thought this is a very light price to pay for such a great game and readily agreed. However, he was not at all pleased to learn that the total quantity of grain required could not be supplied by the entire world output of grain for several years to come. Some accounts of the story claim that he had the inventor beheaded for imposing such a mathematical joke upon royalty! Re-telling this story in his mammoth work A History of Chess, H. J. R. Murray says that the quantity of grain needed is such as to cover England to a uniform depth of 38.4 feet. The actual *number* of grains needed to fulfil the stated conditions is $2^{64}-1$, a figure which also occurs in connection with the story woven around the Tower of Hanoi.

Another form of the story involves either the sale of a horse, or the shoeing of one. In either case the price is fixed at a farthing (over a hundred years ago) or a penny for the first nail in its shoes, doubled-up for the second nail, doubled again for the third nail and so on. The only serious disagreement appears to be concerning the total number of nails (I have stories giving 6, 7, and 8 nails per shoe).

A story can also be woven around the telling of a secret to two friends, each of whom tells it to two other (different) friends, each of whom... Assuming that the actual telling occupies just one minute, and that another minute is lost in scurrying off to find someone else to tell the secret to-how many people will know after one hour has elapsed from the initial telling? By now of course you will have some idea of what is happening and won't be too surprised to learn that by the end of the hour 2,147,483,647 people would know the secret. Since this is just over one-half of the present total world population, it hardly could be called a secret any more! The same story has been presented differently by asking, under the above conditions, in a village of a given number of inhabitants, how soon would it be before everyone knew the secret?

There is a surprising growth rate in the simple matter of doubling at every stage of the sequence. Just think of it next time you fold a piece of paper in half, and don't go on for too long lest you should fall off the top!

For the curious, the exact value of $2^{100}$ is—1,267,650,600,228,229,401,496,703,205,376.

Reprinted from *Games & Puzzles,* December 1974. Copyright 1974 *Games & Puzzles,* 11 Tottenham Court Road, London W1A 4XF, England.

# Magic Squares

| 52 | 61 | 4  | 13 | 20 | 29 | 36 | 45 |
|----|----|----|----|----|----|----|----|
| 14 | 3  |    |    |    |    | 30 | 19 |
| 53 |    |    |    |    |    |    | 44 |
| 11 |    |    |    |    |    |    | 22 |
| 55 |    |    |    |    |    |    | 42 |
| 9  |    |    |    |    |    | 5  | 24 |
| 50 | 63 |    |    |    |    | 34 | 47 |
| 16 | 1  | 64 |    |    | 33 | 32 | 17 |

## on the Computer

Donald T. Piele
Assistant Professor of Mathematics
University of Wisconsin-Parkside
Kenosha, Wisconsin 53140

Magic squares? Humbug! I've never been able to get excited over someone's special arrangement of numbers that total up to the same sum whether you add across a row or down a column or diagonally. Benjamin Franklin when first confronted with them wrote,

"... it is perhaps a mark of good sense of our (English) mathematicians that they would not spend their time in things that were merely *difficiles nugae* incapable of any useful applications."(1)

Franklin had to confess, however,

"In my younger days, having once more leisure time (which I still think I might of employed more usefully) I had amused myself in making these kind of magic squares, and, at length acquired such a knack at it, that I could fill the cells of any magic square of reasonable size with a series of numbers as fast as I could write them, disposed in such a manner that the sum of every row, horizontal, perpendicular or diagonal, should be equal; but not being satisfied with these, which I looked on as common and easy things, I imposed on myself more difficult tasks, and succeeded in making other magic squares with a variety of properties, and much more curious."(2)

In spite of the fact that I knew Benjamin Franklin had been a statesman, a scientist, a politician, a philosopher, and a writer, I was surprised to discover that playing with magic squares was also among his lengthy list of avocations. Reading further, I discovered that there are ways of testing magic squares, besides the usual rows, columns, or diagonals, that I had never seen before. For example, there are generalized diagonals, broken diagonals, corner diagonals, horizontal zig-zags, vertical zig-zags, just to name a few. Next, I found that algorithms existed for generating magic squares which looked relatively easy to program. Maybe magic squares aren't so bad after all? Besides, the computer can be programmed to do all the arithmetic and print out a listing of magical properties for each square. That did it.

I began with a generalized version of the algorithm of De la Loubère.(3) This method fills an n x n square matrix with consecutive integers from 1 to $n^2$ by putting the ith integer in the matrix position $P_i$ as follows:

1. Place the number 1 in any initial position, $P_i = (i,j)$. The standard initial position is the middle of the top row, $(1,(n+1)/2)$ for n odd.
2. Place the successive integers in vacant cells separated by jumps (A,B), $P_i = P_{i-1} + (A,B)$.
3. If $P_i$ moves outside the square n x n matrix, adjust the coordinates modulo (1,2,3, ... ,n) so that $P_i$ moves back into the square. e.g. For n = 3, (2,4) = (2,1) and (0,3) = (3,3).
4. If you encounter a position $P_j$, $j \leq n^2$, that has already been filled, switch for one move to the rule $P_j = P_{j-1} + (C,D)$ and continue as in 2.

---

1. Originally from *Letters and Papers on Philosophical Subjects* by Benjamin Franklin, LL.D., F.R.S., London, 1769. See [1] p. 89.
2. See footnote 1 pp. 89-90.
3. "De la Loubere was the envoy of Louis XIV to Siam in 1687-1688, and there learnt his method." See [3] p. 195.

De la Loubère's original method specified that 1 be placed in the middle of the top row, $P_1 = (1,(n+1)/2)$, and that (A,B,C,D) be fixed at (1,1,0,−1). This is illustrated in the first sample run of the De la Loubère program. But what happens when you try different starting positions $P_1$ and other step values (A,B,C,D)? Will any choice of (A,B,C,D) generate a square? De la Loubère used his algorithm only for odd order squares, what happens for even order squares? Given a De la Loubere magic square, can you tell how it was generated? The second sample run shows the 5 x 5 magic square of Backet de Meziriac which was generated by choosing $P_1 = (3,4)$ and (A,B,C,D) = (1,1,2,0). Originally it was constructed by a completely different method (see [1] p. 17). Can you find other magic squares in books or magazines that can be generated with the De la Loubère program?

For an n x n square of numbers to be considered magic it must at least have the same sum for each row and column. If the square is filled with the consecutive numbers 1 through $n^2$ then each row and column must add up to $n(n^2 + 1)/2$ (why?). All other ways of finding n numbers, symmetrically arranged, that add up to this sum, improves the magic square and makes it more unique. For example, a square may be summed along generalized diagonals as illustrated for a 3 x 3 square in Fig. 1. For an n x n square there are 2n generalized diagonals. The De la Loubère program checks them all in addition to the rows and columns. The best you can do, with this program, is find 4n magical properties for an n x n square.

Benjamin Franklin's magic squares are entirely different and cannot be generated by the De la Loubère algorithm. The best ones are of order 8 and 16, known as the Franklin Magic Squares. The largest one is considered among the most ingenious ever developed. It was impossible even for Franklin to be modest about it.

"... you will readily allow the square of 16 to be the most magically magic of any magic square ever made by any magician."(4)

The Franklin squares are characterized by magical sums along broken diagonals that change direction halfway through the square as illustrated in Fig. 2. They can be constructed to point in four different directions; North, South, East and West. In each direction an n x n square has n broken diagonals, so it is possible to have a total of 4n magic broken diagonals. Franklin Magic Squares have the maximum number. Two other special arrangements that characterize the Franklin order 8 squares are illustrated in Fig. 3.

It is not known how Franklin generated his squares, although it is very likely that they were geometrically motivated. Several investigations have found unique symmetries in the way the numbers are arranged (see [1] p. 93 and [4]). However, there exists an analytical algorithm for reconstructing his squares called the method of *alternation with binate transposition* (see [1] pp. 100-106). It sounds difficult but it is really not. In fact the method can be easily generalized to construct much more than just the Franklin Magic Square. Since there is no difference in the algorithm for squares of order 8 or 16, I will describe, for convenience, the order 8 scheme.

---

4. See footnote 1 p. 93.

Begin with the *plan of construction matrix* (Fig. 4). Number the rows and columns, as usual, and let RC stand for the number at the intersection of row R and column C. The magic square is created as follows:

1. Choose a permutation of the row values, 1 through 8, and denote it by $R_1, R_2, \ldots, R_8$. For the Franklin Square choose 7,8,1,2,3,4,5,6. Let $\overline{R}_i$ be the complementary row $9 - R_i$.
2. Choose an arrangement of the column values 1 through 8, and denote it by $C_1, \underline{C}_2, \overline{C}_1, \overline{C}_2, C_3, C_4, \overline{C}_3, \overline{C}_4$, such that $C_i + \overline{C}_i = 9$. For the Franklin Square choose 4,6,5,3,7,1,2,8. Notice that $4+5, 6+3, 7+2$, and $1+8$ all equal 9.
3. Rearrange the numbers in the *plan of construction matrix* as shown in Fig. 5. For example, using the row and column sequence given in 1 and 2 for the Franklin Magic Square, $R_1 = 7$ and $C_1 = 4$. Thus, $R_1 C_1 = 52$ and is found in the *plan of construction matrix* at the intersection of row 7 and column 4.

Notice the repetition in C values as you move across the columns and in R values as you move down the rows. This characterizes the Franklin squares and makes the computer algorithm relatively short (see program listing).

The original Franklin Magic Square is generated in the first sample run and has a total of 50 magical sums. It has a few other nice properties too, but they are not tested for here (see [1] p. 96). What happens when you try other permutations? If you ignore the restriction on the column permutations, the computer still generates a square but some numbers will be repeated. Try it! Each new row and column permutation will generate a Franklin-like magic square. But, given a Franklin type magic square, can you find a row and column permutation that will generate it?

It probably never occurred to Franklin that anyone would want to, much less be able to, improve upon his "... most magically magic of any magic square." But magic square buffs are a tenacious lot and they should never be underestimated. The most obvious weakness with Franklin's square exists on the main diagonals which are not magic. Many devotees of the subject have tried in vain to remove this imperfection. In 1945, Andrew S. Anema succeeded by constructing, for the first time, a magic square that has all the Franklin properties and in addition is magic along the main diagonals and generalized diagonals (see [2]). His method uses complementary pairs and takes three pages to describe. It turns out that you can generate Anema's improved Franklin square, and many others like it, with the Franklin program described here (see sample run 2). Can you generate other improved Franklin Magic Squares? There are lots of them.

The literature on magic squares is enormous. Probably no other single recreational topic has had more written about it. With very limited experience, my impression is that many of the special methods that have been devised to construct magic squares are merely special cases of more general algorithms. Students who are interested and have a little knowledge of BASIC should be able to step into this area and, with the computer, perform a little magic of their own.

Happy hunting!

## POSTSCRIPT

I'm not sure exactly how many magical properties there are in a Franklin Magic Square, but I do know that the number is much larger than I, or Franklin, ever dreamed. This became apparent one evening when I discovered, or perhaps rediscovered, 139 additional magical arrangements already present and waiting to be counted in a Franklin Square of order 8.

It is a relatively easy exercise to add three subroutines to the Franklin program to check these arrangements for magical sums. Can you do it? Franklin squares appear to have magical properties almost everywhere you look. Can you find other arrangements that sum to 260?

Again, Happy Hunting!

*References*

1. Andrews, W.S. *Magic Squares And Cubes.* The Open Court Publishing Co. 1908.
2. Anema, Andrew S. "Franklin Magic Squares." *Scripta Mathematica* 11:88-96; 1945.
3. Ball, W.W. Rouse. *Mathematical Recreations and Essays.* The Macmillan Co. New York. 1947.
4. Bragdon C. "The Franklin 16 x 16 Magic Square." *Scripta Mathematica* 4:158-60; 1936.

Photograph: Benjamin Franklin 1706-1790
Oval P.M. Alix, c. 1790, after painting by C.P.A. Van Loo, c. 1777-1785, at the American Philosophical Society, New York.

**Donald Piele**

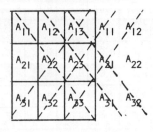

Fig. 1 Generalized diagonals for a 3 x 3 square.

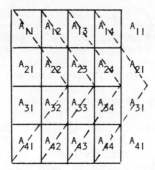

Fig. 2 Broken diagonals in one direction for a 4 x 4 square.

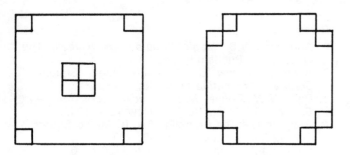

Fig. 3 Special arrangements in the Franklin Magic Square that are magic.
  a. Center 4 squares and four corner squares.
  b. The 4 near-corner squares.

|  | Columns | | | | | | | |
|---|---|---|---|---|---|---|---|---|
|  | 1 | 2 | 3 | 4 | 5 | 6 | 7 | 8 |
| 1 | 1 | 2 | 3 | 4 | 5 | 6 | 7 | 8 |
| 2 | 9 | 10 | 11 | 12 | 13 | 14 | 15 | 16 |
| 3 | 17 | 18 | 19 | 20 | 21 | 22 | 23 | 24 |
| 4 | 25 | 26 | 27 | 28 | 29 | 30 | 31 | 32 |
| 5 | 33 | 34 | 35 | 36 | 37 | 38 | 39 | 40 |
| 6 | 41 | 42 | 43 | 44 | 45 | 46 | 47 | 48 |
| 7 | 49 | 40 | 51 | 52 | 53 | 54 | 55 | 56 |
| 8 | 57 | 58 | 59 | 60 | 61 | 62 | 63 | 64 |

(ROWS)

Fig. 4 The Plan of Construction Matrix.

| $R_1C_1$ | $R_2\overline{C}_1$ | $R_3C_1$ | $R_4\overline{C}_1$ | $R_5C_1$ | $R_6\overline{C}_1$ | $R_7C_1$ | $R_8\overline{C}_1$ |
|---|---|---|---|---|---|---|---|
| $\overline{R}_1C_2$ | $\overline{R}_2\overline{C}_2$ | $\overline{R}_3C_2$ | $\overline{R}_4\overline{C}_2$ | $\overline{R}_5C_2$ | $\overline{R}_6\overline{C}_2$ | $\overline{R}_7C_2$ | $\overline{R}_8\overline{C}_2$ |
| $R_1\overline{C}_1$ | $R_2C_1$ | $R_3\overline{C}_1$ | $R_4C_1$ | $R_5\overline{C}_1$ | $R_6C_1$ | $R_7\overline{C}_1$ | $R_8C_1$ |
| $\overline{R}_1\overline{C}_2$ | $\overline{R}_2C_2$ | $\overline{R}_3\overline{C}_2$ | $\overline{R}_4C_2$ | $\overline{R}_5\overline{C}_2$ | $\overline{R}_6C_2$ | $\overline{R}_7\overline{C}_2$ | $\overline{R}_8C_2$ |
| $R_1C_3$ | $R_2\overline{C}_3$ | $R_3C_3$ | $R_4\overline{C}_3$ | $R_5C_3$ | $R_6\overline{C}_3$ | $R_7C_3$ | $R_8\overline{C}_3$ |
| $\overline{R}_1C_4$ | $\overline{R}_2\overline{C}_4$ | $\overline{R}_3C_4$ | $\overline{R}_4\overline{C}_4$ | $\overline{R}_5C_4$ | $\overline{R}_6\overline{C}_4$ | $\overline{R}_7C_4$ | $\overline{R}_8\overline{C}_4$ |
| $R_1\overline{C}_3$ | $R_2C_3$ | $R_3\overline{C}_3$ | $R_4C_3$ | $R_5\overline{C}_3$ | $R_6C_3$ | $R_7\overline{C}_3$ | $R_8C_3$ |
| $\overline{R}_1\overline{C}_4$ | $\overline{R}_2C_4$ | $\overline{R}_3\overline{C}_4$ | $\overline{R}_4C_4$ | $\overline{R}_5\overline{C}_4$ | $\overline{R}_6C_4$ | $\overline{R}_7\overline{C}_4$ | $\overline{R}_8C_4$ |

Fig. 5 Alternation with permutation scheme.

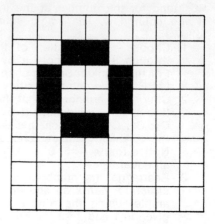

Figure 6. Magic Octagons. Each octagon arrangement sums to 260 wherever it is placed on the 8 x 8 square. There are 25 Magic Octagons.

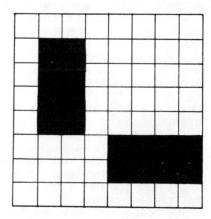

Figure 7. Magic "2 x 4's". Each 2 x 4 rectangle standing up or lying down sums to 260 wherever it is placed on the 8 x 8 square. There are 70 magic "2 x 4's".

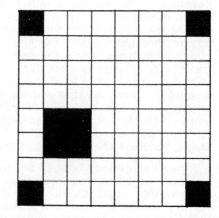

Figure 8. Any 2 x 2 square sums to 130. When it is combined with the four corner squares the sum is 260. There are 45 such arrangements but one has already been counted.

# DE LA LOUBERE PROGRAM

```
10   REM  PROGRAM WRITTEN BY D.T.PIELE 7/15/75
20   REM       ****** THE DE LA LOUBERE PROGRAM  ******
30   PRINT "THIS PROGRAM TESTS NXN SQUARES OF ODD ORDER FOR "
40   PRINT "THEIR MAGICAL PROPERTIES. YOU CAN EITHER ENTER YOUR"
50   PRINT "OWN SQUARE OR LET THE COMPUTER GENERATE ONE FOR YOU "
60   PRINT "USING THE ALGORITHM OF DE LA LOUBERE."
70   PRINT "FOR THE COMPUTER SQUARE TYPE 1. FOR YOUR OWN TYPE 2."
80   INPUT X
90   PRINT "HOW MANY ROWS DO YOU WANT?"
100  PRINT "PICK AND ODD NUMBER BETWEEN 3 AND 11."
110  INPUT N
120  DIM A[12,12],C[12,12],B[12]
130  MAT A=ZER[N,N]
140  MAT C=ZER[N,N]
150  MAT B=ZER[N]
160  IF X=2 THEN 1190
170  K=0
180  PRINT "PICK THE POSITION I,J FOR 1. (MIDDLE OF ROW 1 IS STANDARD)"
190  INPUT I,J
200  PRINT "CHOOSE A,B,C,D (1,1,0,-1 IS STANDARD)"
210  INPUT A,B,C,D
220  REM      ****** THE ALGORITHM FOR GENERATING THE SQUARE ******
230  K=K+1
240  A[I,J]=K
250  IF K=INT(N*N+.5) THEN 370
260  I=I-B
270  J=J+A
280  IF I=0 THEN 700
290  IF I<0 THEN 740
300  IF I>N THEN 700
310  IF J=0 THEN 720
320  IF J<0 THEN 760
330  IF J>N THEN 720
340  IF A[I,J]>0 THEN 670
350  GOTO 230
360  REM ****** END OF ALGORITHM ******
370  PRINT "HERE IS YOUR SQUARE OF ORDER"N
380  PRINT
390  PRINT
400  MAT  PRINT A;
410  S=INT(N*(N*N+1)/2+.5)
420  MAT C=A
430  GOSUB 790
440  R1=C
450  MAT C=TRN(A)
460  GOSUB 790
470  C1=C
480  PRINT
490  PRINT "HERE IS A LIST OF ITS MAGICAL PROPERTIES."
500  PRINT
510  PRINT "ROWS AND COLUMNS:"R1+C1
520  GOSUB 920
530  D1=C+D
540  PRINT "GENERALIZED DIAGONALS:"D1
550  PRINT
560  PRINT "TOTAL MAGICAL SUMS: "R1+C1+D1
570  PRINT
580  PRINT "DO YOU WISH TO TRY AGAIN? TYPE 1 FOR YES, 0 FOR NO."
590  INPUT Y
600  PRINT
610  IF Y=1 THEN 70
620  PRINT "GOODBYE. SEE YOU AT THE FRANKLIN FESTIVAL"
630  PRINT "OCT. 5 TO 11 AT UW-PARKSIDE."
640  PRINT
650  PRINT
660  STOP
670  I=I+B-D
680  J=J-A+C
690  GOTO 280
700  I=ABS(I-N)
710  GOTO 290
720  J=ABS(J-N)
730  GOTO 320
740  I=N+I
750  GOTO 300
760  J=J+N
770  GOTO 330
780  REM  ****** TEST ROWS AND COLUMNS ******
790  C=0
800  FOR I=1 TO N
810  E=0
820  FOR J=1 TO N
830  E=E+C[I,J]
840  NEXT J
850  IF E=S THEN 880
860  GOTO 890
870  GOTO 890
880  C=C+1
890  NEXT I
900  RETURN
910  REM ****** TEST GENERALIZED DIAGONALS ******
920  C=0
930  D=0
940  FOR J=0 TO N-1
950  E=0
960  F=0
970  FOR I=1 TO N
980  R=I+J
990  T=N+1-I-J
1000 IF R <= N THEN 1060
1010 R=ABS(R-N)
1020 GOTO 1060
1030 IF T >= 1 THEN 1080
1040 T=T+N
1050 GOTO 1080
1060 E=E+A[I,R]
1070 GOTO 1030
1080 F=F+A[T,I]
1090 NEXT I
1100 IF E=S THEN 1130
1110 IF F=S THEN 1150
1120 GOTO 1160
1130 C=C+1
1140 GOTO 1110
1150 D=D+1
1160 NEXT J
1170 RETURN
1180 REM  ****** ENTER YOUR OWN SQUARE ******
1190 PRINT "LIST THE MEMBERS OF EACH ROW SEPARATED BY A COMMA."
1200 FOR R=1 TO N
1210 PRINT "ROW"R
1220 MAT  INPUT B
1230 FOR I=1 TO N
1240 A[R,I]=B[I]
1250 NEXT I
1260 PRINT
1270 NEXT R
1280 PRINT
1290 GOTO 370
1300 END
```

# SAMPLE RUN

```
THIS PROGRAM TESTS NXN SQUARES OF ODD ORDER FOR
THEIR MAGICAL PROPERTIES. YOU CAN EITHER ENTER YOUR
OWN SQUARE OR LET THE COMPUTER GENERATE ONE FOR YOU
USING THE ALGORITHM OF DE LA LOUBERE.
FOR THE COMPUTER SQUARE TYPE 1. FOR YOUR OWN TYPE 2.
?1
HOW MANY ROWS DO YOU WANT?
PICK AND ODD NUMBER BETWEEN 3 AND 11.
?3
PICK THE POSITION I,J FOR 1. (MIDDLE OF ROW 1 IS STANDARD)
?1,2
CHOOSE A,B,C,D (1,1,0,-1 IS STANDARD)
?1,1,0,-1
HERE IS YOUR SQUARE OF ORDER 3

     8    1    6
     3    5    7
     4    9    2


HERE IS A LIST OF ITS MAGICAL PROPERTIES.

ROWS AND COLUMNS: 6
GENERALIZED DIAGONALS: 2

TOTAL MAGICAL SUMS:  8

DO YOU WISH TO TRY AGAIN? TYPE 1 FOR YES, 0 FOR NO.
?1
```

**1**

```
FOR THE COMPUTER SQUARE TYPE 1. FOR YOUR OWN TYPE 2.
?1
HOW MANY ROWS DO YOU WANT?
PICK AND ODD NUMBER BETWEEN 3 AND 11.
?5
PICK THE POSITION I,J FOR 1. (MIDDLE OF ROW 1 IS STANDARD)
?3,4
CHOOSE A,B,C,D (1,1,0,-1 IS STANDARD)
?1,1,2,0
HERE IS YOUR SQUARE OF ORDER 5

     3   16    9   22   15
    20    8   21   14    2
     7   25   13    1   19
    24   12    5   18    6
    11    4   17   10   23


HERE IS A LIST OF ITS MAGICAL PROPERTIES.

ROWS AND COLUMNS: 10
GENERALIZED DIAGONALS: 2

TOTAL MAGICAL SUMS:  12

DO YOU WISH TO TRY AGAIN? TYPE 1 FOR YES, 0 FOR NO.
?0

GOODBYE. SEE YOU AT THE FRANKLIN FESTIVAL
OCT. 5 TO 11 AT UW-PARKSIDE.
```

**2**

# FRANKLIN PROGRAM

```
10   REM PROGRAM WRITTEN BY D.T.PIELE 7/15/75
20   PRINT "THIS IS THE FRANKLIN MAGIC SQUARE PROGRAM."
30   PRINT "IT WILL GENERATE AND TEST 8X8 SQUARES. YOU "
40   PRINT "CAN ALSO ENTER AND TEST YOUR OWN 8X8 SQUARES."
50   PRINT
60   DIM A[8,8],C[8],G[8],D[8,8],B[8],E[8,8],F[8]
70   MAT READ A,C,G
80   PRINT "FOR THE COMPUTER GENERATED SQAURE TYPE 1."
90   PRINT "TO ENTER YOUR OWN TYPE 2."
100  PRINT
110  INPUT A
120  IF A=2 THEN 1860
130  PRINT "FIRST PERMUTE THE ROWS 1 THROUGH 8 AND SEPARATE"
140  PRINT "WITH COMMAS. FOR THE FRANKLIN SQUARE CHOOSE"
150  PRINT " 7,8,1,2,3,4,5,6"
160  PRINT
170  MAT INPUT F
180  PRINT
190  PRINT "NEXT SUPPLY A PERMUTATION OF THE COLUMNS 1 THROUGH 8"
200  PRINT "IN THE FORM A,B,C,D,E,F,G,H SUCH THAT A+C=9, B+D=9,"
210  PRINT "E+G=9, AND F+H=9.FOR THE FRANKLIN MAGIC SQUARE CHOOSE"
220  PRINT " 4,6,5,3,7,1,2,8"
230  PRINT
240  MAT INPUT B
250  REM ****** ALGORITHM TO GENERATE SQUARES ******
260  J=1
270  FOR N=1 TO 7 STEP 2
280  FOR I=1 TO 7 STEP 2
290  K=I+1
300  R=B[I]
310  L=F[N]
320  S=B[K]
330  M=9-L
340  D[I,J]=A[L,R]
350  D[K,J]=A[M,S]
360  NEXT I
370  J=J+2
380  NEXT N
390  J=2
400  FOR N=2 TO 8 STEP 2
410  FOR I=1 TO 7 STEP 2
420  K=I+1
430  R=9-B[I]
440  S=9-B[K]
450  L=F[N]
460  M=9-L
470  D[I,J]=A[L,R]
480  D[K,J]=A[M,S]
490  NEXT I
500  J=J+2
510  NEXT N
520  PRINT
530  PRINT "YOUR 8X8 SQUARE IS"
540  PRINT
550  MAT PRINT D;
560  REM ****** END OF THE ALGORITHM ******
570  PRINT
580  FOR I=1 TO 8
590  IF F[I] <> G[I] THEN 670
600  NEXT I
610  FOR I=1 TO 8
620  IF B[I] <> C[I] THEN 670
630  NEXT I
640  PRINT " THIS IS THE BENJAMIN FRANKLIN MAGIC SQUARE OF ORDER 8."
650  PRINT
660  REM ****** TABULATION OF THE MAGICAL PROPERTIES ******
670  PRINT "HERE IS A LIST OF ITS MAGICAL PROPERTIES."
680  PRINT
690  MAT E=D
700  GOSUB 1420
710  R1=C
720  MAT E=TRN(D)
730  GOSUB 1420
740  C1=C
750  PRINT "ROWS AND COLUMNS:"R1+C1
760  M1=0
770  M=D[1,1]+D[2,2]+D[3,3]+D[4,4]+D[5,5]+D[6,6]+D[7,7]+D[8,8]
780  IF M <> 260 THEN 800
790  M1=1
800  M=D[8,1]+D[7,2]+D[6,3]+D[5,4]+D[4,5]+D[3,6]+D[2,7]+D[1,8]
810  IF M <> 260 THEN 830
820  M1=M1+1
830  PRINT "MAIN DIAGONALS:"M1
840  G=0
850  FOR J=1 TO 7
860  E=0
870  F=0
880  FOR I=1 TO 8
890  R=I+J
900  T=9-I-J
910  IF R <= 8 THEN 930
920  R=R-8
930  E=E+D[I,R]
940  IF T >= 1 THEN 960
950  T=T+8
960  F=F+D[T,I]
970  NEXT I
980  IF E <> 260 THEN 1000
990  G=G+1
1000 IF F <> 260 THEN 1020
1010 G=G+1
1020 NEXT J
1030 PRINT "GENERALIZED DIAGONALS:"G
1040 MAT E=D
1050 GOSUB 1520
1060 B1=C
1070 MAT E=TRN(D)
1080 GOSUB 1520
1090 B2=C
1100 MAT E=D
1110 GOSUB 1660
1120 B3=C
1130 MAT E=TRN(D)
1140 GOSUB 1660
1150 Y=B1+B2+B3+C
1160 PRINT "BROKEN DIAGONALS:"Y
1170 C=0
1180 D=0
1190 E=D[1,1]+D[1,8]+D[8,1]+D[8,8]+D[4,4]+D[5,4]+D[4,5]+D[5,5]
1200 C=1
1210 PRINT "SPECIAL CASES:"
1220 PRINT "          CENTER FOUR SQUARES PLUS FOUR CORNER SQUARES."
1230 E=D[1,2]+D[2,1]+D[1,7]+D[2,8]+D[7,1]+D[7,8]+D[8,2]+D[8,7]
1240 IF E <> 260 THEN 1270
1250 PRINT "          THE FOUR CORNER DIAGONAL PAIRS."
1260 D=1
1270 PRINT
1280 W=R1+C1+Y+C+D+M1+G
1290 PRINT "TOTAL MAGICAL SUMS:"W
1300 PRINT
1310 PRINT "DO YOU WANT TO TRY AGAIN? TYPE 1 FOR THE COMPUTER SQUARE,"
1320 PRINT "TYPE 2 TO ENTER YOUR OWN SQUARE, AND TYPE 0 TO STOP."
1330 PRINT
1340 INPUT Z
1350 IF Z=1 THEN 130
1360 IF Z=2 THEN 1860
1370 PRINT "GOODBYE. SEE YOU AT THE BENJAMIN FRANKLIN FESTIVAL"
1380 PRINT "AT UW-PARKSIDE, OCTOBER 5 TO 11."
1390 PRINT
1400 STOP
1410 REM ****** SUBROUTINES TO CHECK FOR MAGICAL SUMS ******
1420 C=0
1430 FOR I=1 TO 8
1440 E=0
1450 FOR J=1 TO 8
1460 E=E+E[I,J]
1470 NEXT J
1480 IF E <> 260 THEN 1500
1490 C=C+1
1500 NEXT I
1510 RETURN
1520 C=0
1530 FOR J=0 TO 7
1540 E=0
1550 FOR I=1 TO 4
1560 R=I+J
1570 T=9-I
1580 IF R <= 8 THEN 1600
1590 R=R-8
1600 E=E+E[I,R]+E[T,R]
1610 NEXT I
1620 IF E <> 260 THEN 1640
1630 C=C+1
1640 NEXT J
1650 RETURN
1660 C=0
1670 FOR J=0 TO 7
1680 E=0
1690 FOR I=1 TO 4
1700 R=9-I
1710 T=9-I-J
1720 IF T >= 1 THEN 1740
1730 T=T+8
1740 E=E+E[T,I]+E[T,R]
1750 NEXT I
1760 IF E <> 260 THEN 1780
1770 C=C+1
1780 NEXT J
1790 RETURN
1800 DATA 1,2,3,4,5,6,7,8,9,10,11,12,13,14,15,16
1810 DATA 17,18,19,20,21,22,23,24,25,26,27,28,29,30,31,32
1820 DATA 33,34,35,36,37,38,39,40,41,42,43,44,45,46,47,48
1830 DATA 49,50,51,52,53,54,55,56,57,58,59,60,61,62,63,64
1840 DATA 4,6,5,3,7,1,2,8,7,8,1,2,3,4,5,6
1850 REM ****** ENTER YOUR OWN SQUARE ******
1860 PRINT
1870 PRINT "LIST THE MEMBERS OF EACH ROW SEPARATED BY COMMAS."
1880 FOR R=1 TO 8
1890 PRINT "ROW"R
1900 PRINT
1910 MAT INPUT B
1920 FOR I=1 TO 8
1930 D[R,I]=B[I]
1940 NEXT I
1950 PRINT
1960 NEXT R
1970 GOTO 650
1980 END
```

# SAMPLE RUN

## 1

```
THIS IS THE FRANKLIN MAGIC SQUARE PROGRAM.
IT WILL GENERATE AND TEST 8X8 SQUARES. YOU
CAN ALSO ENTER AND TEST YOUR OWN 8X8 SQUARES.

FOR THE COMPUTER GENERATED SQAURE TYPE 1,
TO ENTER YOUR OWN TYPE 2.

?1
FIRST PERMUTE THE ROWS 1 THROUGH 8 AND SEPARATE
WITH COMMAS. FOR THE FRANKLIN SQUARE CHOOSE
  7,8,1,2,3,4,5,6

?7,8,1,2,3,4,5,6

NEXT SUPPLY A PERMUTATION OF THE COLUMNS 1 THROUGH 8
IN THE FORM A,B,C,D,E,F,G,H SUCH THAT A+C=9, B+D=9,
E+G=9, AND F+H=9. FOR THE FRANKLIN MAGIC SQUARE CHOOSE
  4,6,5,3,7,1,2,8

?4,6,5,3,7,1,2,8

YOUR 8X8 SQUARE IS

52    61    4     13    20    29    36    45
14    3     62    51    46    35    30    19
53    60    5     12    21    28    37    44
11    6     59    54    43    38    27    22
55    58    7     10    23    26    39    42
9     8     57    56    41    40    25    24
50    63    2     15    18    31    34    47
16    1     64    49    48    33    32    17

THIS IS THE BENJAMIN FRANKLIN MAGIC SQUARE OF ORDER 8.

HERE IS A LIST OF ITS MAGICAL PROPERTIES.

ROWS AND COLUMNS: 16
MAIN DIAGONALS: 0
GENERALIZED DIAGONALS: 0
BROKEN DIAGONALS: 32
SPECIAL CASES:
        CENTER FOUR SQUARES PLUS FOUR CORNER SQUARES.
        THE FOUR CORNER DIAGONAL PAIRS.

TOTAL MAGICAL SUMS: 50

DO YOU WANT TO TRY AGAIN? TYPE 1 FOR THE COMPUTER SQUARE,
TYPE 2 TO ENTER YOUR OWN SQUARE, AND TYPE 0 TO STOP.
```

## 2

```
?1
FIRST PERMUTE THE ROWS 1 THROUGH 8 AND SEPARATE
WITH COMMAS. FOR THE FRANKLIN SQUARE CHOOSE
  7,8,1,2,3,4,5,6

?1,2,8,7,3,4,6,5

NEXT SUPPLY A PERMUTATION OF THE COLUMNS 1 THROUGH 8
IN THE FORM A,B,C,D,E,F,G,H SUCH THAT A+C=9, B+D=9,
E+G=9, AND F+H=9. FOR THE FRANKLIN MAGIC SQUARE CHOOSE
  4,6,5,3,7,1,2,8

?1,2,8,7,5,6,4,3

YOUR 8X8 SQUARE IS

1     16    57    56    17    32    41    40
58    55    2     15    42    39    18    31
8     9     64    49    24    25    48    33
63    50    7     10    47    34    23    26
5     12    61    52    21    28    45    36
62    51    6     11    46    35    22    27
4     13    60    53    20    29    44    37
59    54    3     14    43    38    19    30

HERE IS A LIST OF ITS MAGICAL PROPERTIES.

ROWS AND COLUMNS: 16
MAIN DIAGONALS: 2
GENERALIZED DIAGONALS: 14
BROKEN DIAGONALS: 32
SPECIAL CASES:
        CENTER FOUR SQUARES PLUS FOUR CORNER SQUARES.
        THE FOUR CORNER DIAGONAL PAIRS.

TOTAL MAGICAL SUMS: 66

DO YOU WANT TO TRY AGAIN? TYPE 1 FOR THE COMPUTER SQUARE,
TYPE 2 TO ENTER YOUR OWN SQUARE, AND TYPE 0 TO STOP.
?0
GOODBYE. SEE YOU AT THE BENJAMIN FRANKLIN FESTIVAL
AT UW-PARKSIDE, OCTOBER 5 TO 11.
```

## 0

---

## Pocket Calculator *sʞɔᴉɹ⊥* !

Punch these problems into your pocket calculator, then turn it around (180°) to read the answer. For loads more of calculator problems, see the four calculator books in the *Creative Computing* Library advertisement.

### 1

*The Stock Market Is Dropping!*
$(508^2 - 16^2 + 5^2 + 2) \times 0.03$

*Familiar Principle:*
$(.844561)^{0.5}$

*That's A Big One!*
$50 \times 125^2 - 269^2 + 120$

### 6

*An Ancient Arab Proverb:*
$0.1283 \times 3 + 47 \times 15$

*Where?*
$71 \times 2 + 0.15469 \times 5$

*And Then What?*
$121 \times 57 + 0.25 \times 16 \div 2$

John Jackobs<br/>Heidelberg College

### POOR HOUSE

If you buy 100,000 shares of IBM stock (ENTER 100000) on margin at $148.18 per share (ENTER x 148.18), pay $472 commission (ENTER + 472), and the price goes down 25% (ENTER x 0.25), what do you find yourself in?

David Ahl

# Circular Functions

This is a reprint of one of the original Project Solo curriculum modules developed at the University of Pittsburgh. Project Solo was supported in part by the National Science Foundation, and it was directed by Tom Dwyer and Margot Critchfield. The modules were authored by various persons, including project staff, teachers, and students.

It should be kept in mind that Project Solo began in 1969 (which is probably before **some** of Creative's readers were born). Undoubtedly, many of the modules would be done differently today. There are also surely errors to be found, and neither Creative Computing, the authors, or NSF can warrant the accuracy of the reprints. But as a starting point for your own explorations, they should make a good (albeit slightly ancient) set of shoulders to stand upon. We hope you enjoy the view.

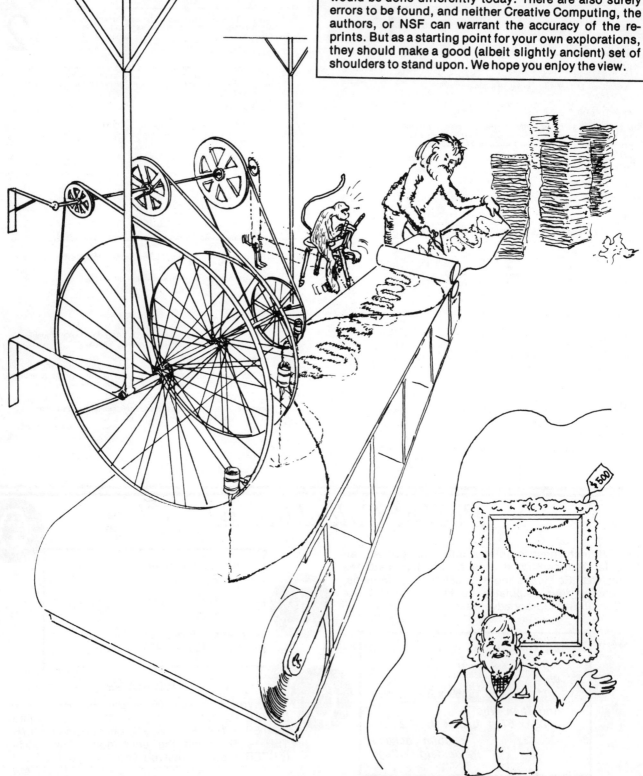

Consider an angle A with its initial side placed along the X-axis, with its vertex at the origin P:(0,0), and with a terminal side that passes through the point P:(X,Y). Let us call the distance from (0,0) to (X,Y) R, so that $R^2 = X^2 + Y^2$. Then the six circular functions of the angle A are defined as:

Sine of the angle = y/r   Contangent of the angle = x/y
Cosine of the angle = x/r   Secant of the angle = r/x
Tangent of the angle = y/x   Cosecant of the angle = r/y

**Sample problem**

Find the circular functions of an angle (A) whose terminal side passes through the point P: (5,12).

**Solution**

x = 5 and y = 12
therefore r = SQR (5*5 + 12*12) = SQR (169) = 13
thus   sin A = 12/13 = .9231

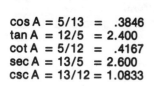
cos A = 5/13 = .3846
tan A = 12/5 = 2.400
cot A = 5/12 = .4167
sec A = 13/5 = 2.600
csc A = 13/12 = 1.0833

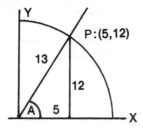

**Using the Computer**

Figure 1 shows the flow chart for a program that asks the user to supply values for X and Y. The program then calculates SIN A, COS A, and TAN A.

Notice that care is taken to avoid dividing by zero. (For which circular functions can this happen???)

Here is a program based on the flow chart in Figure 1:

```
LIST
50 PRINT "THIS PROGRAM CALCULATES THE SIN, COS, AND TAN FUNCTIONS"
60 PRINT "OF AN ANGLE A DEFINED BY THE RATIOS Y/R, X/R, Y/X"
70 PRINT "RESPECTIVELY."
80 PRINT
100 PRINT "TYPE IN THE COORDINATES OF YOUR POINT IN X,Y ORDER USING"
110 PRINT "DECIMAL FORM."
120 INPUT X,Y
200 R=SQR(X*X+Y*Y)
300 S=Y/R
400 C=X/R
450 IF X=0 THEN 900
500 T=Y/X
600 PRINT "SIN(A)=";S,"COS(A)=";C,"TAN(A)=";T
700 PRINT "DO YOU WISH TO ENTER ANOTHER POINT (ANSWER YES OR NO)";
710 INPUT R$
720 IF R$="YES" THEN 100
730 IF R$="NO" THEN 800
740 PRINT "ANSWER YES OR NO"
750 GOTO 700
800 STOP
900 PRINT "SIN(A)=";S,"COS(A)=";C,"TAN UNDEFINED"
910 GOTO 700

RUN
THIS PROGRAM CALCULATES THE SIN, COS, AND TAN FUNCTIONS
OF AN ANGLE A DEFINED BY THE RATIOS Y/R, X/R, Y/X
RESPECTIVELY.

TYPE IN THE COORDINATES OF YOUR POINT IN X,Y ORDER USING
DECIMAL FORM.
? 5,12
SIN(A)= .923077         COS(A)= .384615         TAN(A)= 2.4
DO YOU WISH TO ENTER ANOTHER POINT (ANSWER YES OR NO)? YES
TYPE IN THE COORDINATES OF YOUR POINT IN X,Y ORDER USING
DECIMAL FORM.
? 2,2
SIN(A)= .707107         COS(A)= .707107         TAN(A)= 1
DO YOU WISH TO ENTER ANOTHER POINT (ANSWER YES OR NO)? NO
```

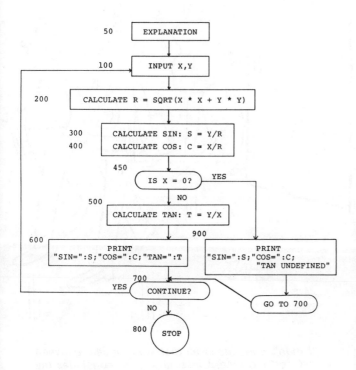

Figure 1. Flowchart for a Program to Calculate Sin, Cos, and Tan.

NOTE: You must type in two decimal numbers. If you don't know a number (say the square root of 3), you can use an expression.

EXAMPLE: Suppose you want to find the circular functions for P : ($\sqrt{3}$, $\sqrt{2}$).

```
TYPE IN THE COORDINATES OF YOUR POINT IN X,Y ORDER USING
DECIMAL FORM.
? SQR(3),SQR(2)  ← WRONG!!


RUN
THIS PROGRAM CALCULATES THE SIN, COS, AND TAN FUNCTIONS
OF AN ANGLE A DEFINED BY THE RATIOS Y/R, X/R, Y/X
RESPECTIVELY.

TYPE IN THE COORDINATES OF YOUR POINT IN X,Y ORDER USING
DECIMAL FORM.
? ^C
Break in 120
Ok
PRINT SQR(3)
 1.73205
Ok
PRINT SQR(2)          CORRECT
 1.41421
Ok
CONT
? 1.73205,1.41421
SIN(A)= .632455         COS(A)= .774597         TAN(A)= .816495
DO YOU WISH TO ENTER ANOTHER POINT (ANSWER YES OR NO)? NO
```

## Computer Problems

1. Use the program above for P:(1,1), P:(0,1) P:(1,0), P:(-3,3), P:(-2,3), P:(-3,-3), P:(3,-3)

2. Modify the program given on Page 4 so that all six circular functions are calculated.
   a. Use your computer program to find the circular functions of the following:
      i. An angle A whose terminal side passes through P:(2*SQR (2), SQR (7)) NOTE: You must use an expression to find 2*SQR (2) etc.
      ii. An angle A whose terminal side passes through P:(-2.386, 7.590)
      iii. An angle A whose terminal side passes through P:(0,2.7)
      iv. An angle A whose terminal side passes through P:(2.7,0)
   b. Run your program again, supplying data on several points in all four quadrants. Can you make a general rule about the signs of the circular functions in each quadrant?
   c. Investigate angles whose terminal sides are coincident with the coordinate axes; that is, circular functions of 0, 90, 180, 360°. Which functions are undefined?

## Advanced Problems

3. Write a program that automatically generates the six circular functions

   FOR X = 1.0, .9, .8, .7, .6, .5, .4, .3, .2, .1, 0.
   -.1, -.2, -.3, -.4, -.5, -.6, -.7, -.8, -.9, -1.0

   WITH R FIXED AT R = 1 (UNIT CIRCLE).

   Here is **part** of a program to do this with some output:

```
100 A$="+#.## +#.## +#.### +#.### +##.### +##.### +##.### +##.###"
110 B$="+#.## +#.## +#.### +#.### UNDEF +##.### UNDEF +##.###"
120 C$="+#.## +#.## +#.### +#.### +##.### UNDEF +##.### UNDEF "
130 R=1
140 PRINT " X    Y    SIN   COS   TAN    COT    SEC    CSC"
150 FOR X=1 TO -1.1 STEP -.1
160 IF ABS(X)>=1 THEN 280
170 Y = ABS(SQR(R*R-X*X))
180 IF ABS(X)<5E-03 THEN 250
190 IF ABS(Y)<5E-03 THEN 280
200 PRINT USING A$;X,Y,Y/R,X/R,Y/X,X/Y,R/X,R/Y
210 NEXT X
230 END
```

```
RUN
X      Y      SIN     COS     TAN      COT     SEC      CSC
+1.00 +0.00 +0.000 +1.000 +0.000   UNDEF  +1.000   UNDEF
+0.90 +0.44 +0.436 +0.900 +0.484  +2.065 +1.111  +2.294
+0.80 +0.60 +0.600 +0.800 +0.750  +1.333 +1.250  +1.667
+0.70 +0.71 +0.714 +0.700 +1.020  +0.980 +1.429  +1.400
+0.60 +0.80 +0.800 +0.600 +1.333  +0.750 +1.667  +1.250
+0.50 +0.87 +0.866 +0.500 +1.732  +0.577 +2.000  +1.155
+0.40 +0.92 +0.917 +0.400 +2.291  +0.436 +2.500  +1.091
+0.30 +0.95 +0.954 +0.300 +3.180  +0.314 +3.333  +1.048
+0.20 +0.98 +0.980 +0.200 +4.899  +0.204 +5.000  +1.021
+0.10 +0.99 +0.995 +0.100 +9.950  +0.101 +10.000 +1.005
-0.00 +1.00 +1.000 -0.000  UNDEF  -0.000  UNDEF  +1.000
-0.10 +0.99 +0.995 -0.100 -9.950  -0.101 -10.000 +1.005
-0.20 +0.98 +0.980 -0.200 -4.899  -0.204 -5.000  +1.021
-0.30 +0.95 +0.954 -0.300 -3.180  -0.314 -3.333  +1.048
-0.40 +0.92 +0.917 -0.400 -2.291  -0.436 -2.500  +1.091
-0.50 +0.87 +0.866 -0.500 -1.732  -0.577 -2.000  +1.155
-0.60 +0.80 +0.800 -0.600 -1.333  -0.750 -1.667  +1.250
-0.70 +0.71 +0.714 -0.700 -1.020  -0.980 -1.429  +1.400
-0.80 +0.60 +0.600 -0.800 -0.750  -1.333 -1.250  +1.667
-0.90 +0.44 +0.436 -0.900 -0.484  -2.065 -1.111  +2.294
-1.00 +0.44 +0.436 -1.000 -0.436   UNDEF  -1.000   UNDEF
```

## Man Against Machine

4. Try to make the computer blow its mind by calculating TAN(A), with R = 1, and X approaching zero.

```
10 LET X=1
20 IF X<1E-13 THEN 60
30 PRINT "X=";X;TAB(35);"TAN(X)=";SQR(1-X*X)/X
40 LET X=X/2
50 GOTO 20
60 END
```

5. The picture on the opening page shows that the paint dripping from the paint dispensers on the three wheels (rotated by monkey power) traces out "graphs" of the circular functions on the canvas (moved by mad artist power).

   Write a program that:

   (a) Graphs the SIN or COS function.
   OR (b) Graphs the SIN and COS function on the same axis, using two different values of R.
   OR (c) Graphs SIN and COS functions with different values of R and different starting points, possibly making "Computer generated" art as your goal.

"I didn't understand all that stuff he said between 'Good Morning, Class' and 'That concludes my lecture for today'."

# Trigonometric Functions and Tchebychev Approximations

**(Or—Can a humble computer that only knows how to do arithmetic calculate transcendental functions?)**

> This is a reprint of one of the original Project Solo curriculum modules developed at the University of Pittsburgh. Project Solo was supported in part by the National Science Foundation, and it was directed by Tom Dwyer and Margot Critchfield. The modules were authored by various persons, including project staff, teachers, and students.
>
> It should be kept in mind that Project Solo began in 1969 (which is probably before some of Creative's readers were born). Undoubtedly, many of the modules would be done differently today. There are also surely errors to be found, and neither Creative Computing, the authors, or NSF can warrant the accuracy of the reprints. But as a starting point for your own explorations, they should make a good (albeit slightly ancient) set of shoulders to stand upon. We hope you enjoy the view.

Problem 1. Below you see three representations of the "sine" function. Which is the right one?

☐ 1. (a)
☐ 2. (b)
☐ 3. (c)
☐ 4. All of the above.
☐ 5. None of the above.

(a) The SINE function:

| Y | Y/R |
|---|---|
| 0.0 | 0.000 |
| 0.1 | 0.100 |
| 0.2 | 0.200 |
| 0.3 | 0.300 |
| 0.4 | 0.400 |
| etc. | |

(b) The SINE function:

| X | $\pm\sqrt{(R^2-X^2)}/R$ |
|---|---|
| 1.0 | 0.000 |
| 0.9 | 0.436 |
| 0.8 | 0.600 |
| 0.7 | 0.714 |
| 0.6 | 0.800 |
| etc. | |

(c) The SINE function:

| A | SIN(A) |
|---|---|
| 0.0 | 0.000 |
| 0.1 | 0.099 |
| 0.2 | 0.198 |
| 0.3 | 0.295 |
| 0.4 | 0.389 |
| etc. | |

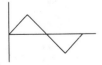

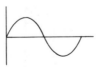

Note: We are using the letters X, Y, A, and R to represent the same quantities as in the module "Circular Functions."

Answer to Problem 1: All of the representations are correct.

Here are three programs which you can try to convince yourself that you can get three different-looking "graphs" of the sine function. The shape of these graphs depends on which variable (X, Y, or A) you decide to increment in equal steps.

After looking over these programs, you may get the feeling that something has been 'put over' on you. Question: Has a little mathematical "hanky-panky" crept into these demonstration programs??? Which one??? (For the answer, read on!!!)

Demonstration Program (a)

```
200 FOR Y=0 TO .95 STEP .1
210 PRINT TAB(30); "+"; TAB(30+30*Y);"2"
220 NEXT Y
230 FOR Y=1 TO .05 STEP -.1
240 PRINT TAB(30); "+"; TAB(30+30*Y);"2"
250 NEXT Y
260 FOR Y=0 TO -.95 STEP -.1
270 PRINT TAB(30+30*Y); "2"; TAB(30); "+"
280 NEXT Y
290 FOR Y=-1 TO .05 STEP .1
295 PRINT TAB(30+30*Y); "2"; TAB(30);"+"
296 NEXT Y
299 END
```

# Trigonometric, con't...

Demonstration Program (b)

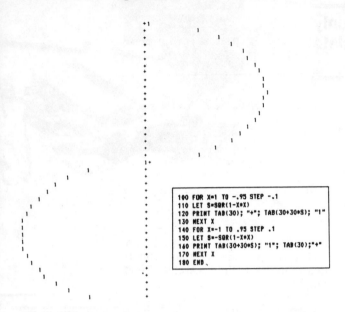

Demonstration Program (c)

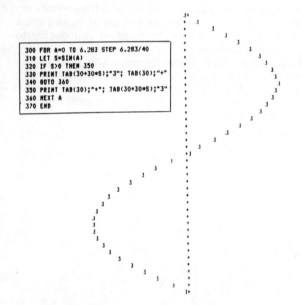

If you haven't spotted the hanky-panky by now, check line 310 above. Assuming computers can only do arithmetic (which is true), an expression like 30 + 30 * S in line 330 makes sense, but it isn't clear that arithmetic is being used in line 310. Where is the arithmetic being done? Answer: SECRETLY! The computer actually goes into a "library" sub-routine at line 310, which tells it what arithmetic operation must be done on the angle A (expressed in radians) in order to calculate SIN(A).

## The Secret Unveiled

The library functions in digital computers used to evaluate complicated functions almost always employ expertly designed polynomials as approximations to these functions. These approximations can be made quite accurate, using techniques originated by the Russian mathematician Tchebychev (also spelled Chebyshev), and put into useful form by Hastings.[1]

Here is one of the approximations given by Hastings:

(Hastings Sheet 14)

Function: $\sin\left(\frac{\pi}{2} X\right)$

Range: $-1 \leq X \leq 1$

Approximation: $\sin\left(\frac{\pi}{2} X\right) \approx C_1 X + C_3 X^3 + C_5 X^5$

where $C_1 = 1.5706268, C_3 = -.6432292,$ and $C_5 = .0727102$.

Before using this approximation, it will be useful to make two changes. First, we will change the range of the independent variable by the substitution:

$Y = (\pi/2) * X$ so that when: $-1 \leq X \leq 1$
we will have: $-\pi/2 \leq Y \leq \pi/2$

Note: Both Y and X are **angles**, measured in radians. Don't confuse these variables with the Y and X coordinates used in Problem 1. The variable Y in the rest of this discussion is identical to the variable A shown in Problem 1. cover.

After making this substitution, the Hastings Sheet 14 approximation becomes:

$$\sin(Y) \approx K_1 Y + K_3 Y^3 + K_5 Y^5$$

where $K_1 = (2/\pi) * C_1; \quad K_3 = (2^3/\pi^3) * C_3; \quad K_5 = (2^5/\pi^5) * C_5$.

The second change we will make is to rearrange the right side of the preceding formula into "nested multiplication" form:

$$\sin(Y) \approx (((K_5 * Y * Y + K_3) * Y) * Y + K_1) * Y$$

The advantage of this form is that it only takes five multiplications, whereas the original polynomial takes nine.

Here is a program which uses this form to calculate sin(Y), and compares it with the library routine (which uses a higher degree approximation—see problem 3).

```
run
Program to compare library SIN function with
Hastings 14.

Radians         Library           Hastings 14
-------------------------------------------------
-1              -.841471          -.841534
-.9             -.783327          -.783407
-.8             -.717356          -.717433
-.7             -.644218          -.644278
-.6             -.564643          -.564679
-.5             -.479425          -.479439
-.4             -.389418          -.389413
-.3             -.29552           -.295505
-.2             -.198669          -.198653
-.1             -.0998335         -.0998233
 0               0                 0
 .1              .0998334          .0998233
 .2              .198669           .198653
 .3              .29552            .295505
 .4              .389418           .389413
 .5              .479426           .479439
 .6              .564643           .564679
 .7              .644218           .644278
 .8              .717356           .717433
 .9              .783327           .783407
 1               .841471           .841534
Ok
```

## Trigonometric, con't...

```
list
290 PRINT "Program to compare library SIN function with"
295 PRINT "Hastings 14.": PRINT: PRINT
300 LET P=3.14159265#
310 LET P2=P*P
315 LET P3=P*P2
320 LET C1=(1.5706268#*2)/P
330 LET C3=(-.643229*8)/P3
340 LET C5=(.0727102*32)/(P2*P3)
350 PRINT "Radians";TAB(20);"Library";TAB(40);"Hastings 14"
355 PRINT "------------------------------------------------"
360 FOR Y=-1 TO 1 STEP .1
365 Y=INT(10*Y+.5)/10              'correct roundoff errors
370 S1=(((C5*Y*Y+C3)*Y*Y+C1)*Y
380 PRINT " ";Y;TAB(20);SIN(Y);TAB(40);S1
390 NEXT Y
400 END
Ok
```

Problem 2. Change the FOR loop in the above program to cover the range for Y = 1 to 3 radians, and see what happens. Also try the range for Y = -1 to -3. If something "blows up," can you modify your program so that **any** value of Y can be used in the Hastings' approximation? (Hint: recall sin(Y) = sin( -Y) = sin(2 + Y) = sin(3 -Y) = etc.)

Problem 3. Here is the Hastings Sheet 15 approximation which ought to come a lot closer to the values produced by the library function. Modify your program to try it out, again comparing it with the library function.

$$\sin((\pi/2) * X) \simeq C_1 * X + C_3 * X^3 + C_5 * X^5 + C_7 * X^7$$

where
$C_1 = 1.570794852$   $C_3 = -0.645920978$
$C_5 = 0.079487663$   $C_7 = -0.004362476$

and $-1 \leq X \leq 1$

Problem 4. (Optional) If you find the idea of super-precision interesting, ask your teacher for information on your computer's double precision arithmetic and library functions. One of the best handbooks to use as a standard for "correct" values is "The Handbook of Mathematical Functions, Applied Math Series No. 55" available from the Superintendent of Documents, U.S. Government Printing Office, Washington, D.C. 20402 ($6.50). A paperback version published by Dover Co. may be available at book stores for about $5.00.

Problem 5. (Easy) Use the library functions TAN and SQRT to find the lengths of the guy-wires X and Y shown below. (Use of direct mode as a desk calculator is a good way to solve this problem.)

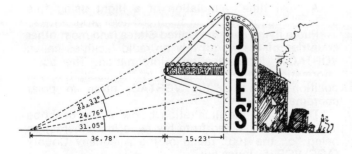

[1] A book your library ought to have is Hastings, Cecil, **Approximations for Digital Computers**, Princeton University Press, 1955.

# Puzzles and Problems For Fun

▶ The number $153 = 1^3 + 5^3 + 3^3$ Find all other 3-digit numbers that have the same property. How about 4-digit numbers? To the 4th?

Bill Morrison<br>Sudbury, Mass.

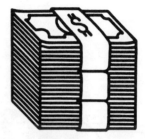

▶ Mr. Karbunkle went to the bank to cash his weekly paycheck. In handing over the money, the cashier, by mistake, gave him dollars for cents and cents for dollars.

He pocketed the money without examining it and spent a nickel on candy for his little boy. He then discovered the error and found he possessed exactly twice the amount of the check.

If he had no money in his pocket before cashing the check, what was the exact amount of the check? One clue: Mr. Karbunkle earns less than $50 a week.

▶ Can you find the missing number for each diagram? You first have to figure the pattern which may be horizontal or vertical with a relationship between every number, every second or third number. You may have to add, subtract, multiply, divide, invert or do a combination of these things. Have fun!

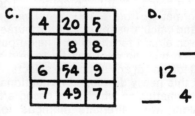

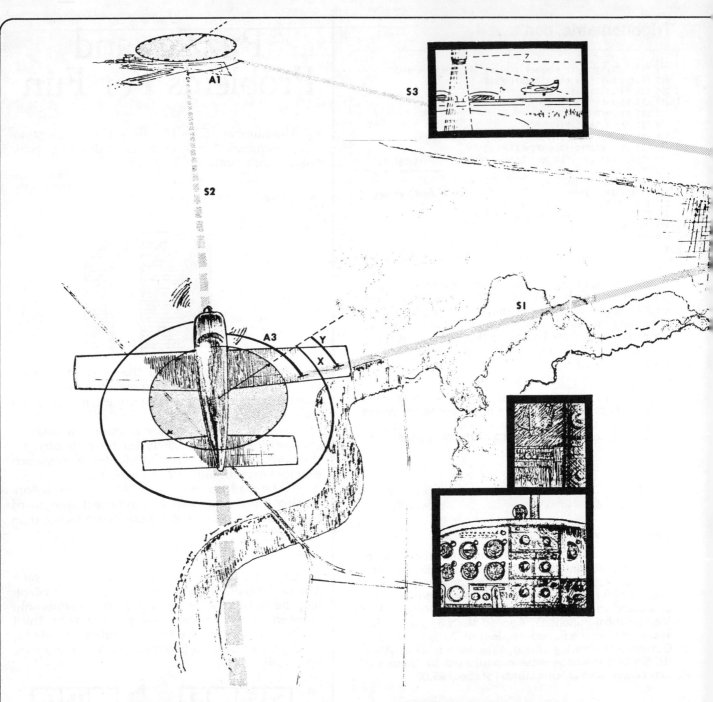

# Phantom VORTAC

This module will guide you in preparing the master program for an on-board flight computer. The computer takes information from a VHF omni range system (VORTAC), and calculates the magnetic course and distance to a given airport for the pilot. The output of the program is such that the pilot can fly toward a "phantom" VORTAC located at any airport he selects.

A description of this newly developed navigational system and the mathematics on which it is based are contained below along with alternate methods for handling the computation, and suggestions for ways in which previous programs you have written might be incorporated as sub-routines.

A "real time" simulation of a flight using this system is suggested as an advanced level program.

Pilots flying over the United States (and most other countries of the world) rely on radio facilities called VORTACs[1] for navigational information. The basic information the pilot receives in the cockpit is his position relative to the VORTAC, given in polar coordinates.

The pilot in the illustration below would describe his position (obtained from his radio instruments) as being "on the 100° radial of the Allegheny County (AGC) VOR, 50 miles out."

It is easy for this pilot to note that he can get to

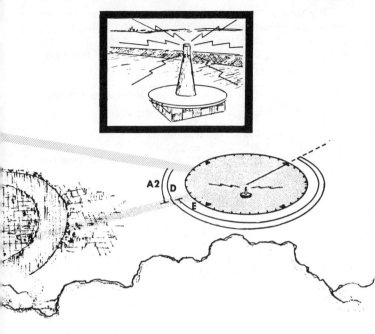

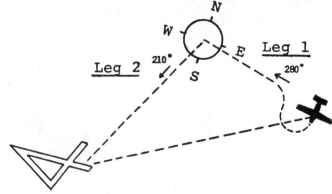

Before writing a program for such an on-board computer, it would be useful for a programmer to solve by hand typical problems arising in this situation.[3] At this point, let's look at three such solutions.

**Problem 1.** The aircraft is 150 miles out on the 135° radial and the destination airport is 100 miles out on the 210° radial. That is, in the diagram, S1 = 150, E = 135°, S3 = 100, D = 210°.

Let A2 be the angle determined by S1 and S3. Hence, A2 = 210° - 135° = 75°. By the Law of Cosines,
S2 = $\sqrt{100^2 + 150^2 - 2 \cdot 100 \cdot 150 \cdot \cos(75°)}$ = 157.3 miles. Now by the Law of Sines, sin A3 = (100 · .966)/157.3 = .614 and cos A3 = $\sqrt{1 - .614^2}$ = .789. From this we can get A3 = arctan (.614/.789) = 38°. (Why wasn't A3 computed directly from sin A3?) Therefore x = 135° + 180° - 38° = 277°. Output to the pilot is 157.3 miles, 277°--is this answer reasonable?

AGC by turning right, and flying a course[2] of 280° (Why 280°?). If he is going 200 miles per hour, it is also easy for him to estimate that he will arrive at AGC in 15 minutes (Is this exactly true?--go back over the article on vector addition if you are not sure.)

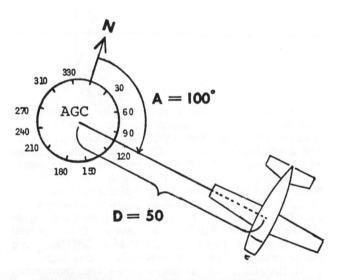

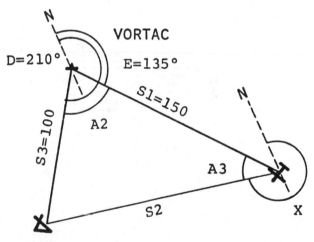

The catch to all of this is that the location of the VOR is usually different from the location of the destination airport. This difficulty can be handled by flying two legs, the first from the present position to the VOR, and the second from the VOR to the airport. This is obviously an inefficient route.

A new navigational system uses an on-board, special-purpose computer to tell the pilot what course and distance to fly in order to go directly from his present position to the airport. Let's first examine three ways of analyzing the mathematics involved in such a computation.

**Problem 2.** The airplane is 250 miles out on the 45° radial and the destination airport is 100 miles out on the 280° radial. In terms of the diagram, S1 = 250, E = 45°, S3 = 100, D = 280°.

Converting the positions of the aircraft and destination to rectangular coordinates with the VORTAC as origin we have

$x_p$ = 250 sin 45° = 176.8 miles,
$y_p$ = 250 cos 45° = 176.8 miles,
$x_d$ = 100 sin 280° = -98.48 miles,
$y_d$ = 100 cos 280° = 17.36 miles.

A = arctan $\frac{176.8 - 17.36}{176.8 - (-98.48)}$ = arctan (.5792) = 30.1°

# VORTAC, con't...

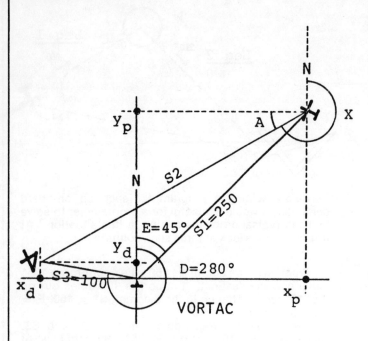

Therefore x = 270° - 30.1° = 239.9°. By the Pythagorean Theorem

$$S2 = \sqrt{(176.8 - 17.36) + (176.8 - (-98.48))^2} = 318.1.$$

Hence the output to the pilot is : 318.1 miles, 239.9°.

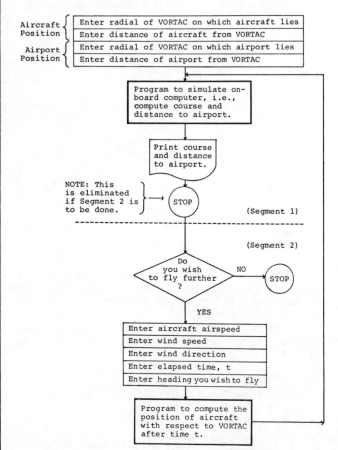

**Problem 3.** The aircraft is 150 miles out on the 120° radial and the destination airport is 400 miles out on the 150° radial.[4] In the diagram, E = 120°, S1 = 150 miles, D = 150°, S3 = 400 miles. Using the same reasoning as in Problem 1 yields:

A2 = 30°
$S2 = \sqrt{400^2 + 150^2 - 2 \cdot 150 \cdot 400 \cdot \cos 30°}$ = 290.3 miles
sin A3 = 400 (.5236/280.3) = .7135
$\cos A3 = \sqrt{1 - .7135^2}$ = .7007
A3 = arctan(.7135/.7007) = 45.52°
X = 120° + 180° - 45.52° = 254.48°

Output to the pilot is 280.3 miles, 254.48°--does this answer look reasonable? What went wrong?

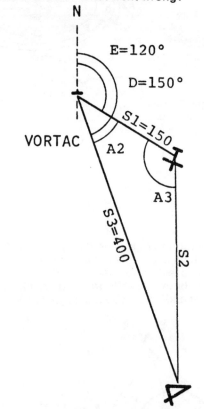

**Assignment.** Write a program for an on-board navigational computer which will accept as input the radial and distance from a VORTAC of both an aircraft and a destination, and which will compute a course and distance to the destination. The trigonometric subroutines in the computer require arguments in radians, but pilots think in terms of degrees, so it will be necessary for you to convert degrees to radians and back (see the module on converting to radians). Other possibly useful topics are inverse trigonometric functions, the Laws of sines and Cosines, and transformation of polar to rectangular coordinates[5].

**Optional Section.** During most flights, the aircraft moves through the air and the air moves as well. Write an addition to your program which will:

1. Accept as additional input, specified by the 'pilot,' (a) the aircraft's speed, (b) the speed of the wind, (c) the direction of the wind, (d) an elapsed time, t, and (e) a heading[6] for the aircraft. Any heading should be acceptable: the 'pilot' should be able to fly wherever he likes.

2. Compute the new position of the aircraft on the basis of the above information.

3. Repeat steps 1 and 2 as often as desired or necessary to reach the destination airport.

1. Very high frequency **O**mni Range and **TACAN**, where TACAN is an older military system of distance measuring equipment. Most civilian pilots call these facilities VOR-DME stations.

2. The angular direction of the intended flight path, measured clockwise from N.

3. This is another way of saying that computers do not remove the responsibility of analyzing a problem before solving it--quite the contrary; they demand more thought than ever. This may be, in fact, one of the most important contributions computing systems can make to learning.

4. By now you should have noticed that the word radial is used to designate the angular position of a line segment that starts at the VORTAC.

5. See previous articles.

6. Heading is defined as the angular direction of the longitudinal axis of the aircraft with respect to North. In the picture below, the pilot is flying a course of 90°, but his heading is 65°. Why?

> This is a reprint of one of the original Project Solo curriculum modules developed at the University of Pittsburgh. Project Solo was supported in part by the National Science Foundation, and it was directed by Tom Dwyer and Margot Critchfield. The modules were authored by various persons, including project staff, teachers, and students.
> It should be kept in mind that Project Solo began in 1969 (which is probably before some of Creative's readers were born.) Undoubtedly, many of the modules would be done differently today. There are also surely errors to be found, and neither Creative Computing, the authors, or NSF can warrant the accuracy of the reprints. But as a starting point for your own explorations, they should make a good (albeit slightly ancient) set of shoulders to stand upon. We hope you enjoy the view.

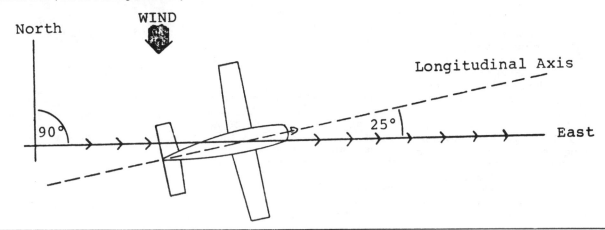

An APL and Basic approach to one of the oldest, and most interesting, programming problems.

# Pascal's Triangle: What's It All About?

### Jordan Mechner

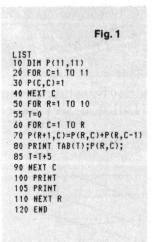

This triangle has quite a few interesting properties.

```
      1
     1 1
    1 2 1
   1 3 3 1
  1 4 6 4 1
```

Each row is symmetrical. Each row also happens to contain the coefficients for a binomial expansion. The descending diagonals are the same as the columns. The sums across the ascending diagonals form the Fibonacci sequence. The sums across the rows are all powers of two. Each row corresponds to the digits of a power of 11. Every element is the sum of the one above it and the one to the left of the one above it. All the elements are identities in combinatorial theory.

Fig. 1

```
LIST
10 DIM P(11,11)
20 FOR C=1 TO 11
30 P(C,C)=1
40 NEXT C
50 FOR R=1 TO 10
55 T=0
60 FOR C=1 TO R
70 P(R+1,C)=P(R,C)+P(R,C-1)
80 PRINT TAB(T);P(R,C);
85 T=T+5
90 NEXT C
100 PRINT
105 PRINT
110 NEXT R
120 END
```

Fig. 2

```
RUN
 1
 1  1
 1  2   1
 1  3   3    1
 1  4   6    4    1
 1  5  10   10    5    1
 1  6  15   20   15    6    1
 1  7  21   35   35   21    7    1
 1  8  28   56   70   56   28    8    1
 1  9  36   84  126  126   84   36    9    1
 1 10  45  120  210  252  210  120   45   10    1
 1 11  55  165  330  462  462  330  165   55   11   1
```

This is, of course, Pascal's Triangle, a favorite programming problem. The ways it can be generated are as varied and interesting as its properties, though often more difficult to figure out. The powers-of-eleven idea, for instance, which seemed pretty simple, conks out when we get to higher powers ($11^6$ = 161051, which doesn't look like it belongs in the triangle) because the digits carry over and make a mess.

How else can the triangle be generated? Let's look at it closely:

```
1
1  1
1  2  1
1  3  3  1
1  4  6  4  1
```

Any element of any column can be found by adding the previous element of that column to the previous element of the previous column. Let's see if we can pack this into a BASIC program. Fig. 1 is a listing of the program and Fig. 2 is a sample run.

Lines 20 through 40 set the diagonal at 1. Lines 50 through 110 do the actual calculating and printing out of the triangle. The variable T simply contains the number of tabs the computer should space over to make the output look nice. It's a simple enough program.

Fig. 3 is another BASIC program which uses a different approach and, incidentally, does not use arrays. (Fig. 4 is a sample run.) It generates the triangle one element at a time. Can you figure out what makes this tick? Line 60 is the crucial one.

If you're familiar with APL you may have seen that this could be a beautiful demonstration of its power and conciseness. Fig. 5 is an APL version of the program. Line 1 sets X, which contains only one row at a time, at 1. Line 2 prints out X. Line 3 catenates a zero onto the end, then onto the beginning, and adds the two together. Line 4 loops back to line 2. Fig. 6 is a run of the program.

It looks like line 3 is the interesting part of the function.
Let's examine its action more closely:

```
   1  0
 + 0  1
 ───────                  X = 1
   1  1
                          X = 1 1
   1  1  0
 + 0  1  1
 ──────────
   1  2  1
                          X = 1 2 1
   1  2  1  0
 + 0  1  2  1
 ─────────────
   1  3  3  1
                          X = 1 3 3 1
```

This is a much simpler and more elegant way of looking at the adding up of elements, but it would be nearly impossible to work out in BASIC.

There are other ways to generate Pascal's Triangle (there will always be other ways). In fact, here are a few challenges:

1 - Write a BASIC program to print out a specified row of Pascal's Triangle without wasting memory by storing all the others.
2 - Write an AF function to do the same thing. (Can you do it with eight characters?)
3 - Write programs in BASIC and APL to prove that when you sum across the rows, you get powers of two and that when you sum across the ascending diagonals, you get the Fibonacci sequence.
4 - Use the powers-of-eleven idea to generate a triangle. Find some way of catching the digits when they carry over. ■

**Fig. 3**
```
LIST
10 FOR N=0 TO 11
20 LET T=0
30 FOR R=0 TO N
40 LET C=1
45 IF N<N-R+1 THEN 80
50 FOR X=N TO N-R+1 STEP -1
60 LET C=C*X/(N-X+1)
70 NEXT X
80 PRINT TAB(T); C;
90 LET T=T+5
100 NEXT R
110 PRINT
120 PRINT
130 NEXT N
140 END
```

**Fig. 4**
```
RUN
1
1  1
1  2  1
1  3  3  1
1  4  6  4  1
1  5  10 10  5  1
1  6  15 20 15  6  1
1  7  21 35 35 21  7  1
1  8  28 56 70 56 28  8  1
1  9  36 84 126 126 84 36 9 1
```

```
∇PASCAL1[□]∇

    ∇ PASCAL1
[1]   X←1
[2]   X
[3]   X←(X,0)+0,X
[4]   →2
    ∇

    PASCAL1
1
1 1
1 2 1
1 3 3 1
1 4 6 4 1
```

# Art, Graphics and Mathematics

# ART & MATHEMATICAL STRUCTURES

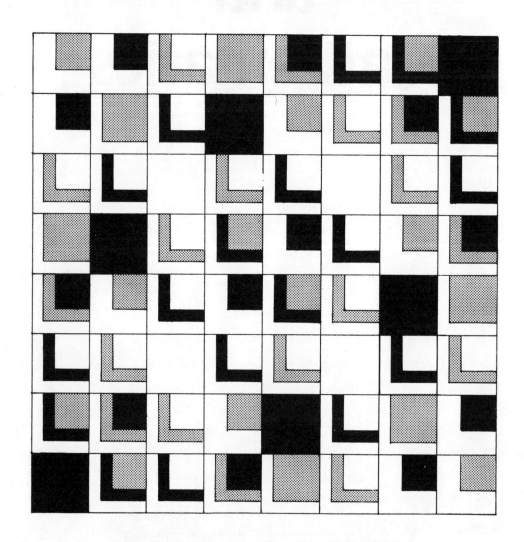

## FLIPS AND SPINS
### by R. Chandhok and M. Critchfield

Reprinted from Soloworks Module #2160 — PROJECT SOLO, Dept. of Computer Science, University of Pittsburgh, Pgh., PA 15260

## Polygons: The Algebra of Symmetry

There is an important mathematical system that is not based on numbers, but on changing the position of a given polygon.

Imagine a rectangle that is lettered on both sides with the same letter in the same corner. In how many ways can you pick up the rectangle lettered ABCD and then "spin" it or "flip" it so that when you put it down the shape looks the same but the letters are in different positions.

Exercise: Make yourself a paper rectangle like the one above, use it to try out the manipulations discussed below.

One way is to flip it on its vertical axis of symmetry:

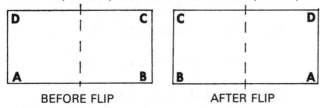

BEFORE FLIP          AFTER FLIP

This is the "V" (flip on vertical axis) configuration. The other possible motions (and their resulting configurations) are:

"H" (flip on horizontal axis)

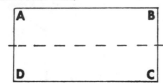

"R" (rotate rectangle 180°)

"I" (itself, a rotation of 0° or no flip—the identity)

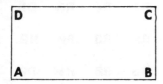

These are the elements of the system:

I, R, V, H

The binary operation of following one motion by another will allow us to create a table like this:

| "followed by" | I | R | V | H |
|---|---|---|---|---|
| I | I |   | V | H |
| R | R |   |   | V |
| V | V | H | F |   |
| H | H | V |   |   |

Exercise: Finish the table by manipulating your rectangle. What are the properties of this table?

## Regular Polygons

From here on we will be considering only the regular polygons.

Exercise: Make a paper equilateral triangle. Instead of lettering the corners, number them. Can you do the same motions with it as with the rectangle?

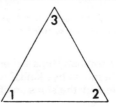

For the sake of simplicity, we will start using the following notations (which can be used for all regular polygons).

$R_0$ will be a rotation of 0 degrees (replaces I)
$R_k$ will be a rotation of $k * (360/n)$ degrees
where n is the number of sides of the polygon

So, for n = 3, a triangle, there are three possible rotations or spins.

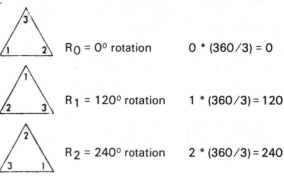

$R_0 = 0°$ rotation     $0 * (360/3) = 0$

$R_1 = 120°$ rotation   $1 * (360/3) = 120$

$R_2 = 240°$ rotation   $2 * (360/3) = 240$

There are also three flips, and a new notation for them.

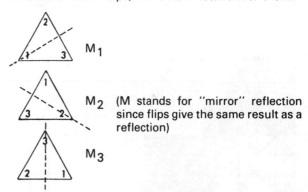

$M_1$

$M_2$ (M stands for "mirror" reflection since flips give the same result as a reflection)

$M_3$

A table for the 3-sided regular polygon (alias equilateral triangle) would look like this:

| "followed by" | $R_0$ | $R_1$ | $R_2$ | $M_1$ | $M_2$ | $M_3$ |
|---|---|---|---|---|---|---|
| $R_0$ |   |   |   |   |   |   |
| $R_1$ |   |   |   |   |   |   |
| $R_2$ |   |   |   |   |   |   |
| $M_1$ |   |   |   |   |   |   |
| $M_2$ |   |   |   |   |   |   |
| $M_3$ |   |   |   |   |   |   |

Exercise: You guessed it! Use your paper triangle to fill in the above table. Look for its properties.

You can see that it would get quite tedious to try to uncover the properties of all the regular polygons by hand this way.

However, try one more, since there is a slight variation when the number of sides is *even*.

### Symmetries of the Square

Using the formula previously given, with n = 4 you can find that the number of rotations of the square are 4—$0°$, $90°$, $180°$, and $270°$.

Exercise: Make a paper square. How many flips (reflections) does it have? Can you classify them into two kinds? Make a table like the one for the triangle with the spins and flips in the following order:

$$R_0, R_1, R_2, R_3, D_1, D_2, M_1, M_2$$

where D stands for "diagonal flip" and M, for a flip that bisects opposite sides.

Note: It is possible to write a computer program to produce this, and all other tables from the symmetries of regular polygons. However, it is usually a good idea to go through the construction of tables, by hand, before trying to write a program. An annotated listing and a run of one such program is included at the end of this module.

Exercise: (optional) Make tables for the pentagon, hexagon, octagon, etc.

### Designs from Symmetries

Designs based on these tables can be quite surprising and beautiful.

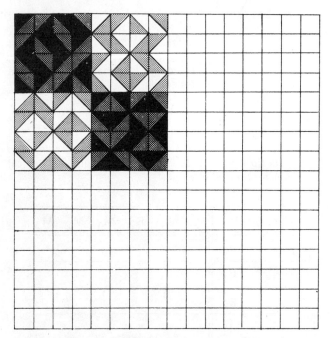

Exercise: (NOT OPTIONAL) Fill in the rest of the above design using colored markers. You decide whether to repeat, reflect, or rotate the original design.

Exercise: (optional) There are other plane figures which have symmetry but are not regular polygons, such as rectangles, parallelograms, trapezoids, and rhombuses. Devise a computer program to produce the table for one of these figures.

## ENTER THE COMPUTER

Here's a RUN of the computer program we promised earlier. The program listing is on the following page.

```
? 3

R0  =  1   2   3
R1  =  2   3   1
R2  =  3   1   2

M1  =  1   3   2
M2  =  3   2   1
M3  =  2   1   3

R0   R1   R2   M1   M2   M3

R1   R2   R0   M2   M3   M1

R2   R0   R1   M3   M1   M2

M1   M3   M2   R0   R2   R1

M2   M1   M3   R1   R0   R2

M3   M2   M1   R2   R1   R0

? 4

R0  =  1   2   3   4
R1  =  2   3   4   1
R2  =  3   4   1   2
R3  =  4   1   2   3

D1  =  1   4   3   2
D2  =  3   2   1   4
M1  =  2   1   4   3
M2  =  4   3   2   1

R0   R1   R2   R3   D1   D2   M1   M2

R1   R2   R3   R0   M2   M1   D1   D2

R2   R3   R0   R1   D2   D1   M2   M1

R3   R0   R1   R2   M1   M2   D2   D1

D1   M1   D2   M2   R0   R2   R1   R3

D2   M2   D1   M1   R2   R0   R3   R1

M1   D2   M2   D1   R3   R1   R0   R2

M2   D1   M1   D2   R1   R3   R2   R0
```

**FURTHER READING:**

*A First Course in Abstract Algebra*, by John S. Fraleigh, Addison-Wesley, 1967.

*Mathematical Reasoning*, Anita Harnadek, Midwest Publications, 1972.

```
LIST
POLY4    04:26 PM        16-DEC-75
10 PRINT"THIS PROGRAM CALCULATES THE ROTATIONS AND REFLECTIONS
11 PRINT"OF THE REGULAR POLYGONS .
12 PRINT "WHAT NUMBER OF SIDES (>2) DO YOU WANT?"
15 F$="! # "
16 R$="R"
17 D$="D"
18 M$="M"
20 INPUT N
30 PRINT
40 IF 360/N=INT(360/N) THEN 80
50 PRINT "THERE IS NO REGULAR POLYGON OF";N;"SIDES WITH INTEGRAL ANGLES"
60 GO TO 20
70 DIM C(20,10)
80 D=0
85 REM CALCULATES THE ROTATIONS
90 FOR V=0 TO N-1
100 FOR I= 1 TO N
110 LET C(V+1,I)=I
120 NEXT I
130 FOR L= 1 TO N
140 Q=L+V
150 IF Q>N THEN Q=Q-N
160 C(V+1,L)=Q
170 NEXT L
180 PRINTUSING F$,R$,V;:PRINT"=";
190 FOR I= 1TO N
200 PRINT C(V+1,I);
210 NEXT I
220 PRINT
230 NEXT V
240 PRINT
245 REM CALCULATES THE FLIPS
250 IF N/2 = INT(N/2) THEN D=0 ELSE D=1
260 IF N/2 = INT(N/2) THEN A=N/2 ELSE A=N
270 FOR I= 1 TO N
280 P= N+2-I
290 IF P>N THEN P=P-N
310 C(N+1,I)=P
320 NEXT I
330 V=1
335 K=N
340 GO SUB 450
350 FOR K=N+1 TO 2*N-1
360 FOR L= 1 TO N
370 C(K+1,L)= C(K,L)+2
380 IF K=3*N/2 THEN C(K+1,L)=C(K+1,L)+1
390 IF C(K+1,L)>N THEN C(K+1,L)=C(K+1,L)-N
400 NEXT L
410 V=V+1
420 GOSUB 450
430 NEXT K
440 GO TO 540
450 REM PRINTS OUT D'S AND M'S
460 IF V>A THEN D=D+1
470 IF V>A THEN V=1
480 IF D>1 THEN 540
490 IF D>0 THEN PRINTUSING F$,M$,V; :PRINT"=";:GO TO 510
500 PRINTUSING F$,D$,V;:PRINT"=";
510 FOR I=1 TO N: PRINT C(K+1,I);:NEXT I:PRINT
530 RETURN
540 PRINT
550 REM CALCULATES THE PRODUCT USING FIRST 2 NUMBERS
560 FOR K=1 TO 2*N
570 FOR K1= 1 TO 2*N
580 FOR I=1 TO 2
590 J=C(K,I)
600 T(I)=C(K1,J)
610 NEXT I
620 REM NEXT PART RECOGNIZES AND PRINTS RESULT
630 IF T(1)<>N-1 THEN 650
640 IF T(2)=T(1)+1 THEN 720
650 IF T(2)=(T(1)+1)-INT(((T(1)+1)/N))*N THEN 720
660 IF T(1)/2=INT(T(1)/2) THEN T(1)= (T(1)+1+N)/2 ELSE T(1)=(T(1)+1)/2
670 IF N/2<>INT(N/2) THEN 700
680 IF T(1)<=N/2 THEN PRINTUSING F$,D$,T(1);:GO TO 730
690 PRINTUSING F$,M$,INT(T(1)-N/2);: GO TO 730
700 PRINTUSING F$,M$,T(1);
710 GO TO 730
720 PRINTUSING F$,R$,T(1)-1;
730 NEXT K1
740 PRINT:PRINT
750 NEXT K
760 GO TO 20
770 END
```

# Patterns
## William Games

Patterns in nature are aesthetically pleasing as well as a key to understanding processes and events. So it is with functions of two variables. They, too, may exhibit wondrous patterns and symmetries that help one appreciate the order and beauty of mathematics. Here is a program that generates contour maps of two-variable functions over domains of the user's choice. The results can be both beautiful and educational.

When one first inspects the function z=cos(xy), he is probably left cold. A question that first arises might be, "What does it look like?" The function can be analyzed for critical points. Points can be evaluated and plotted. Only after much time, abstract imagination, and artistic effort, may that object of one's curiosity be seen. Unfortunately, many of us do not have such mathematical training or the perseverance to behold such sublime splendor. It is for the impatient and the lazy that computerized graphics are so useful. Unfortunately, though, the cost of graphics systems and terminals are well above the means of most hobbyists and schools. Let us improvise, using BASIC, the ASCII character set, and a Teletype-like printer.

Our objective is to graphically represent functions of two variables such as z=cos(xy). Mathematically, this involves plotting in space a function whose domain is a subset of the xy-plane. In other words, the ordered pair (x,y) is mapped onto z=f(x,y). If f(x,y) is continuous, the result is a "surface" suspended in space where each point is of the form (x,y, f (x,y)).

The best way to "see" these functions or any surface on a two-dimensional piece of paper is with a contour map. In the case of many contour maps, equal elevations are represented by a continuous curve through those points. Another approach is to color or shade the map according to elevation. It is the second technique that is used in this program. Since there is the constraint of the discontinuous Teletype, equal elevations, or values of the function, must be represented by ASCII characters. In this program, the greater the value of the function, the more dense (darker) the combination of characters printed. When viewed very closely, such output makes little sense. When viewed as a whole, though, the discontinuities tend to blend together creating the overall affect of gradual darkening of greater and greater values of the function and thus the curvature of the surface itself.

The following program is an efficient tool for "seeing" what functions of two variables look like and/or creating beautiful patterns. The program is designed to allow easy manipulation of parameters for discovery of their effect on the whole. To change functional parameters or the function itself, simply redefine the function in line 110. Upon execution, the program is designed to first interrogate the user. The user is asked to specify the domain of interest, first the x-axis interval and then the y-axis interval. In effect, a rectangular area of the xy-plane is defined for plotting. It is recommended that the intervals be the same length to minimize scale distortion. Once the domain of interest is specified, the user specifies the number of pages of output. A response of one results in exactly the specified domain being printed. A response less than or greater than one results in a corresponding fraction or multiple of the original domain being printed. Initially, a response of one page is recommended. Next, the user is questioned whether or not to automatically compute the range of the function. Since the function in line 210 defines a linear one-to-one correspondence between the range and all characters in A$ and B$, accuracy in specifying the range is important. A NO response to the query allows the user to input the range. To the YES response, the program responds by automatically searching all plottable points for the absolute minimum and maximum values. Finding the range automatically may take more than a few seconds. Note that the range (R1, R2) is widened at both ends by .00001 to compensate for round-off errors.

Once all parameters are set, the output phase begins. Simply, the output section consists of a horizontal printing loop nested in a vertical advancement loop. The inner x-loop is responsible for the printing of a single line representing the value of the function across the entire x-interval for a fixed-y. The x-loop is incremented by the length of the interval divided by the number of print positions. Thus, a greater number of print positions per fixed interval increases the sense of continuity. The function defined in line 210 converts the numeric value of the function in line 110 into a position on A$ and B$. These strings list the output characters arranged by increasing density. Note that each line is printed twice. First, the determined position in A$ is output in each print position. The carriage is returned to the beginning of the same line where characters from B$ are then printed. By printing each line twice, the range of discrete densities is increased. The result is a smoother surface in appearance. In effect, each position (P) returned by the function in line 210 is graphically represented by the "sum" of the given position in A$ and B$. The characters assigned to A$ and B$ give the best result for a Decwriter II terminal. Modification of the character strings may be necessary for other terminals. If one is working with a CRT, or desires only one sweep of each line, then change line 60 to read: 60 N=1 (n is the number of sweeps of a given line). The I-loop determines the number of times each line is printed and is controlled by the assignment in line 60. The Y-loop sweeps the domain one line at a time, beginning with the greatest value of y. The Y-loop is stepped by the same increment used in the X-loop. A scale factor is introduced to compensate for discrepancies in the number of columns and lines per inch. It is assumed that 10 columns=6 rows=1 inch. After a map has been printed, one may want to repeat the map but extended above and below. This may be achieved by responding to the page prompt with a number greater than one. By increasing the number of pages to two, the length of the Y-interval will be doubled as will the length of the output. Changing this parameter has no effect on the scale or output of the initial domain. Caution should be exercised here as the original range may be exceeded. Whenever the value of the function is outside the specified range, the letter E prints to indicate the error. Repeat the program with widened range when this happens. ■

William Games, 8357 Alexe Ct., Stockton, CA 95209.

## Experimental Functions and Activities

Try these functions for interesting results:
1. cos(x)*sin (y)
2. cos(x)+cos(y)
3. exp(sqr(x↑2+y↑2))-int(exp(sqr(x↑2+y↑2)))
4. cos(x*y/sqr(x↑2+y↑2))
5. cos((x+y)/(log(abs(x*y)+.5)))
6. cos((abs(x)+.5)↑y)
7. cos(y/(abs(x)+.5))
8. sin(x-y)/(1.5+cos(y))

An interesting assignment for high-school students might be to investigate the effect of manipulating constants in arguments. For example, how is the map of cos(x)*sin(y) transformed when the function is changed to cos(x)*sin(2*y) ? Another inquiry might illustrate various trigonometric identities. An example is sin(x+y)=sin(x)*cos(y)+cos(x)*sin(y).

**Six Sample Patterns** from the program will be found on the next two pages.

## Table of Program Variables

- A$: output characters ordered by increasing density
- B$: output characters for second sweep
- E$: error indicator
- L: the number of characters in A$
- N: the number of times each line is printed
- W: width or number of print columns
- X1,X2: domain interval (X1, X2) along x-axis
- Y1,Y2: domain interval (Y1, Y2) along y-axis
- X$: dummy string
- R1,R2: range of FNZ (X, Y)
- X, Y: the coordinates (X, Y) being printed

```
1    REM ****************************************************************
2    REM ***                                          12/26/77    ***
3    REM ***      WRITTEN BY: BILL GAMES                          ***
4    REM ***                  8357 ALEXA CT.                      ***
5    REM ***                  STOCKTON, CALIFORNIA    95209       ***
6    REM ****************************************************************
10   DIM A$[100],B$[100]
20   A$=" '-~!<=;[kX&*#X*"
30   B$="              HO"
35   E$="E"
40   L=LEN(A$)
50   REM *** N: # OF TIMES EACH LINE IS PRINTED ***
60   N=2
100  REM ***FUNCTION TO BE GRAPHED***
110  DEF FNZ(X)=COS(X*Y)
200  REM ***CONVERTS VALUE OF FNZ(X,Y) INTO A POSITION ON A$ AND B$***
210  DEF FNP(X)=INT(L*(FNZ(X)-R1)/(R2-R1))+1
1000 REM ***INPUT PARAMETERS***
1010 PRINT "HOW MANY PRINT POSITIONS";
1020 INPUT W
1030 PRINT "SPECIFY INTERVALS AS FOLLOWS: LEAST#,GREATEST#"
1040 PRINT "INPUT DOMAIN INTERVAL OF X-AXIS:";
1050 INPUT X1,X2
1060 PRINT "INPUT DOMAIN INTERVAL OF Y-AXIS:";
1070 INPUT Y2,Y1
1080 PRINT "HOW MANY PAGES OF OUTPUT";
1090 INPUT M
1100 M0=(M-1)*(Y1-Y2)/2
1110 PRINT "SET RANGE OF FNZ(X,Y) ATOMATICALLY? ";
1120 LINPUT X$[1,1]
1130 IF X$="Y" THEN 1170
1140 PRINT "INPUT RANGE OF FNZ(X,Y):";
1150 INPUT R1,R2
1160 GOTO 1300
1170 REM ***AUTOMATIC RANGE FINDER***
1180 PRINT "***THIS WILL TAKE AWHILE.   PLEASE HOLD ON."
1190 R1=999999.
1200 R2=-999999.
1210 FOR Y=Y1 TO Y2 STEP -(Y1-Y2)/(.6*W)
1220 FOR X=X1 TO X2 STEP (X2-X1)/W
1230 IF FNZ(X)>R1 THEN 1250
1240 R1=FNZ(X)
1250 IF FNZ(X)<R2 THEN 1270
1260 R2=FNZ(X)
1270 NEXT X
1280 NEXT Y
1290 PRINT "LOWER BOUND=";R1,"UPPER BOUND=";R2
1300 R1=R1-.00001
1310 R2=R2+.00001
2000 REM ***OUTPUT ROUTINE***
2010 PRINT LIN(2)
2020 FOR Y=Y1+M0 TO Y2-M0 STEP -(Y1-Y2)/(.6*W)
2030 FOR I=1 TO N
2040 FOR X=X1 TO X2 STEP (X2-X1)/W
2050 P=FNP(X)
2060 IF P >= 1 AND P <= L THEN 2090
2070 PRINT  USING "#,A";E$
2080 GOTO 2130
2090 IF I=2 THEN 2120
2100 PRINT  USING "#,A";A$[P,P]
2110 GOTO 2130
2120 PRINT  USING "#,A";B$[P,P]
2130 NEXT X
2140 PRINT LIN(0);
2150 NEXT I
2160 PRINT
2170 NEXT Y
2180 PRINT LIN(10)
9999 END
```

This statement dimensions strings A$ and B$ to hold 100 characters each. The strings are scalars, not arrays.

Just a fancy INPUT statement which puts the first character typed into X$. It does not print a ? as a prompt.

PRINT LIN(2) results in the printing of three blank lines: two because of the LIN(2) function, and a third from the PRINT statement itself.

The PRINT USING causes the printing of the leftmost character of E$ with no carriage return or linefeed after printing. The # character is used for carriage control.

Likewise. Note that A$(P,P) is a substring—the character at position P in A$.

PRINT LIN(0); prints a carriage return but no line feed, so that a line may be overprinted. Use PRINT CHR$(13); in some other BASICs.

Prints 11 blank lines at the end of the printout.

```
110 DEF FNZ(X)=COS((X+Y)/(LOG(ABS(X*Y+.5))))
RUN
XYZ2

HOW MANY PRINT POSITIONS?74
SPECIFY INTERVALS AS FOLLOWS: LEAST#,GREATEST#
INPUT DOMAIN INTERVAL OF X-AXIS:?-5,5
INPUT DOMAIN INTERVAL OF Y-AXIS:?-5,5
HOW MANY PAGES OF OUTPUT?2
SET RANGE OF FNZ(X,Y) ATOMATICALLY? NO
INPUT RANGE OF FNZ(X,Y):?-1,1
```

```
110 DEF FNZ(X)=COS((X+Y)/(LOG(X^2+Y^4)))
RUN
XYZ2

HOW MANY PRINT POSITIONS?75
SPECIFY INTERVALS AS FOLLOWS: LEAST#,GREATEST#
INPUT DOMAIN INTERVAL OF X-AXIS:?-200,200
INPUT DOMAIN INTERVAL OF Y-AXIS:?-200,200
HOW MANY PAGES OF OUTPUT?2
SET RANGE OF FNZ(X,Y) ATOMATICALLY? NO
INPUT RANGE OF FNZ(X,Y):?-1,1
```

```
110 DEF FNZ(X)=COS(Y/(ABS(X)+.5))
RUN
XYZ2

HOW MANY PRINT POSITIONS?74
SPECIFY INTERVALS AS FOLLOWS: LEAST#,GREATEST#
INPUT DOMAIN INTERVAL OF X-AXIS:?-10,10
INPUT DOMAIN INTERVAL OF Y-AXIS:?-10,10
HOW MANY PAGES OF OUTPUT?2
SET RANGE OF FNZ(X,Y) ATOMATICALLY? NO
INPUT RANGE OF FNZ(X,Y):?-1,1
```

110 DEF FNZ(X)=COS(X*EXP(-(Y/10)))

RUN
XYZ2

HOW MANY PRINT POSITIONS?75
SPECIFY INTERVALS AS FOLLOWS: LEAST*,GREATEST*
INPUT DOMAIN INTERVAL OF X-AXIS:?-10,10
INPUT DOMAIN INTERVAL OF Y-AXIS:?-10,10
HOW MANY PAGES OF OUTPUT?2
SET RANGE OF FNZ(X,Y) ATOMATICALLY N
INPUT RANGE OF FNZ(X,Y):?-1,1

110 DEF FNZ(X)=COS(X*Y/SQR(X^2+Y^2))

RUN
XYZ2

HOW MANY PRINT POSITIONS?76
SPECIFY INTERVALS AS FOLLOWS: LEAST*,GREATEST*
INPUT DOMAIN INTERVAL OF X-AXIS:?-20,20
INPUT DOMAIN INTERVAL OF Y-AXIS:?-20,20
HOW MANY PAGES OF OUTPUT?2
SET RANGE OF FNZ(X,Y) ATOMATICALLY Y
***THIS WILL TAKE AWHILE. PLEASE HOLD ON.
LOWER BOUND=-.999996          UPPER BOUND= 1.

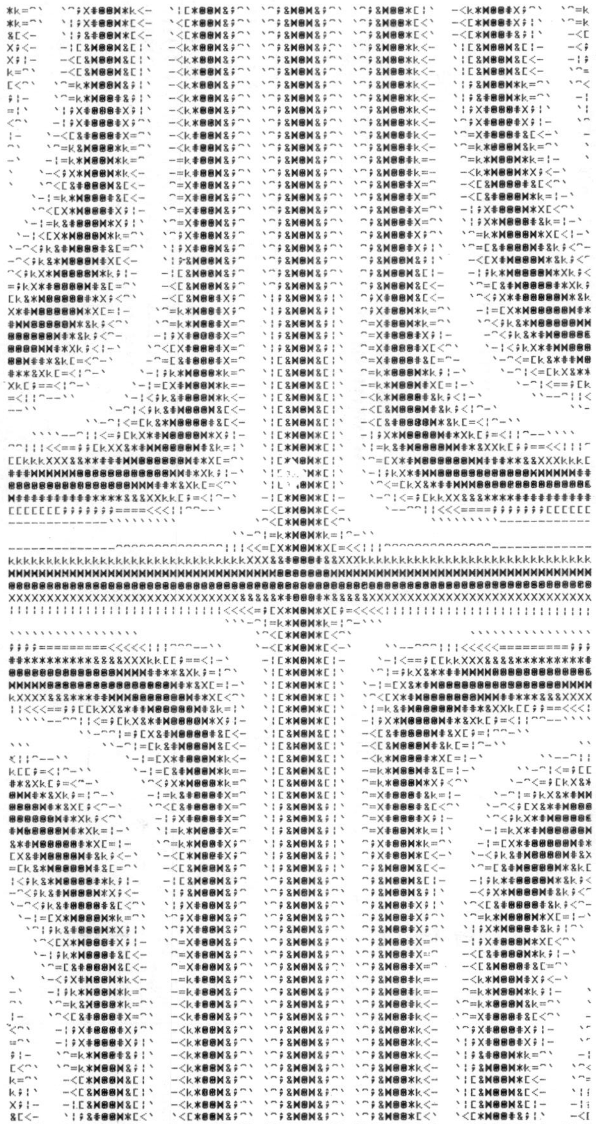

110 DEF FNZ(X)=SIN(X)*SIN(Y)

RUN
XYZ2

HOW MANY PRINT POSITIONS?75
SPECIFY INTERVALS AS FOLLOWS: LEAST*,GREATEST*
INPUT DOMAIN INTERVAL OF X-AXIS:?-5,5
INPUT DOMAIN INTERVAL OF Y-AXIS:?-5,5
HOW MANY PAGES OF OUTPUT?2
SET RANGE OF FNZ(X,Y) ATOMATICALLY NO
INPUT RANGE OF FNZ(X,Y):?-1,1

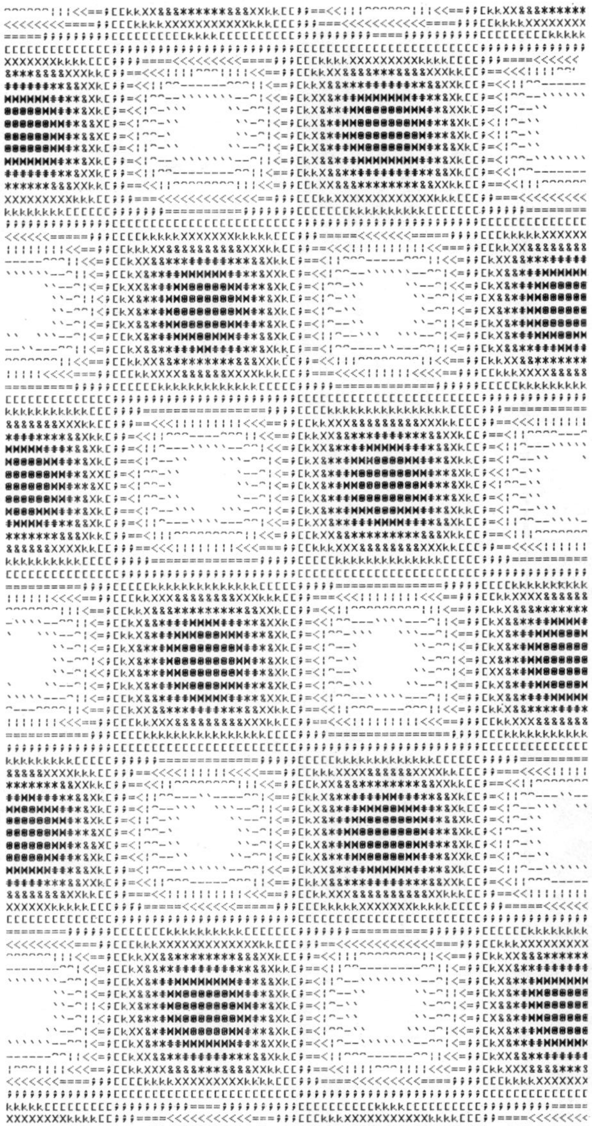

# Inkblot

INKBLOT is a program that creates "inkblots" similar to those used in the famous Rorschach Inkblot Test. The program generates these inkblots randomly so that literally millions of different patterns can be produced. Many of these patterns are quite interesting and serve not only as conversation pieces, but also as good examples of computer "art."

In addition, INKBLOT is interesting from a mathematical point of view. This is because INKBLOT actually creates inkblots by plotting ellipses on the left side of the page and their mirror-images on the right side. The program first chooses the ellipses to be plotted by randomly selecting the values a, b, j, k and θ in the equation for a rotated ellipse:

$$\frac{[(x-j)\cos\theta + (y-k)\sin\theta]^2}{a^2} + \frac{[(y-k)\cos\theta - (x-j)\sin\theta]^2}{b^2} = 1$$

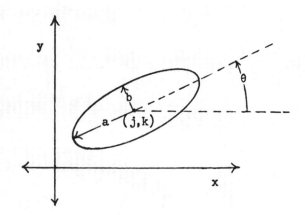

where a = the horizontal radius of the ellipse
b = the vertical radius of the ellipse
j = the distance from the ellipse center to the y-axis
k = the distance from the ellipse center to the x-axis
θ = the angle of rotation in radians

Since the actual method by which the program plots the ellipses is quite complicated, it won't be discussed here.

INKBLOT could be enhanced in several ways, for example allowing the user to specify which character is to be used in printing the inkblot. It could have an option to print the "negative" of an inkblot by filling in the area around the ellipses rather than the ellipses themselves. Finally, it is possible to build in a "repeatable randomness" feature so that exceptional outputs could be reproduced at any time. These enhancements are left for the ambitious programmer to make.

Program and description are by Scott Costello.

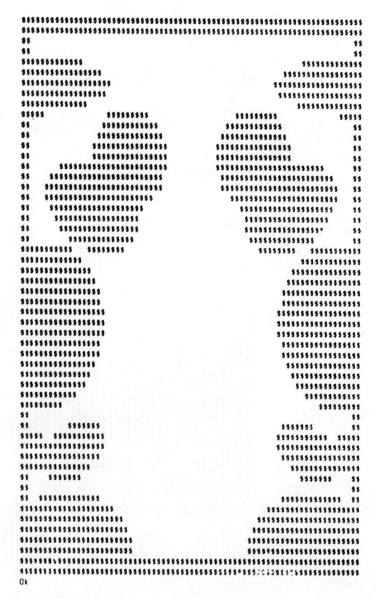

```
LIST
100 PRINT TAB(26);"INKBLOT"
105 PRINT TAB(20);"CREATIVE COMPUTING"
110 PRINT TAB(18);"MORRISTOWN, NEW JERSEY"
115 PRINT:PRINT:PRINT
120 REM *** WORKS BY PLOTTING ELLIPSES AND THEIR MIRROR IMAGES
130 DIM A (12,13),B$(36),A$(36)
140 REM *** CHOOSE FROM 5 TO 12 ELLIPSES
150 M=INT(8*RND(1))+5
160 REM *** CREATE SIZE, LOCATION AND ANGLE OF M ELLIPSES
170 FOR L=1 TO M
180 A(L,1)=34*RND(1)
190 A(L,2)=80*RND(1)
200 A(L,3)=(15*RND(1)+2)^2
210 A(L,4)=(15*RND(1)+2)^2
220 T=3.14159*RND(1)
230 A(L,5)=COS(T)
240 A(L,6)=SIN(T)
250 A(L,7)=A(L,5)*A(L,6)
260 A(L,5)=A(L,5)*A(L,5)
270 A(L,6)=A(L,6)*A(L,6)
280 A(L,8)=A(L,1)*A(L,1)*A(L,6)
290 A(L,9)=A(L,1)*A(L,1)*A(L,5)
300 A(L,10)=A(L,1)*A(L,7)
310 A(L,11)=-2*A(L,1)*A(L,6)
320 A(L,12)=-2*A(L,1)*A(L,5)
330 A(L,13)=A(L,6)/A(L,4)+A(L,5)/A(L,3)
340 NEXT L
350 REM *** PRINT TOP BORDER;  B$ CONTAINS 36 DOLLAR SIGNS
360 B$="$$$$$$$$$$$$$$$$$$$$$$$$$$$$$$$$$$$$"
370 PRINT B$;B$
380 PRINT B$;B$
390 REM *** LOOP Y IS Y-COORDINATE OF PLOT; EACH TIME Y LOOP
400 REM *** IS EXECUTED, A LINE IS PRINTED
410 FOR Y=79.9 TO 0 STEP -1.6
420 A$="  $$                                "
430 REM *** LOOP E CHECKS THE EQUATION OF EACH ELLIPSE TO SEE
440 REM *** IF IT INTERSECTS THE LINE TO BE PRINTED
450 FOR E=1 TO M
460 Y1=Y-A(E,2)
470 Y2=Y1*Y1
480 Y3=Y1*A(E,10)
490 Y4=Y1*A(E,7)
500 B=(A(E,12)+Y4)/A(E,3)+(-Y4+A(E,11))/A(E,4)
510 C=(Y2*A(E,6)+A(E,9)-Y3)/A(E,3)+(Y2*A(E,5)+A(E,8)+Y3)/A(E,4)-1
520 REM *** R IS THE RADICAL IN THE STANDARD QUADRATIC FORMULA
530 R=B*B-4*A(E,13)*C
540 IF R<0 THEN 690
550 R=SQR(R)
560 REM *** FIND WHERE THE LINE INTERSECTS THE ELLIPSE
570 R1=INT(-(B+R)/2/A(E,13)+1)
580 IF R1>34 THEN 690
590 R2=INT((R-B)/2/A(E,13))
600 IF R2<1 THEN 690
610 IF R2<35 THEN 630
620 R2=34
630 IF R1>0 THEN 660
640 R1=1
650 REM *** FILL IN THE LINE WHERE IT CROSSES THE ELLIPSE
660 FOR J=R1+2 TO R2+2
670 A$=LEFT$(A$,J-1)+"$"+RIGHT$(A$,LEN(A$)-J)
680 NEXT J
690 NEXT E
700 REM *** PRINT LINE
710 PRINT A$;
720 FOR K=36 TO 1 STEP -1
730 PRINT MID$(A$,K,1);
740 NEXT K
750 NEXT Y
760 REM *** PRINT BOTTOM BORDER
770 PRINT B$;B$
780 PRINT B$;B$
790 END
Ok
```

Inkblot is reprinted from the new book, **More Basic Computer Games** which contains 90 new, fascinating games in Microsoft Basic. Reserve your copy by sending $7.50 plus $1.00 shipping and handling in USA ($2.00 foreign) to Creative Computing, P.O. Box 789-M, Morristown, N.J. 07960. Visa, MasterCharge and American Express okay.

# A Picture In 20 Lines

by E. Young
Beavercreek High School
Xenia, Ohio

We've all heard that a picture is worth 1000 words. Well what kind of picture can be produced in a 20-line BASIC program (approx. 1000 characters).

My assignment to my first-semester computer science class was simply "to produce a picture with a 20-line BASIC program with no PRINT quote formats allowed."

The variety of programming methods surprised me. They ranged from 3 data codes for what, where, and how many characters to read — to single numerical data that was sectioned algebraically to code a whole line.

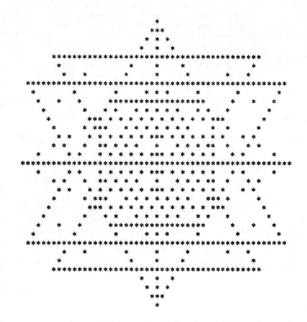

Star of Beaver Creek by Dave Triwush

Susan Gordon

Dottie Dimiduk

Lowell VonRuden

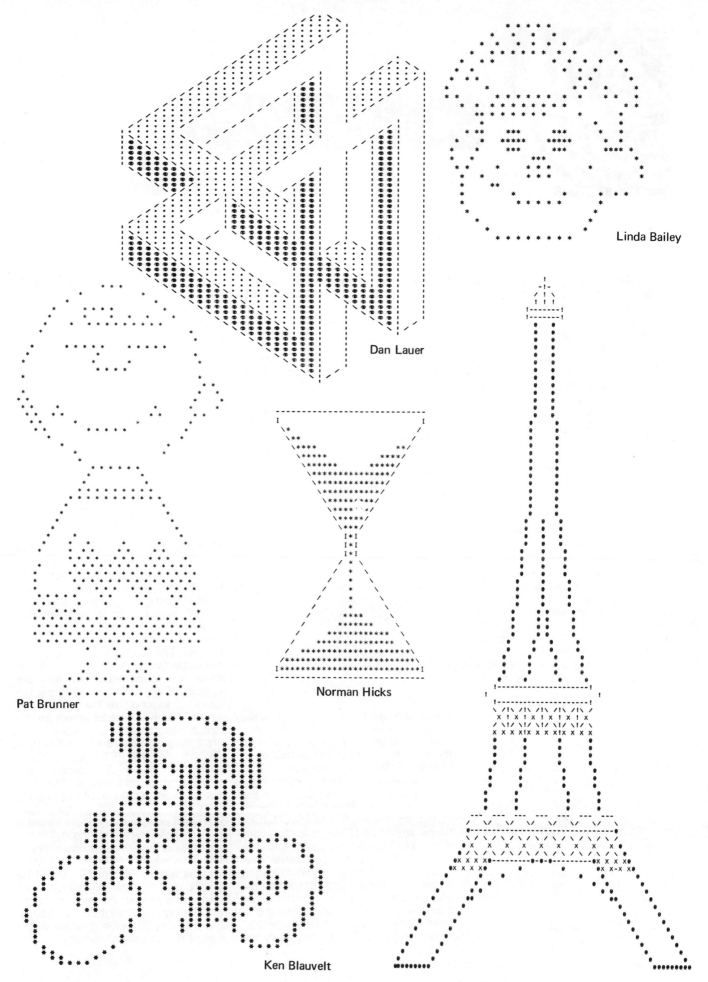

# SUPEROSE

### Michael D. Zorn

I was inspired by the article "Gumowski," in the Sept/Oct *Creative Computing*, to find a similar routine for my new PET. The limiting factor of a 25 by 40 display precludes anything approaching the subtle tracery produced by that remarkable program. This routine, however, has several advantages (besides that of running on a PET!): it is easily adaptable to any BASIC computer driving a character or high-resolution display — such as the TRS-80, the Apple, the new Exidy and all the rest. It can be used with a line printer (though storage space and time will be a consideration here), and if you have an x-y plotter, you're really in luck, because you can generate the figures in their greatest detail. The accompanying figures show examples of line printer output, on a 40 by 70 grid, and x-y plotter output.

## HOW IT WORKS

The routine itself is quite simple, yet provides a great deal of flexibility. The essence of the algorithm is in line 107. Think of a pen mounted on a rotating arm of varying length. Let the radius change as a function of the angle, and the pen will trace out a curved line. In particular, if R doesn't change at all, we have a circle centered on the pivot of the arm. Line 107, then, defines the function, in this case, R = Z*SIN(TH*T). Line 100 lets the arm go once around a circle (to $2\pi$ radians) in increments of two degrees.

That's all there is to it. However, unless you're writing to a radar screen, you need to get from the land of (r,theta) to the land of (x,y). This is done by the next two lines: 110 and 115. In the PET, X% and Y% automatically convert the results to integer. Be sure not to omit this conversion — it's an essential part of the process.

We're still not there, though, because while we do have the x,y coordinates,

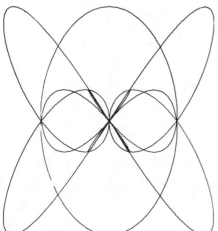

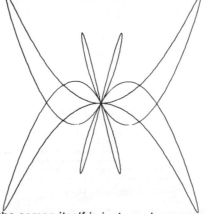

the screen itself is just one long array (1000 bytes in the PET, for example). Line 120 takes care of this problem. (Notice the similarity to the routine in the Short Programs section in that same issue.) Let's take the transformation in two steps: first, consider the screen as a 40 by 25 grid, with the point (1,1) at the lower left. The upper left, or home, would be (1,25), and the upper right, (40,25). So we can find the location of (x,y) in the screen array by P = (25 − y)*40 + x. Since we'd like the origin to be in the center of the screen, all we need to do is translate the origin from the lower left, 20 units right and 13 units up: P = (25−(y+13))*40+(x+20). This formula then reduces to line 120

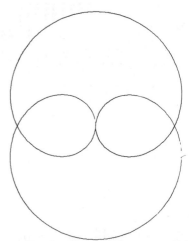

(actually, the sign of Y is reversed, but all that does is invert the figures, and I like them better that way).

## CONVERTING TO OTHER SYSTEMS

Now let's consider modifying the routine for your particular computer. In line 120, the 40 represents the number of characters or points across the display, and the 20 is just half that number. The 12 is half the number of lines — each "half" truncated to integer. The Z in line 37 sets the maximum radius, and should be set to about the same value as half the number of lines. If Z is too large, the figure will extend past the edges of the screen. The K in line 38 allows for the screen aspect ratio. It insures that a circle comes out circular, rather than elliptical. For the PET, 1.25 seems to work quite nicely. To calculate the value for your particular system, first plot (or print) a column of 10 symbols. Then find out (by marking the length on a card) how many symbols across it takes to make the same length. Then K = number across/number down. The Q = 81 tells which character is POKEd onto the screen. For the PET, this is the big dot, shift-Q. Asterisks seem to work nicely on a printer. For higher resolution displays, you might prefer points. The S in line 39 is the location of the start of the screen in memory.

Michael D. Zorn, 833 S. Peck Rd. #4, Monrovia, CA 91016

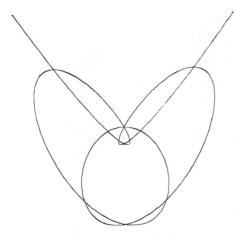

## THE ANT

For a line printer, the setup is a little different. In this case, we start with an M by N array of blanks. As each X,Y is generated, set ARRAY (X,Y) to DOT. Then, when the figure is completed, print the entire array at one time. To figure the size of the array, start with the numbers of lines on a page — let's suppose it's 40, as in my case. Multiply this by the printer's aspect ratio. Most printers I've seen are 10/6. This gives 66.6666, which we round up to the next convenient round number — say, 70. Then dimension the array 40 by 70, and proceed as before. If the 70 is larger than the number of print positions on your printer, start with that number and work backwards.

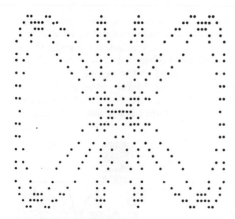

## THE BUTTERFLY

For a plotter, take out the integer conversion, replace line 120 with the equivalent of CALL PLOT(X,Y) (that is, move the pen to the next x,y), and set Z according to the plotter's specifications. If the figure is too granular, decrease the step size in line 100, to 1 or even .5.

## CHECKOUT

To check out the routine for a non-PET system, first put a REM into line 107 to hold the radius constant, and input A=1 and B=1. This should generate a circle of maximum radius. (You might want to take out the integer conversion at lines 110 and 115 at this point to see what happens.) If that works, you're home free. Take out the REM, input A=1 and B=1, and you should see the 3-leaf rose. If the circle is too big, however, reduce Z. If the circle isn't circular, adjust the value of K. If the dots seem randomly sprinkled over the screen, check out line 120. Copy it into another line, say 5000, followed by a PRINT,P. Then feed it x=0,y=0, and P should be at the center of the screen (500 for the PET). Feed it the coordinates of the corners, and see if they map correctly into the screen array. The upper left coordinate (-19,12 for the PET) should give 0 or 1.

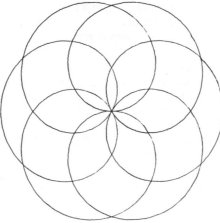

## THE LOTUS

For a plotter, make sure that the pen starts out at the middle of the page. Also, make sure that the pen doesn't come down until the first (x,y) is reached (it may not always be at the origin). Finally, make sure that the value for Z gives a maximum size circle during the initial checkout.

## APPLICATIONS

The fundamental shape for a given value of T is drawn when A=B=1. In general, when T is odd, there are T leaves (use them later to have your fortune told) and when T is even, there are 2T leaves. Then let A and B vary as integers in the range 1 through 10 or so to produce the variations on the

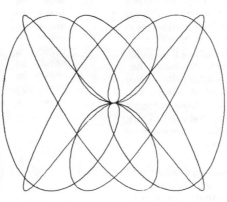

fundamental shape. Some of the figures are quite striking, and deserve to be given names, such as the Glider (A=1,B=5), the Ant (4,5), the Butterfly (1,6), and the Lotus (4,4). All of these are for T=3, as are the figures on these pages. After you've gone through these few hundred possibilities, here are some directions for opening up the investigation:

— let A and B be non-integers
— holding A constant, produce a series of figures by slowly varying B in fractional steps from one integer to the next
— let T be a non-integer (interesting values for A, B, and T would be numbers of the form p/q, where p and q are small integers)
— use the X and Y values to drive a digital-to-analog converter which in turn drives an oscilloscope
— run the routine in a "warp drive" mode by setting A=B= a very large number (between 30000 and 100000)

All of this has come from the simple formula r = a sin n. There are any number of other formulas you might use. The only restriction is that r should not be allowed to grow without limit, as would be the case with r = a tan. If you do use a formula like that, be sure to check the limits on X and Y, or you'll be POKEing into the operating system, into your hi-fi set, and into the PET next door.

There's an almost infinite variability to these figures, yet they're not random — they're symmetrical, mathematical, and beautiful.

```
37   Z=12 : Q=81
38   K=1.25
39   S=32767
40   T=3
41   INPUT "A<B";A,B : PRINT "⌂"
100  FOR TH=0 TO 2*π STEP 2*π/180
107      R=Z*SIN(TH*T)
110      X% =K*R*COS(A*TH)
115      Y% =R*SIN(B*TH)
120      P=(12+Y%)*40+X%+20
130      POKE S+P,Q
140  NEXT TH
150  END
```

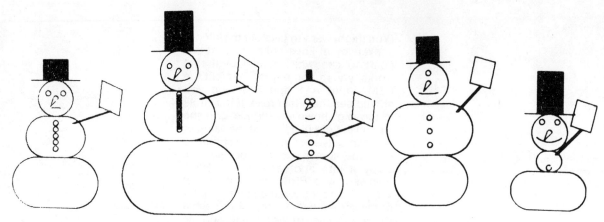

# Computer-Planned Snowmen

Robert S. McLean
Ontario Institute for Studies in Education
Toronto, Canada

Computer models are useful for testing theories. If you have a model of a process or structure expressed as a computer program, you have a powerful tool. A computer can then produce many different instances of the modeled process through the use of a range of parameter values. It will produce the results of the model without being influenced by extraneous notions of what the outcome should be. In addition to investigating the range of applicable parameter values, one can even push those values beyond "reasonable" limits and observe the results. The computer tirelessly shows the results of the chosen conditions without requiring very many input resources. If the result of the model is a picture, so much the better; its adequacy can often be judged visually by the user very rapidly. What is more, any pictorial outputs can be very entertaining when they illustrate weaknesses of the model or data, thereby producing very unusual pictures.

Computer modeling at first seems like such a high-powered idea that it would be hard to apply for fun. Maybe it ought to be reserved for sending men to the moon, finding petroleum resources, or managing large construction projects. Although these projects use modeling, simple things that an individual does can also benefit from these techniques. Consider building a snowman, for instance. Here is an important problem for the individual which can be solved with the power of modern computing.

By choosing the proper aspects of a snowman as parameters for our model, we can use the computer to draw pictures of the resulting snowmen. We can observe the shape of hundreds of combinations of parameter values and select the most pleasing one for implementation with real materials. In addition to increasing the range of choice available, this procedure has several other advantages. Most of the design process as well as some of the construction effort is no longer at the mercy of the weather. Much of the work can be done in locations not previously suitable for this activity (home, office, school room, etc. were not suitable places for snowman construction activity previously). Valuable resources are not squandered in making real prototypes since one will have a better idea of the outcome of the process before starting to use these materials. Thus, computer-planned snowmen become feasible in areas where snow is in short supply. One can appreciate that the benefits are many.

What are the parameters of use in modeling a snowman? The reader may wish to propose his own set; for purposes of illustration, we provide a suggested set that were used to design the accompanying illustration. Our standard snowman will consist of three balls of snow, referred to as B1 (on bottom), B2 (middle), and head. In addition to the features of these components, we will add a hat and a broom. This is the economy version, since it could be given many more accessories, such as scarves, cigars, glasses, etc. These are left as an exercise for the reader.

Snowballs 1 and 2 are not necessarily round; they often become somewhat flattened by the load above. Thus, two parameters specify these balls: the radius of the two half circles that form the curved part and the length of the flat parts on top and bottom of the ball, RAD and SIZ respectively, giving the first four parameters, B1SIZ, B1RAD, B2SIZ, and B2RAD. The middle snowball has, in addition, some number of buttons down its front (NBUT). These are spaced out over the vertical extent of B2.

In the world of economy snowmen, heads are always round; hence, there is one head size parameter, HEAD, the radius of the head. Five other parameters are used to place the eyes, nose and mouth on the head. In this version of the model, the nose is always placed in the center of the head and always has the peculiar (carrot-like) shape shown. The sole parameter available here is the length of the nose, NOSE. The eyes are located symmetrically above the center

```
BISIZ   200
B1RAD   100
B2SIZ   150
B2RAD   100
N BUT     4
HEAD    100
NOSE     30
SMILE    20
SMIL2   200
EYEHT    20
EYEWD    40
HATWD   120
HATHT    20
BROOM    60
BDISP   200
```

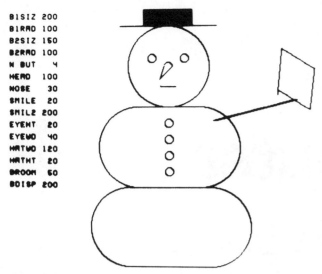

of the head along a horizontal line EYEHT units above the center of the head. They are separated by EYEWD units, the interoccular distance.

The mouth is specified by two parameters, where SMILE gives the length of an arc used to denote mouth, and SMIL2 gives the radius of curvature of that arc. Since models should strive for generality in their parameters, SMIL2 may be negative as well as positive. If positive, the center of the arc is above the mouth, and if negative, it is below the mouth. Thus, we obtain the relationship that a frown is a negative smile! The mouth is always placed halfway between the nose and the chin, again in aid of simplicity.

The remaining four parameters specify the two accessories, the hat and the broom. HATWD and HATHT specify the width and height of the hat; the brim is always twice as wide as the hat. The broom is perhaps the most difficult part of the model and as can be seen looks more like a shovel (a kind interpretation) or a strike sign (less kind). It surely leaves room for the reader to improve the model. The parameters used here are BROOM, giving the length of the handle, and BDISP, giving the distance that the broom handle upper end is displaced to the right from the lower end. Some very crude modeling results in the funny quadrilateral that looks more like a hoe in some cases.

The reader can, no doubt, suggest refinements and is encouraged to do so. The inspiration for this program came from Chernoff's faces (1971) and an adaptation of them that is used to adjust the parameters of oral surgery and to illustrate the various ways in which facial features might be rebuilt (Eisenfeld, et al, 1974). But that's pretty serious business. For the amateur simulator, there are many other possibilities.

### REFERENCES

Chernoff, H. The use of faces to represent points in n-dimensional space graphically. Tech. Report No. 71. Department of Statistics, Standford University, Stanford, California, December 1971.

Eisenfeld, J., Barker, D. R., and Mishelevich, D. J. Iconic representation of the human face with computer graphics. *Computer Graphics* (SIGGRAPH–ACM) 1974, 8(3), 9–15.

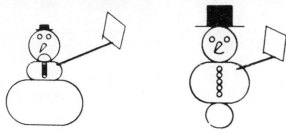

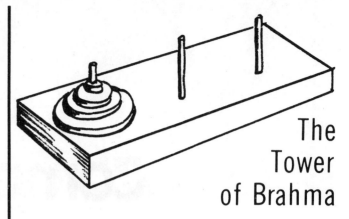

# The Tower of Brahma

In the great temple at Benares beneath the dome which marks the center of the world, rests a brass plate in which are fixed three diamond needles, each a cubit high and as thick as the body of a bee. On one of these needles, at the creation, God placed sixty-four discs of pure gold, the largest disc resting on the brass plate and the others getting smaller and smaller up to the top one. This is the Tower of Brahma. Day and night unceasingly, the priests transfer the discs from one needle to another, according to the fixed and immutable laws of Brahma, which require that the priest on duty must not move more than one disc at a time and that he must place this disc on a needle so that there is no smaller disc below it. When the sixty-four discs shall have been thus transferred from the needle on which, at the creation, God placed them, to one of the other needles, tower, temple, and Brahmans alike will crumble into dust, and with a thunderclap, the world will vanish.

If the priests were to effect one transfer every second, and work twenty-four hours a day for each day of the year, it would take them 58,454,204,609 decades plus slightly more than 6 years to perform the feat, assuming they never made a mistake—for one small slip would undo all their work.

How many transfers are required to fulfill the prophecy?

A. Set up a program which allows the user to move disks by hand. You can try your ingenuity at drawing the result by some sort of plot or graph.

B. At least verbally, indicate how one would proceed in any arbitrary case (5 disks, 6 disks, etc. 64 is too much to try!).

C. The monks, like monks everywhere, never eat, sleep, rest or die. If they have been moving one disk per second since the world began, how long will the total age of the universe be on Thunderclap Day?

Note: Prove that a game of N disks can be played in 2-1 (2-to-the-N, minus 1) moves.

D. Analyze the problem in this way: to move 5 disks, what kind of 4-disk moves are required? How do the "from" and "to" of these moves relate to the "from" and "to" of the 5-disk level?

| DISCS | MOVES | DISCS | MOVES |
|---|---|---|---|
| 1 | 1 | 6 | 63 |
| 2 | 3 | 7 | 127 |
| 3 | 7 | 8 | 255 |
| 4 | 15 | 9 | 511 |
| 5 | 31 | 10 | 1023 |

# Computer Assisted Instruction

**Computer Assisted Instruction (CAI) was one of the earliest and most successful uses of computers in education. In this series of 5 articles we'll show how you can produce and use CAI software on your home or school computer.**

# CAI: Mathematics Drill and Practice

**David H. Ahl**

In its most elemental form, CAI presents drill and practice exercises to a student on a subject that he or she has already learned in class or elsewhere. On larger systems this is refined to the point where the computer keeps track of each student and presents proportionately more material of the type with which the student is having difficulty.

For example, in second grade arithmetic a student may receive drill and practice on horizontal addition, vertical addition, horizontal subtraction, and vertical subtraction in equal doses, i.e., 25% of each type of problem. However, over time the student may miss more of the horizontal type problems, particularly subtraction. In this case after several sessions the ratio of problems might be 30% horizontal addition, 15% vertical addition, 40% horizontal subtraction, and 15% vertical addition.

These problem categories are sometimes known as "strands" and a student may progress along each of the strands independently of other strands and independently of his or her overall grade level. Thus, in an extreme case, a third grader could be at sixth grade level in vertical addition and first grade level in fractions.

This is the type of drill and practice that has proved so successful in Chicago, Compton and numerous other places using large-scale computers or dedicated time-sharing mega-minis. However, there's no reason that we can't produce a similar system for micros and minis, or a non-dedicated time sharing system.

Before we produce a relatively elaborate record-keeping system, it's important to understand some of the basics of writing drill and practice for any computer. For example, consider the following problem:

3 + ___ = 17

Where does the student type the answer? With "normal" high-level languages you can request input to the right of the problem or on the next line (a or b). With cursor addressing you could request input at the more desirable location where it actually belongs (c).

3 + **c** ___ = 17   **a**
**b**

On a problem like this:

32
− 17
───

do you require the answer as 15 or do you allow the student to work from right to left, first inputting a 5 and then having the cursor back up for the 1?

Initially, we'll assume the only language available is Basic with no extended capabilities and with no cursor addressing. However, the principles of writing CAI are the same no matter what language you're dealing with.

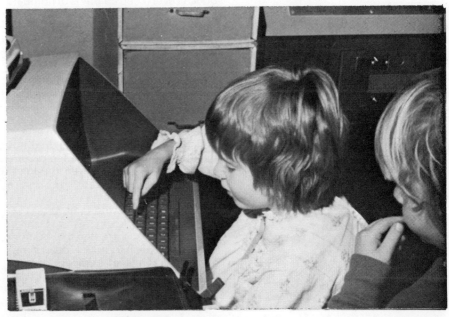

## EXAMPLE 1

```
10 RANDOMIZE
20 N=10
30 W=0
40 A=INT(N*RND(0))      ← Numbers in problems
50 B=INT(N*RND(0))        will be between 0
100 PRINT                 and 9.
110 PRINT "  ";A
130 PRINT "+ ";B
140 R=A+B               ← R = Right answer
200 PRINT "------"        G = "Guess" or student
210 INPUT G               input
220 IF G=R THEN 300
230 W=W+1
240 IF W>1 THEN 270
250 PRINT "WRONG, TRY AGAIN."
260 GOTO 100
270 PRINT "YOU MISSED THAT ONE TWICE."
280 PRINT "THE CORRECT ANSWER IS ";R
290 GOTO 310
300 PRINT "CORRECT !!"
310 PRINT "HERE'S ANOTHER..."
320 GOTO 30
999 END
```

## EXAMPLE 2

```
25 P=0
60 P=P+1
70 IF B<=A THEN 100    ← Make sure that a
80 C=A                   smaller number (B) is
85 A=B                   subtracted from the larger (A).
90 B=C
120 IF P/2=INT(P/2) THEN 160   ← Alternate between
150 GOTO 200                     addition and sub-
160 PRINT "- ";B                 traction problems.
170 R=A-B
```

```
     6
  +  2
  -----
? 8
CORRECT !!
HERE'S ANOTHER...

     8
  -  2
  -----
? 6
CORRECT !!
HERE'S ANOTHER...

     7
  +  2
  -----
? 10
WRONG, TRY AGAIN.

     7
  +  2
  -----
? 9
CORRECT !!
HERE'S ANOTHER...

     9
  -  3
  -----
? 6
CORRECT !!
HERE'S ANOTHER...

     7
  +  5
  -----
? 12
CORRECT !!
HERE'S ANOTHER...

     7
  -  4
  -----
? 3
CORRECT !!
```

```
     2
  +  0
  -----
? 2
CORRECT !!
HERE'S ANOTHER...

     1
  +  7
  -----
? 8
CORRECT !!
HERE'S ANOTHER...

     1
  +  7
  -----
? 8
CORRECT !!
HERE'S ANOTHER...

     6
  +  2
  -----
? 9
WRONG, TRY AGAIN.

     6                 ← Two chances to
  +  2                   get the correct
  -----                  answer seems about
? 10                     right with young children.
YOU MISSED THAT ONE TWICE.
THE CORRECT ANSWER IS 8
HERE'S ANOTHER...

     9
  +  1
  -----
? 10
CORRECT !!
HERE'S ANOTHER...

     7
  +  8
  -----
? STOP
PROGRAM HALTED
```

Example 1 generates and presents vertical addition problems. It doesn't keep score, it doesn't use cursor addressing, it doesn't have timing, it doesn't even keep columns of numbers lined up, but it's a starting point. And, incidentally *it is useful.* Children are incredibly adaptable and it's frequently easier to get a child to accept a less-than-beautiful format on the computer than to go through the programming gyrations to get everything "just right." The *really important* reasons that CAI is so successful is that it is personal, it is self-paced, it is not critical (in an ego deflating or destructive way), and it is infinitely patient. All these factors are present in Example 1 even though it lacks the niceties of more sophisticated programs.

Notice the following features:

• *Problem difficulty.* This is set in Statement 20. Currently the number range is between 0 and 9. N determines the upper range of numbers used in problems.

• *Number of trials allowed.* Statement 230 counts the number of times a problem is gotten wrong. Statement 240 allows two trials; if you wish to allow 3 trials before giving the correct answer, then Statement 240 should be IF W 2 THEN 270.

By adding 10 statements to Example 1 (see Example 2) we can present addition and subtraction problems alternately. Statement 60 is a problem counter; Statement 120 branches to subtraction problems on even numbers. Statements 70 through 90 simply make sure that a smaller number is being subtracted from a larger one (not necessary, of course, if the student understands the concept of negative numbers).

Example 3, for multiplication problems adds two additional features not in Examples 1 or 2:

• *Personal feedback.* The child's name, input in Statement 10, is used liberally in comments throughout the exercise (Statements 114, 145, 220).

• *Scoring.* Variable R counts the number of problems right on the first or second trial and Statements 210-220 compute the total score.

# EXAMPLE 3

```
2 RANDOMIZE
5 PRINT "MULTIPLICATION PRACTICE."
10 PRINT "YOUR NAME";
12 INPUT A$
14 PRINT\PRINT "HI ";A$;". TO STOP, TYPE 999 FOR YOUR ANSWER."
15 R=0\P=-1
20 P=P+1
25 Q=0
30 A=INT(12*RND(0)+1)
40 B=INT(10*RND(0)+1)
50 PRINT
60 PRINT "      ";A
70 PRINT "   X  ";B
80 PRINT "   ------"
90 PRINT "  ";
100 INPUT G
110 IF G=A*B THEN 140
111 IF G=999 THEN 200
112 IF Q<1 THEN 122
114 PRINT "YOU MISSED THAT ONE TWICE,";A$
115 PRINT "THE CORRECT ANSWER IS ";A*B
117 PRINT
120 GOTO 20
122 Q=Q+1
125 PRINT "NO. TRY AGAIN."
130 GOTO 50
140 R=R+1
145 PRINT "RIGHT, ";A$
150 GOTO 20
200 PRINT
210 PRINT "YOU GOT";R;"RIGHT OUT OF";P;"PROBLEMS."
215 S=INT(100*R/P)
220 PRINT "SCORE IS";S;"PERCENT, ";A$
230 END
```

```
MULTIPLICATION PRACTICE.
YOUR NAME? DEREK

HI DEREK. TO STOP, TYPE 999 FOR YOUR ANSWER.

       9
    X  7
    ------
  ? 54
NO. TRY AGAIN.

       9
    X  7
    ------
  ? 64
YOU MISSED THAT ONE TWICE,DEREK
THE CORRECT ANSWER IS  63

       2
    X  3
    ------
  ? 6
RIGHT, DEREK

      11
    X  7
    ------
  ? 77
RIGHT, DEREK

       5
    X  4
    ------
  ? 20
RIGHT, DEREK

      10
    X  8
    ------
  ? 80
RIGHT, DEREK

       4
    X  6
    ------
  ? 24
RIGHT, DEREK

       2
    X  3
    ------
  ? 6
RIGHT, DEREK

       1
    X  1
    ------
  ? 999

YOU GOT 6 RIGHT OUT OF 7 PROBLEMS.
SCORE IS 85 PERCENT, DEREK
```

Let's now take a bigger jump to Example 4 which presents 9 different types of horizontal and vertical addition and subtraction problems Starting with the same basics, we've added some additional features:

• *Digit alignment* in vertical problems. Statements 114, 115, and 401 determine how many spaces to tab over (Statements 210, 220, etc.) so that the digits are right justified.

• *Different reinforcement messages.* Problem counter Y coupled with Statements 750-795 alternates between 4 reinforcement messages. More could be used, of course.

Notice that scoring is not in this program. Scoring is most valuable when it is an internal variable used to alter the ratio of different types of problems in response to what the child is getting right or wrong. However many children feel threatened by scores (like grades) so it may not be desirable to print it out.

Next issue we'll look at how the scores on different types of problems can be used to vary the ratio of problem types presented and we'll also look at keeping records from one session to the next.

# EXAMPLE 4

```
10 RANDOMIZE
20 N=15
30 PRINT "HI. WHAT'S YOUR NAME";
40 INPUT A$
50 PRINT
60 PRINT "OK, ",A$,", WE'RE GOING TO DO SOME ARITHMETIC PROBLEMS."
80 E=0
85 Y=0
90 FOR P=1 TO 18
100 A=INT(N*RND(0)+1)
105 B=INT(N*RND(0)+1)
110 IF A>B THEN 114
111 D=B
112 B=A
113 A=D
114 S=3-INT(LOG(A)/2.302585+1)
115 T=2-INT(LOG(B)/2.302585+1)
120 Q=P
130 IF P<10 THEN 150
140 Q=P-9
150 ON Q GOTO 200,250,300,350,400,450,500,550,600
200 R=A+B
210 PRINT TAB(S);A
220 PRINT TAB(T);"+";B
225 PRINT "------"
230 INPUT G
235 GOSUB 700
240 IF E>0 THEN 210
245 GOTO 680
250 R=A-B
260 PRINT A;" - ";B;" = ";
270 INPUT G
280 GOSUB 700
290 IF E>0 THEN 260
295 GOTO 680
```

```
300 R=A-B
310 PRINT TAB(T)" "B
320 PRINT "+"
325 PRINT "------"
330 PRINT TAB(S);A,,
332 INPUT G
335 GOSUB 700
340 IF E>0 THEN 310
345 GOTO 680
350 R=A-B
360 PRINT A" - "" = ";B,,
370 INPUT G
380 GOSUB 700
390 IF E>0 THEN 360
395 GOTO 680
400 C=INT(N*RND(0)+1)
401 U=3-INT(LOG(C)/2.302585+1)
402 R=A+B+C
410 PRINT TAB(S);A
415 PRINT TAB(U);C
420 PRINT TAB(T);"+";B
425 PRINT "------"
430 INPUT G
435 GOSUB 700
440 IF E>0 THEN 410
445 GOTO 680
450 R=A+B
460 PRINT A" + "B" = ";
470 INPUT G
480 GOSUB 700
490 IF E>0 THEN 460
495 GOTO 680
500 R=A-B
510 PRINT TAB(S);A
520 PRINT TAB(T)" -"B
525 PRINT "------"
530 INPUT G
535 GOSUB 700
540 IF E>0 THEN 510
545 GOTO 680
550 R=A-B
560 PRINT B" + "" = "A,,
570 INPUT G
580 GOSUB 700
590 IF E>0 THEN 560
595 GOTO 680
600 R=A-B
610 PRINT TAB(S);A
615 PRINT "-"
620 PRINT "------"
625 PRINT TAB(T)" "B,,
630 INPUT G
635 GOSUB 700
640 IF E>0 THEN 610
645 GOTO 680
680 NEXT P
690 GOTO 900
700 IF G=R THEN 750
705 E=E+1
710 IF E>2 THEN 800
720 PRINT "WRONG.  TRY AGAIN."
725 PRINT
730 RETURN
750 Y=Y+1
752 E=0
755 ON Y GOTO 760,770,780,790
760 PRINT "VERY GOOD "A$
765 GOTO 725
770 PRINT "SUPER !"
775 GOTO 725
780 PRINT "THAT'S RIGHT "A$
785 GOTO 725
790 PRINT "CORRECT !"
792 Y=0
795 GOTO 725
800 PRINT "YOU MISSED THAT ONE 3 TIMES "A$"."
805 PRINT "THE CORRECT ANSWER IS "R"."
810 PRINT "HERE'S ANOTHER PROBLEM."
815 E=0
820 GOTO 725
900 PRINT
910 PRINT "THAT WAS LOTS OF FUN "A$"."
920 PRINT "DO YOU WANT ANY MORE PROBLEMS TODAY (YES OR NO)";
930 INPUT B$
940 IF B$="YES" THEN 85
950 IF B$="NO" THEN 960
953 PRINT "PLEASE ANSWER 'YES' OR 'NO'."
955 GOTO 920
960 PRINT
970 PRINT "OK.  GOODBYE FOR NOW "A$".  PLEASE TYPE 'BYE' AND"
980 PRINT "HANG UP THE PHONE.  THANKS."
999 END
```

- 310–330: Vertical addition (type 2)  →  `  2` / `+ ` / `---` / ` 7`
- 360: Horizontal subtraction (type 2)  →  `8 - __ = 5`
- 410–425: Vertical addition  →  `10` / ` 3` / `10`
- 460: Horizontal addition  →  `15 + 1 = __`
- 510–525: Vertical subtraction  →  ` 15` / `-15`
- 560: Horizontal addition (type 2)  →  `2 + __ = 11`
- 610–625: Vertical subtraction (type 2)  →  `10` / `- ` / `--` / `10`
- 700: Subroutine to check answer (G) against correct one (R). E counts the number of incorrect trials.
- 760–790: Alternates between 4 reinforcement messages.

---

```
HI.  WHAT'S YOUR NAME? DETTA

OK, DETTA, WE'RE GOING TO DO SOME ARITHMETIC
                                        PROBLEMS.
  11
+  2
------
? 13
VERY GOOD DETTA

  12  -  4  = ? 8
SUPER !

   2
+
------
   7                                         ? 5
THAT'S RIGHT DETTA

  8 -       = 5                              ? 3
CORRECT !

  10
   3
+ 10
------
? 23
VERY GOOD DETTA

  15 + 1 = ? 16
SUPER !

  15
- 15
------
? 1
WRONG.  TRY AGAIN.

  15
- 15
------
? 10
WRONG.  TRY AGAIN.

  15
- 15
------
? 20
YOU MISSED THAT ONE 3 TIMES DETTA.
THE CORRECT ANSWER IS  0 .
HERE'S ANOTHER PROBLEM.

  2 +       = 11                             ? 9
THAT'S RIGHT DETTA

  10
-
------
  10                                         ? 1
WRONG.  TRY AGAIN.

  10
-
------
  10                                         ? 0
CORRECT !

  13
+  6
------
? 19
VERY GOOD DETTA

  5 - 1 = ? 4
SUPER !

  10
+
------
  13                                         ? 3
THAT'S RIGHT DETTA

  13 -       = 8                             ? 5
CORRECT !

  12
   5
+  2
------
? 19
VERY GOOD DETTA
```

Without cursor addressing, the answer required here must be put here.

Answers to problems like this must be input as 23, not 3 then 2 which may be what the student is used to.

The second in this five-part series on Computer Assisted Instruction shows how the score can be used to tailor the type and difficulty of problems to the individual student.

# CAI: Structuring the Lesson to the Student

David H. Ahl

In Part 1 of this series we looked at how various types of arithmetic problems can be generated. In this section we'll look at the use of scoring to vary the ratio or type and difficulty of problems.

Let's look at the performance of 3 students on 4 types of problems. We'll assume that each student has completed 100 problems, 25 of each type. Here are the percentage missed (incorrect) of each type of problem.

| | Student | | |
|---|---|---|---|
| Problem Type | 1 | 2 | 3 |
| Vertical addition | 0% | 8% | 24% |
| Horizontal addition | 4 | 12 | 32 |
| Vertical subtraction | 0 | 12 | 36 |
| Horizontal subtraction | 8 | 20 | 48 |

It is obvious that Student 1 is doing quite well—3 problems wrong out of 100. It is equally obvious that Student 3 is having trouble, with 35 out of 100 problems incorrect. Student 2 is somewhere in between.

Now our situation is that we'd like to alter the ratio of problems presented to each student in order to give them additional practice on the types with which they're having trouble. A straightforward way of doing this is to come up with a ratio based on problems missed previously (or in a pre-test). Let's try this for Student 2:

| Problem Type | % Wrong | Ratio | New Distribution of Problems |
|---|---|---|---|
| Vertical addition | 8% | 2 | 16.7% |
| Horizontal addition | 12 | 3 | 25.0 |
| Vertical subtraction | 12 | 3 | 25.0 |
| Horizontal subtraction | 20 | 4 | 33.3 |
| | | 12 | 100.0% |

Good approach? Well, maybe. Except that Student 1 would get 67% horizontal subtraction and 33% horizontal addition and no other types. How then does the student advance to a higher grade-level in vertical subtraction? Clearly, we've overlooked something. But before discussing this, let's consider another factor.

If a student misses 5 out of 5 fraction problems on Monday and then misses 0 out of 5 on Tuesday, what does that mean? Does this mean:

(1) The overall score is 50% and the student needs more practice.

(2) The student learned the concept after a dismal performance Monday and now needs no further practice.

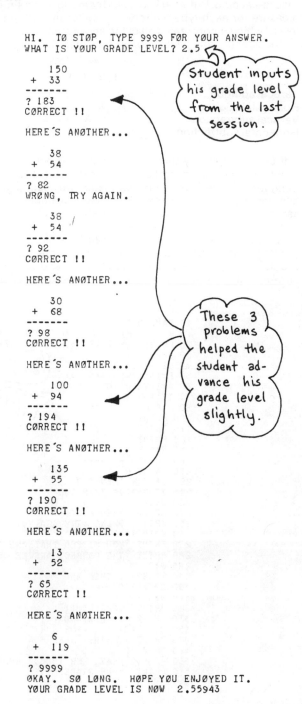

(3) The student lucked out on Tuesday and got a batch of trivial problems.

What would be the explanation if the scoring was reversed; that is 5 correct Monday and 5 missed Tuesday? The point is, and this ties in with the previous scoring situation, that we need a moving-average type of scoring system which meets the following criteria:

(1) It weighs the most recent performance most heavily, but does not ignore previous performance.

(2) It allows a student to advance to more difficult problems than their current mastery level.

(3) It continues to give some minimal practice on problem types the student has already mastered.

(4) It is simple to understand for both student and teacher (or parent or administrator).

If we proceed along traditional lines, we'll have to determine what type of problems a student should be receiving practice in, his score on each in the last session, and his score in the sessions before that—a tricky bit of record-keeping. But what if we could come up with a single measure for each type of problem, say "estimated grade level," which incorporated all of the above measures?

Consider the following "scores" for Derek Carlson:

|  | Grade Level | | | | | |
|---|---|---|---|---|---|---|
|  | 1 | 2 | 3 | 4 | 5 | 6 |
| Vertical addition |  |  | X |  |  |  |
| Horizontal addition |  | X |  |  |  |  |
| Vertical subtraction |  |  |  | X |  |  |
| Horizontal subtraction |  | X |  |  |  |  |

If Derek is halfway through Grade 2, he is behind in 2 problems types, ahead in 1 and on-target with 1. The nice thing for us is we have all the information we need to give him more problems at the "right" level. Naturally this assumes we know what problems of what complexity are being done at every grade level.

So how do we score? Simply by letting the most recent problem done count 10% of the overall score (of that problem type) if the problem was correct and over his current grade level or if it was incorrect and under his current level. The problem is ignored if he got it wrong and it was over his current level, or got it right and it was under his grade level. In other words:

|  | Answer | |
|---|---|---|
|  | Right | Wrong |
| **Problem** Higher than grade level | Raise student grade level | Ignore |
| Lower than grade level | Ignore | Lower student grade level |

At first glance this may look complex and even somewhat goofy, however what it really means is that is that a student is "rewarded" for doing a problem beyond his grade level but he is not penalized if he can't do it. On the other hand he is penalized if he can't do a problem lower than his grade level, but not rewarded for doing one lower.

A flowchart of this process is shown below.

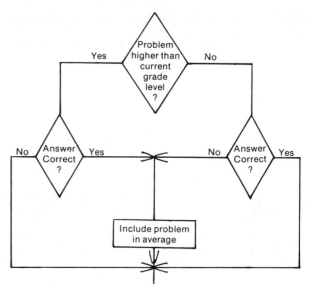

```
10 RANDOMIZE
15 PRINT "HI. TO STOP, TYPE 9999 FOR YOUR ANSWER."
20 PRINT "WHAT IS YOUR GRADE LEVEL";
30 INPUT G1
40 G2=G1-.5+RND(0)
50 R=INT(2*1.73*G2^4)
60 A=INT(100*G2*RND(0))
70 IF A>R THEN 60
80 B=R-A
100 PRINT
110 PRINT "   ";A
120 PRINT " + ";B
140 R=A+B
200 PRINT "-------"
210 INPUT G
220 IF G=R THEN 310
225 IF G=9999 THEN 500
230 W=W+1
240 IF W>1 THEN 270
250 PRINT "WRONG, TRY AGAIN."
260 GOTO 100
270 PRINT "YOU MISSED THAT ONE TWICE."
280 PRINT "THE CORRECT ANSWER IS ";R
285 W=0
290 IF G2>G1 THEN 400
300 GOTO 350
310 W=0
320 PRINT "CORRECT !!"
330 IF G2<G1 THEN 400
350 G1=.9*G1 + .1*SQR(SQR(R/2/1.73))
400 PRINT \ PRINT "HERE'S ANOTHER..."
410 GOTO 40
500 PRINT "OKAY. SO LONG. HOPE YOU ENJOYED IT."
510 PRINT "YOUR GRADE LEVEL IS NOW ";G1
999 END
```

Annotations:
- *G2, the grade level of the problem to be done is ±½ a grade of the student's level, G1.*
- *Statements 50 and 60 calculate the addends of the problem.*
- *Statement 350 gives us a moving grade level average.*

We said that if a problem is to be counted in the student's overall average, it counts 10% of the total. If his current grade level (on a particular problem type) is L and the level of the problem to be averaged in is P, then the averaging formula is simply:

$$L = .9L + .1P$$

A grade level must be kept for the student for each type of problem that he is doing. So if a student is dealing with 9 different math concepts (or standards), he is assigned a grade level for each one.

Our next task is to assign a grade level to each problem presented. This unfortunately will vary depending upon the local school system, the textbooks used, and teaching method. Since we certainly don't want to store a huge data base of problems tagged with a grade level, for each problem type, we probably should try to devise a simple method of determining its grade level. We must also determine what level should be presented to the student. One straightforward approach is to present problems up to one-half a grade level over and under where the student currently is. Thus the overall range of problems for a student at grade level 3.2 would be 2.7 to 3.7.

How do we generate the right problems? Let's consider one type for now: vertical addition. Say it's first introduced in Grade 1 and continues through Grade 4 (actually 4.9). The simplest problem is 1+1 and most difficult 999+999. In other words:

| Grade Level | Addend |
|---|---|
| 1.0 | 1 |
| 4.9 | 999 |

Clearly, learning is not a linear process so we can't use a simple linear formula, hence we need something that represents the exponential process of learning.

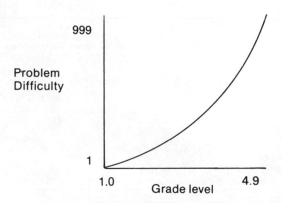

There are lots of exponential representations like logs, powers, etc. A fairly trivial, but workable formula for this problem is:

Addend = 1.73 x Grade Level $^4$

or

Grade level = $\sqrt[4]{\text{Addend}/1.73}$

To fill in a few more values on our tables:

| Grade level | Addend |
|---|---|
| 1.0 | 1 |
| 2.0 | 27 |
| 3.0 | 140 |
| 4.0 | 442 |
| 4.9 | 997 |

Now it's a relatively straightforward, although somewhat tedious matter, to tie all these elements together in a computer program.

A few notes about the program. G2 is a variable that is within one-half a grade level of the current level of the student, G1. The complicated mess in statement 350 determines a fair approximation of grade level based on the answer to the problem. The recording of the grade level and carrying over to the next lesson is a manual process; grade level could just as easily be retained on a file medium such as floppy disk or cassette tape and keyed to the name or number (heaven forbid) of the student. ■

"...T12X31LC, this is A561205,LC4A ... I have to make pee pee..."

# CAI: Further Considerations for Presenting Multiple-Problem Types

Part three in this five-part series on Computer-Assisted Instruction looks at incorporating a sliding grade level into programs that generate several types of problems.

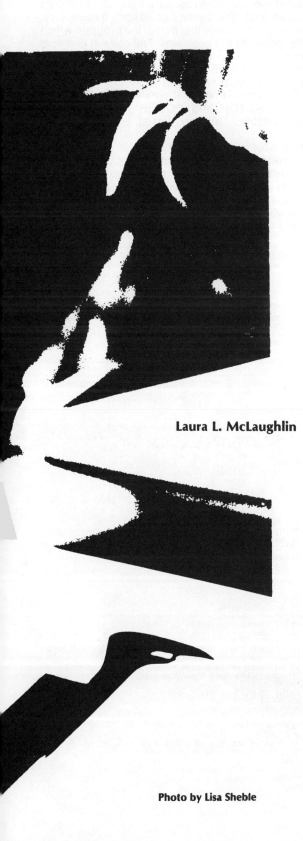

Laura L. McLaughlin

Photo by Lisa Sheble

```
B>RUN24 MATHART
BASIC-E INTERPRETER: VER K1.3

HI! WHAT'S YOUR NAME? TRACY
I GUESS WE'VE NEVER MET BEFORE TRACY
WHAT GRADE ARE YOU IN? 2

OKAY, LET'S PLAY WITH SOME NUMBERS

     1
   + 4
   -----
? 5

THAT'S RIGHT, TRACY

LET'S TRY ANOTHER...

    21
  + 15
  -----
? 36

THAT'S RIGHT, TRACY

LET'S TRY ANOTHER...

    41
  + 15
  -----
? 50

WRONG, TRY AGAIN? 55

WRONG, TRY AGAIN? 56

THAT'S RIGHT, TRACY

LET'S TRY ANOTHER...

    24
  + 14
  -----
? 38

THAT'S RIGHT, TRACY

LET'S TRY ANOTHER...

    54
  + 12
  -----
? 64

WRONG, TRY AGAIN? 54

WRONG, TRY AGAIN? 62

YOU DON'T SEEM TO UNDERSTAND THIS ONE TRACY
THE CORRECT ANSWER IS 66

LET'S TRY ANOTHER...

41 + 7 =
```

EXAMPLE 1

```
10      DEF FNIN$(X)=LEFT$(STR$(X+.5), INT(LOG(X)/2.302585+1))
        DIM G(6)
        NAMES$="NAME.RND"
        NAMES=1
        FILE NAMES$(128)
        IF END #NAMES THEN 300
        INPUT "HI! WHAT'S YOUR NAME"; NAME$
        RANDOMIZE
100     COUNT=COUNT+1
        READ #NAMES; NAMEF$, G(1), G(2), G(3), G(4), G(5), G(6)
        IF NAME$=NAMEF$ THEN 500
        GOTO 100
300     PRINT "I GUESS WE'VE NEVER MET BEFORE "; NAME$
        INPUT "WHAT GRADE ARE YOU IN"; G(1)
        G(1)=G(1)+.3
        FOR I=1 TO 5
        G(I+1) = G(I)
        NEXT I
500     PRINT
        PRINT "OKAY, LET'S PLAY WITH SOME NUMBERS"

7000    PRINT
        PRINT "THAT'S ENOUGH FOR NOW "; NAME$
        PRINT "COME BACK AGAIN SOON"
        PRINT #NAMES, COUNT; NAME$, G(1), G(2), G(3), G(4), G(5), G(6)
9999    END
A>
```

The previous article in this series discussed a method of calculating a sliding grade level based on the range of the numbers presented in a particular problem. But how can we incorporate this concept into a program that handles multiple-problem types and what other factors should we consider?

Let's say we want a program that will generate six types of problems for grades 1-4: addition, subtraction, and multiplication in both the vertical and horizontal formats. We want this program to keep track of a student's grade level for each type of problem separately. We can then give him practice problems, in each area, that fall within a half-grade on either side of his current level for that particular type of problem. The program should constantly adjust the student's grade levels based on the accuracy of his responses.

We would also like the program to be both flexible and concise. Flexibility is important so that we can later expand and/or modify the types of problems to be presented or the grade levels to be covered. The need for conciseness is evident to all (except maybe those with their own personal IBM 370). To accomplish this, we are going to make extensive use of the BASIC FOR/NEXT and GOSUB verbs.

But before we can generate any problems, we must have a means of obtaining the student's grade levels. Since we are now talking about six different levels based on problem type and format, it is no longer practical to expect the student to enter his grade level data. Therefore, we are going to build a file with one record for each student. It will contain his name and a level for each type of problem. Example 1 shows the beginning and end of our program.

(1) At 100, we search through the file for the student's name. When a match is found we can proceed with the session, since we have his current levels.

(2) If there is no record for a student, we must ask him what grade he is in. This will provide us with initial values (see Line 300).

(3) Since his true level is probably somewhere between what he answers and the next higher grade, we will set our initial values up by three-tenths. Remember, the program will adjust these numbers in either direction based on his ability, so the initial grade levels are not absolute.

(4) The end of the program (at 7000) either updates or creates his record with the levels that have been adjusted during this session.

Just a word about the function (FNIN$) defined at the beginning of the program. As you may know, each of the different BASICs available has its own "quirks" (or, more accurately, problems). This program was developed using BASIC-E Version K1.4. One of its "quirks" is that on occasion it will incorrectly convert a numeric integer into a string variable as a non-integer number (for example, 93 becomes "93.00001"). The function FNIN$ is used to correct this problem. When writing a CAI program, it is essential that all such inaccuracies be identified and corrected or it will result in confusing, rather than helping, the student. Check your BASIC, nothing is perfect.

Now, what kinds of common subroutines can we use? Well, Example 2 shows four of the more obvious ones. The routine at 4000 will set up to three random numbers based on the value of G1 (the student's grade level). The formula for determining the maximum value of any one of the numbers is the same as used in the previous article for addition problems (Addend = 1.73 x Grade Level^4.)* However, we have provided the formula with greater flexibility by making the values for both FACTOR and POWER variable. This way they may be changed based on problem type. Note that the value of NUM must have been previously set for both this routine and the two that follow. Routines for printing a problem vertically or horizontally are shown at 5000 and 5500 respectively, with the type of problem determined by the value of OP$. The last routine, at 6000, will get the student's answer and process it. Three incorrect responses are accepted prior to showing the right answer. Note that this could be easily changed to vary depending on grade level, if we later determined that allowing only two mistakes would be more appropriate for the younger student. We then decide if the problem should effect his grade level and do the calculations if necessary.

We are now ready to write that part of the program which will actually generate the problems. The code in Example 3 will generate addition (1000), subtraction (2000), and multiplication (3000) problems. Let's look at it a little more closely and see what it really does:

(1) The values for G1 (student's grade level), OP$ (type of problem), FACTOR and POWER (parameters for calculating variables) are set based on the type of problem.

(2) Five problems of each type are presented in succession (with the exception that no multiplications are generated if his grade level for horizontal multiplication is below 2.5). By presenting the same math concept multiple times in succession, while at the same time limiting the number, we will give him the opportunity to learn from his mistakes but decrease the possibility of boredom or

```
                                EXAMPLE 2

4000    G2=G1-.5+RND
        MAX=INT(FACTOR*G2^POWER)
        A=INT(RND*MAX)
        B=INT(RND*MAX)
        IF NUM=2 THEN C=0:RETURN
        C=INT(RND*MAX)
        RETURN

5000    PRINT
        PRINT "   ";
        IF A=0 THEN PRINT A ELSE PRINT FNIN$(A)
        IF NUM=2THEN 5100
        PRINT "   ";
        IF C=0 THEN PRINT C ELSE PRINT FNIN$(C)
5100    PRINT " ";OP$;" ";
        IF B=0 THEN PRINT B ELSE PRINT FNIN$(B)
        PRINT "------"
        RETURN

5500    PRINT
        IF A=0 THEN PRINT A; ELSE PRINT FNIN$(A);
        PRINT " ";OP$;" ";
        IF NUM = 2 THEN 5600
        IF C=0 THEN PRINT C; ELSE PRINT FNIN$(C);
        PRINT " ";OP$;" ";
5600    IF B=0 THEN PRINT B ELSE PRINT FNIN$(B);
        PRINT " = "
        RETURN

6000    INPUT ANSWER
        IF ANSWER = R THEN 6200
        W=W+1
        IF W>2 THEN 6100
        PRINT
        PRINT "WRONG, TRY AGAIN";
        GOTO 6000
6100    PRINT
        PRINT "YOU DON'T SEEM TO UNDERSTAND THIS ONE ";NAME$
        PRINT "THE CORRECT ANSWER IS ";
        IF R=0 THEN PRINT R ELSE PRINT FNIN$(R)
        IF G2>G1 THEN GOTO 6300 ELSE GOTO 6250
6200    PRINT
        PRINT "THAT'S RIGHT, ";NAME$
        IF G2<G1 THEN 6300
6250    MAX=SQR(MAX/FACTOR)
        IF POWER = 4 THEN MAX=SQR(MAX)
        G1=.9*G1+.1*MAX
6300    W=0
        IF OP$="-" AND G(5)<2.5 AND I=10 THEN RETURN
        IF OP$="X" AND I=10 THEN RETURN
        PRINT
        PRINT "LET'S TRY ANOTHER..."
        RETURN

A>
```

---

*Some terminals use this "upside-down saucer" to indicate exponentiation, rather than an up-arrow.

choice of coding technique should be based upon the extent of modification envisioned to achieve the final iteration.

The addition and subtraction loops also call two routines we have not yet defined. Take a look at Example 4. The routine at 4500 is called from addition and subtraction only for problems of that below a particular level. It will insure that addition problems will not require a "carry" if both numbers are greater than 9, and that subtractions will not need a "borrow." This kind of special editing (independent of the range of the variables) must be considered for each type of math concept presented. Otherwise, we run the risk of completely frustrating the student by expecting him to attempt something which could well be far beyond his capabilities. Similarly, since we do not want to give multiplication too soon, the routine at 6500 is used to set the student's grade level to a value that will allow the generation of multiplication problems only after he has reached a particular level (grade 3) in one of the other types.

Now that we have a basis from which to work, it will be relatively easy to expand upon. Subroutines can be written for generating different formats (mixed operations, fractions) or more operations (division, square roots). Also, additional checks can be included to give a student a larger variety of problems and eliminate those which he has already mastered.

In the next article, we will discuss some of the considerations that should be made concerning the interaction between the student and the computer. Careful thought must be given to such things as edit requirements and presentation formats so that the student does not need to learn a whole new set of rules. Remember, for CAI to be a useful tool, it must be something that is easy and comfortable to use. ∎

frustration.

(3) The value of NUM (number of variables) is set for addition based on the grade level of the problem (G2), while for subtraction and multiplication it is set to 2.

(4) When done with a particular problem type, the student's grade level for that type is updated.

We could have put the generation of all three problem types into one large FOR-NEXT loop. This approach would have saved us close to 50% of the code necessary to produce the problems, but would have made modifications and/or additions much more difficult (the old ease of maintenance/size & efficiency trade-off). The

```
                            EXAMPLE 3
    1000    G1 = G(1)
            OP$ = "+"
            FACTOR = 1.73:POWER = 4
            FOR I = 1 TO 10
            IF G1<4 THEN NUM=2 ELSE NUM=3
            GOSUB 4000
            IF G2<2.5 AND I<6 THEN GOSUB 4500
            IF G2<3 AND I>5 THEN GOSUB 4500
            R=A+B+C
            IF I<6 THEN GOSUB 5000 ELSE GOSUB 5500
            GOSUB 6000
            IF I=5 THEN G(1)=G1:G1=G(2)
            NEXT I
            G(2)=G1
            GOSUB 6500

    2000    G1=G(3)
            OP$="-":NUM=2
            FOR I=1 TO 10
            GOSUB 4000
            IF A<B THEN X=A:A=B:B=X
            IF G2<3 AND I<6 THEN GOSUB 4500
            IF G2<3.5 AND I>5 THEN GOSUB 4500
            R=A-B
            IF I<6 THEN GOSUB 5000 ELSE GOSUB 5500
            GOSUB 6000
            IF I=5 THEN G(3)=G1:G1=G(4)
            NEXT I
            G(4)=G1
            GOSUB 6500

    3000    G1=G(5)
            IF G1<2.5 THEN 7000
            OP$="X"
            FACTOR= 68:POWER=2
            FOR I=1 TO 10
            GOSUB 4000
            R=A*B
            IF I<6 THEN GOSUB 5000 ELSE GOSUB 5500
            GOSUB 6000
            IF I=5 THEN G(5)=G1:G1=G(6)
            NEXT I
            G(6)=G1
            GOTO 7000
    A>
```

```
                            EXAMPLE 4
    4500    IF A<10 OR B<10 THEN RETURN
            TSTA$=RIGHT$(FNIN$(A),1)
            TSTB$=RIGHT$(FNIN$(B),1)
            TSTA=VAL(TSTA$)
            TSTB=VAL(TSTB$)
            IF OP$="-" THEN 4600
            IF TSTA + TSTB < 10 THEN RETURN
            IF TSTA > 4 THEN A=A-5
            IF TSTB > 5 THEN B=B-5
            RETURN
    4600    IF TSTA - TSTB >= 0 THEN RETURN
            IF TSTA < 5 THEN A=A+5
            IF TSTB > 5 THEN B=B-5
            RETURN

    6500    IF G(5) >= 3 THEN RETURN
            IF G1<3 THEN RETURN
            G(5)=3
            G(6)=3
            RETURN
    A>
```

*"Now hear this! I am the programmer. You are the programmee!"*

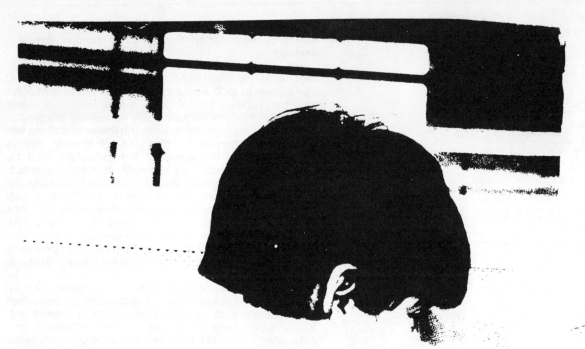

# CAI: Interaction Between Student and Computer

**Part four in this series on Computer Assisted Instruction investigates "some of the more subtle areas that deal with how the student and computer interact."**

Laura L. McLaughlin

The previous article in this series (Nov-Dec 1977, p. 74) gave us an example of a program that would maintain sliding grade levels for a student by problem type, both within a session and from one session to the next. This is a very important concept for good CAI. Now, however, we're going to take a closer look at some of the more subtle areas that deal with how the student and computer interact.

For instance, consider my son Jeff's initial reaction when he sat down at the terminal to try out the program. His first problem was a vertical addition with three numbers of three digits each. "If I could put this in the right direction, it would be easy!", were the first words out of his mouth. And he had a very valid point. Why should he have to put his answer in from left to right just because he's working with a computer?

Another occurrence, in many BASIC programs, is that non-numeric input

---

Laura L. McLaughlin, Computer Mart of Pennsylvania, 550 DeKalb Pike, King of Prussia, PA 19406.

will either cause the program to abort and return to BASIC (worst case) or generate a non-specific error message which could leave the student confused. Why can't it simply state that what he entered was not a number and it would like him to put his answer in again?

The solution to these problems is simple. The program could do these things — it's just a matter of someone deciding they are important enough to spend some additional programming effort to overcome them (of course, the use of a CRT with cursor control instead of a Teletype might make life a little easier).

Let's consider what kinds of routines we might be able to add to a CAI program to make the interaction between student and computer as smooth as possible, from the student's point of view.

(1) *Editing of input.* This would catch any non-numeric or otherwise invalid input (like, let's not assume on a yes/no answer that if it wasn't a "yes" it was a "no"). We could also give the student an out if he is presented with a problem totally beyond his ability by

```
WE ARE GOING TO HELP YOU PRACTICE MULTIPLICATION
WHAT IS YOUR NAME? JEFF

IF YOU DON'T UNDERSTAND A PROBLEM, OR IT IS TOO DIFFICULT,
JUST TYPE RETURN TO GO ON TO THE NEXT ONE -
I WILL SHOW YOU THE ANSWER AND GIVE YOU A CHANCE TO STUDY
IT SO THAT YOU MIGHT BE ABLE TO GET IT THE NEXT TIME

OKAY, JEFF LET'S GET TO IT
THE FIRST SET OF PROBLEMS WILL BE SIMILAR TO THIS:

                    727
                   X  3
                   ----
                   2181

YOU HAVE YOUR CHOICE OF ENTERING YOUR ANSWERS FROM
LEFT TO RIGHT (IF YOU WANT TO FIGURE IT OUT IN YOUR HEAD)
OR FROM RIGHT TO LEFT (IF YOU'D RATHER WORK IT OUT STEP
BY STEP).

WOULD YOU LIKE TO ENTER FROM RIGHT TO LEFT
WORKING IT OUT STEP BY STEP? ■
```

accepting a CR (carriage return) only as a statement that he has given up. Frustration can be very negative and should therefore be avoided.

(2) *Answer positioning.* In many cases the student should have the option of entering his answer in the direction that is easiest for him. We could also, in certain cases, allow him to work out the entire problem (including any intermediate steps in long multiplications and divisions, for example). Although at times, for various reasons, we might want to specify these things ourselves.

(3) *Timing routines.* The length of time it takes for a response to be entered can be very important. Some students are reluctant to admit they do not understand and will therefore not use the "out" we have given them. So maybe, after a certain period of time, we should prompt them with a message. Another use for a timing routine would be to provide speed drills for things like multiplication tables (automatic flash-cards).

The program included in this article, MATHMULT, which is designed to give a student practice in multiplication, incorporates all of these ideas. Let's take a look at it and see how it is done. Three groups of problems are given to the student, with the code for each starting at lines 100, 200, and 300, respectively.

The first set consists of twelve problems that require multiplying a number up to three digits long by a one-digit number. Two screens are presented with six problems on each of them. The student is given a choice between entering his answer from left to right or from right to left. For this level of problem, the option is given to him since for some students it would be easier to do the calculation in their heads while for others it would not. The program will keep track of the time it takes for the student to enter his response and give him a prompting message if he is taking an excessive amount of time. The length of time will be greater if he has elected to enter his answer from left to right, since he must do the entire calculation in his head. The input will be edited to insure that it is numeric. Invalid data is captured immediately on input so it is not even put out to the screen. Instead, the BELL is sounded to indicate that the character entered was not accepted. When he has completed all twelve problems, an appropriate message will be written to indicate how well he did.

Next the student will be given 52 "flash-card" type problems. They will all be put on one screen, and he will have only a few seconds to enter each response. On these problems, as opposed to the previous ones, if an answer is incorrect, no opportunity is

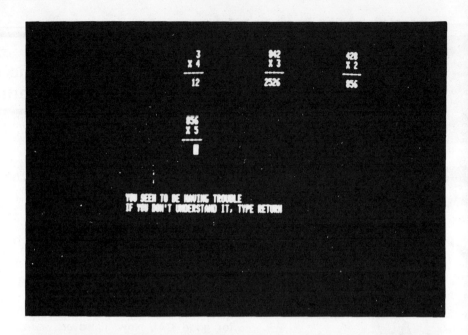

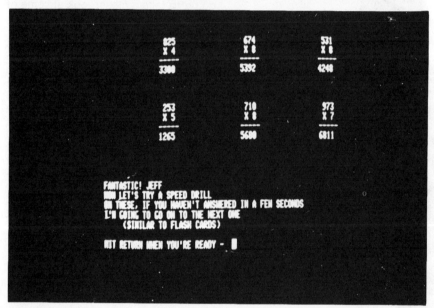

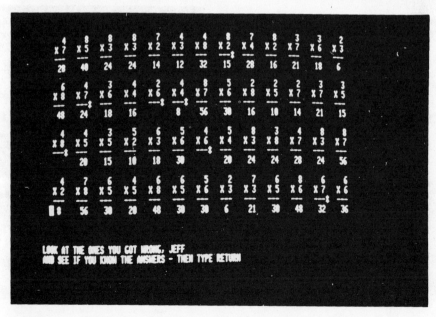

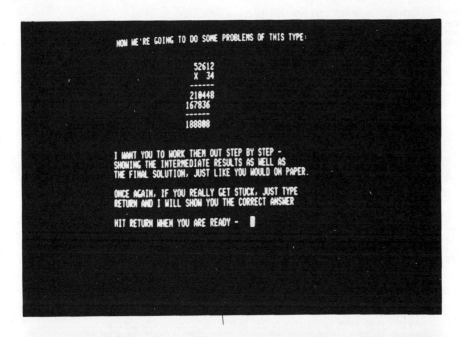

given to re-enter the data since the whole idea is speed. When he is done he will be shown which ones he missed and be given a chance to look them over to see if he knows the answers. When he is ready to go on, the corrected answers will be displayed. This is very important for two reasons: (1) if he simply did not know the right answer it will give him a chance to study it, and (2) it will reinforce his knowledge if he just couldn't get it in the amount of time allowed.

The final set of problems presented will be "long" multiplications: a multiple-digit number times another multiple-digit number. These problems require that the student enter his answers from right to left because we want him to work them out just like he would on paper, including intermediate results. Note that all answers, final as well as intermediate ones, should be edited for non-numerics. In this example, all responses are also checked for a CR only, indicating that the student gave up, and for validity. By providing him with the correct answer on an intermediate result which he cannot get himself, he still has the opportunity to finish the rest of the problem. In other words, all three entries are treated as individual problems, except that only the final answer is actually counted as right or wrong. There are obviously a number of ways to handle this depending on the level of the student. This program is set up for a relative newcomer to this type of problem, and therefore gives him a fair amount of leeway so he has the chance to complete the problems. Remember, this is a practice session, not a test of his knowledge.

This gives us an overall view of what the program is trying to accomplish. It also explains some of the things included for each problem type that will make the student encounter the least amount of frustration in his interaction with the machine. Now let's look at a few of the more general techniques that are incorporated to further increase the student's feeling of comfort with the tool he is using. Many of these points were discussed in the first article of this series. However, to emphasize their importance, we will mention them again. Briefly, some of the considerations should be:

(1) Alignment of numbers
(2) Use of the student's name
(3) Random responses
(4) Informational messages
(5) Variety of layout

A student can be easily confused if the numbers in a problem are not lined up correctly, causing him to make errors on things he would otherwise be able to do correctly. This program makes use of the PRINT USING statement to insure the alignment of numbers, a very convenient method if your BASIC has this statement. If it does not, another means for right justification was shown in the previous article.

Personal feedback in the form of using the student's name is very effective. But remember that anything can be overdone. When you are talking to someone, you do not append their name to every statement. Therefore, the program uses the name somewhat selectively.

Which brings us to the use of random responses. Repetition of the same reply can be very annoying. So the program has been set up with a variety of both positive and negative responses which can be selected from randomly, a much better approach than constantly repeating ourselves.

A liberal number of informational messages is displayed throughout a session. This serves a dual purpose: beyond providing the student with an explanation, they tend to break up the session, giving him a chance to take a breath.

There is no reason to always have problems presented in the same format. Using a CRT, why couldn't we sometimes use the center of the screen, or put up multiple problems at the same time. With this in mind, each group of problems in the program is presented with a different layout on the screen. This avoids the monotony of just displaying one problem after another.

This sample program has shown us how many of the important considerations for a good CAI session can be programmed relatively easily. There are, of course, many variations and extensions of these ideas that could be used. For instance, in the "flash-card" section we might want to allow the student a second chance to go back and correct his own mistakes. Or perhaps we'd want to add the code necessary to eliminate duplicate problems.

Ideally, we would like to combine the sliding-grade concept discussed in the previous article with everything we have considered in this one. Moreover, when we did this, we might also like to use the timing routine to supply response time information as further input to our sliding-grade calculations.

So far, in this series, we have covered much of what is needed to develop a good math drill and practice CAI session. Next we are going to look at some other possibilities for CAI that will still follow the same concepts but which are in areas outside of mathematics. Spelling, grammar, and vocabulary all lend themselves to CAI. And there is no reason why the study of science, social studies, and other subjects cannot be aided by this tool. ∎

```
REM     MATHMULT - A MULTIPLICATION DRILL & PRACTICE SESSION
REM           WRITTEN BY LAURA L MCLAUGHLIN   DECEMBER 1977
REM                   COMPUTER MART OF PENNSYLVANIA
REM              USING CBASIC VERSION 1.0
REM                FOR AN ADM-3A WITH CURSOR CONTROL ENABLED
10      DIM FLASH(4,13)
        FOR I = 1 TO 4
        FOR J = 1 TO 13
        FLASH(I,J)=-1
        NEXT J,I
        CLEAR=26:CR=13:LF=10:BACKSP=8
        ESC=27:DEL=127:BELL=7
        BLANK$="                                                        "
        PRINT CHR$(CLEAR);"WE ARE GOING TO HELP YOU PRACTICE MULTIPLICATION"
        INPUT "WHAT IS YOUR NAME? "; LINE NAME$
        RANDOMIZE
        PRINT
        PRINT "IF YOU DON'T UNDERSTAND A PROBLEM, OR IT IS TOO DIFFICULT,"
        PRINT "JUST TYPE RETURN TO GO ON TO THE NEXT ONE -"
        PRINT "  I WILL SHOW YOU THE ANSWER AND GIVE YOU A CHANCE TO STUDY"
        PRINT "  IT SO THAT YOU MIGHT BE ABLE TO GET IT THE NEXT TIME"
        PRINT
        PRINT "OKAY, ";NAME$;" LET'S GET TO IT"
REM     GROUP 1: 12 PROBLEMS OF 3 DIGITS BY 1 DIGIT - SIX TO A SCREEN
REM             STUDENT OPTION ON DIRECTION OF ENTRY
REM             TWO RETRIES ON INCORRECT ANSWER
100     PRINT "THE FIRST SET OF PROBLEMS WILL BE SIMILAR TO THIS:"
        PRINT
        PRINT TAB(26);"727"
        PRINT TAB(26);"X 3"
        PRINT TAB(25);"-----"
        PRINT TAB(25);"2181"
        PRINT
        PRINT "YOU HAVE YOUR CHOICE OF ENTERING YOUR ANSWERS FROM"
        PRINT "LEFT TO RIGHT (IF YOU WANT TO FIGURE IT OUT IN YOUR HEAD)"
        PRINT "OR FROM RIGHT TO LEFT (IF YOU'D RATHER WORK IT OUT STEP"
        PRINT "BY STEP)."
        PRINT
        PRINT "WOULD YOU LIKE TO ENTER FROM RIGHT TO LEFT"
        INPUT "WORKING IT OUT STEP BY STEP? ";A$
100.1   IF LEFT$(A$,1)="Y" THEN DIREC$="RL":LIMIT=1500:GOTO 100.3
        IF LEFT$(A$,1)="N" THEN DIREC$="LR":LIMIT=3000:GOTO 100.3
        PRINT "I DON'T UNDERSTAND, PLEASE ANSWER YES OR NO ";
        INPUT A$
        GOTO 100.1
100.3   TYPE$="REG":NUM1=998:NUM2=7:NUM3=2
        FOR I = 1 TO 2
        X=15:Y=2
        PRINT CHR$(CLEAR);
        FOR J = 1 TO 2
        FOR K = 1 TO 3
        GOSUB 900.1
        Y1=Y:X1=X+1:GOSUB 900.5
        PRINT USING "###";A
        Y1=Y+1:X1=X+1:GOSUB 900.5
        PRINT USING "X #";B
        Y1=Y+2:X1=X:GOSUB 900.5:PRINT "-----"
        Y1=Y+3:WRONG=0
        IF DIREC$="LR" THEN X1=X ELSE X1=X+3
        GOSUB 400
        X=X+20
        NEXT K
        Y=Y+7:X=15
        NEXT J,I
        GOSUB 900.2
        IF RIGHT=12 THEN PRINT "FANTASTIC! ";NAME$:GOTO 200
        IF RIGHT<8 THEN PRINT "LOOKS LIKE YOU COULD USE MORE PRACTICE":GOTO 200
        PRINT "THAT WAS PRETTY GOOD ";NAME$
REM     GROUP 2: 52 PROBLEMS OF 1 DIGIT BY 1 DIGIT - ALL ON ONE SCREEN
REM             ENTRY - LEFT TO RIGHT
REM             TIME OUT AFTER A FEW SECONDS
REM             NO RETRIES ALLOWED - ERRORS & CORRECTIONS SHOWN AT END
200     PRINT "NOW LET'S TRY A SPEED DRILL"
        PRINT "ON THESE, IF YOU HAVEN'T ANSWERED IN A FEW SECONDS"
        PRINT "I'M GOING TO GO ON TO THE NEXT ONE"
        PRINT "     (SIMILAR TO FLASH CARDS)"
        PRINT
        PRINT "HIT RETURN WHEN YOU'RE READY -";:INPUT " ";LINE A$
        DIREC$="LR":TYPE$="FLASH":NUM1=7:NUM3=2:LIMIT=250:RIGHT=0
        Y=1:PRINT CHR$(CLEAR);
        FOR I = 1 TO 4
        X=4
        FOR J = 1 TO 13
        GOSUB 900.1
        Y1=Y:X1=X+1:GOSUB 900.5:PRINT A
        Y1=Y+1:X1=X:GOSUB 900.5:PRINT "X";B
        Y1=Y+2:X1=X:GOSUB 900.5:PRINT "---"
        Y1=Y+3:X1=X+1
        GOSUB 400
        X=X+6
        NEXT J
        Y=Y+5
        NEXT I
        GOSUB 900.2
        IF RIGHT = 52 THEN PRINT "YOU GOT THEM ALL RIGHT!":GOTO 300
REM     LOOP FOR SHOWING MISTAKES
        PRINT "LOOK AT THE ONES YOU GOT WRONG, ";NAME$
        PRINT "AND SEE IF YOU KNOW THE ANSWERS - THEN TYPE RETURN ";
        Y=1
        FOR I = 1 TO 4
        X=4
```

```
                FOR J = 1 TO 13
                IF FLASH(I,J) = -1 THEN 200.5
                Y1=Y+2:X1=X+3:GOSUB 900.5:PRINT "*"
200.5           X=X+6
                NEXT J
                Y=Y+5
                NEXT I
                INPUT " "; LINE A$
                GOSUB 900.2
REM             LOOP FOR SHOWING CORRECT ANSWERS
                PRINT "HERE ARE ALL THE CORRECT ANSWERS"
                PRINT "STUDY THEM, AND THEN HIT RETURN TO GO ON ";NAME$;
                Y=1
                FOR I = 1 TO 4
                X=4
                FOR J = 1 TO 13
                IF FLASH(I,J)= -1 THEN 200.9
                Y1=Y+3:X1=X+1:GOSUB 900.5
                PRINT USING "##";FLASH(I,J)
200.9           X=X+6
                NEXT J
                Y=Y+5:PRINT
                NEXT I
                INPUT " "; LINE A$
REM             GROUP 3: 4 PROBLEMS OF 4 DIGITS BY 2 DIGITS - ONE PER SCREEN
REM                     INTERMEDIATE & FINAL RESULTS TO BE ENTERED
REM                     ENTRY - RIGHT TO LEFT
REM                     TWO RETRIES ON INCORRECT ANSWER FOR BOTH INTERMEDIATES & FINAL
300             PRINT CHR$(CLEAR);
                PRINT "NOW WE'RE GOING TO DO SOME PROBLEMS OF THIS TYPE:"
                PRINT:PRINT
                PRINT TAB(21);"52612"
                PRINT TAB(21);"X  34"
                PRINT TAB(20);"------"
                PRINT TAB(20);"210448"
                PRINT TAB(19);"167836"
                PRINT TAB(19);"-------"
                PRINT TAB(19);"188808"
                PRINT:PRINT
                PRINT "I WANT YOU TO WORK THEM OUT STEP BY STEP -"
                PRINT "SHOWING THE INTERMEDIATE RESULTS AS WELL AS"
                PRINT "THE FINAL SOLUTION, JUST LIKE YOU WOULD ON PAPER."
                PRINT
                PRINT "ONCE AGAIN, IF YOU REALLY GET STUCK, JUST TYPE"
                PRINT "RETURN AND I WILL SHOW YOU THE CORRECT ANSWER"
                PRINT
                PRINT "HIT RETURN WHEN YOU ARE READY - ";
                INPUT " "; LINE A$
                DIREC$="RL":LIMIT=2000:TYPE$="DIFF":NUM1=9988:NUM2=88:NUM3=11
                X=30:Y=5:RIGHT=0
                FOR I = 1 TO 2
                FOR J = 1 TO 2
                GOSUB 900.1
                PRINT CHR$(CLEAR);
                Y1=Y:X1=X:GOSUB 900.5
                PRINT USING "#####";A
                Y1=Y+1:X1=X:GOSUB 900.5
                PRINT USING "X  ##";B
                Y1=Y+2:X1=X-1:GOSUB 900.5:PRINT "------"
                Y1=Y+3:X1=X+4:WRONG=0
                LEVEL=1:TYPE$="INTERMEDIATE":GOSUB 400
                Y1=Y+4:X1=X+3:WRONG=0
                LEVEL=2:GOSUB 400
                Y1=Y+5:X1=X-2:GOSUB 900.5:PRINT "-------"
                Y1=Y+6:X1=X+4:WRONG=0
                LEVEL=3:TYPE$="DIFF":GOSUB 400
                NEXT J,I
                GOSUB 900.2
                IF RIGHT=4 THEN PRINT NAME$;" THAT WAS EXCELLENT":GOTO 300.9
                IF RIGHT>1 THEN PRINT "NOT TOO BAD, ";NAME$:GOTO 300.9
                PRINT "YOU SEEM TO NEED MORE PRACTICE, ";NAME$
REM             END OF SESSION
300.9           PRINT "HOPE YOU COME BACK SOON"
                PRINT "BYE FOR NOW":STOP
REM             VALIDATE ANSWER ROUTINE
REM                     CHECKS FOR NO ANSWER (CR ONLY) FIRST
REM                     THEN CHECKS IF RIGHT OR WRONG (AND COUNTS EACH)
400             GOSUB 900.6:GOSUB 500
                IF LEN(ANSWER$) = 0 THEN 450
                ANSWER=VAL(ANSWER$)
                IF (TYPE$="INTERMEDIATE") AND (LEVEL=1) AND (ANSWER=ANSW1) THEN RETURN
                IF (TYPE$="INTERMEDIATE") AND (LEVEL=2) AND (ANSWER=ANSW2) THEN RETURN
                IF TYPE$="INTERMEDIATE" THEN 400.07
                IF ANSWER <> ANSW THEN 400.07
                RIGHT=RIGHT+1
REM             SELECTS RANDOM RESPONSE FOR CORRECT ANSWER (EXCEPT FOR FLASH CARDS)
REM                     RETURNS TO GET NEXT PROBLEM
                IF TYPE$ = "FLASH" THEN RETURN
                GOSUB 900.2
                ON (RND*5)+1 GOTO 400.051, 400.052, 400.053, 400.054, 400.055, 400.056
400.051 PRINT "THAT'S RIGHT, ";NAME$:GOSUB 900.7:RETURN
400.052 PRINT "VERY GOOD":GOSUB 900.7:RETURN
400.053 PRINT "KEEP UP THE GOOD WORK ";NAME$:GOSUB 900.7:RETURN
400.054 PRINT "YOU GOT IT":GOSUB 900.7:RETURN
400.055 PRINT "PERFECT ";NAME$:GOSUB 900.7:RETURN
400.056 PRINT "THAT'S CORRECT":GOSUB 900.7:RETURN
REM             SELECTS RANDOM RESPONSE FOR WRONG ANSWER (EXCEPT FOR FLASH CARD)
REM                     IF WRONG MORE THAN TWICE BRANCHES TO GIVE ANSWER ROUTINE
REM                     OTHERWISE RETURNS TO GET NEW ANSWER
400.07  IF TYPE$ = "FLASH" THEN FLASH(I,J)=ANSW:RETURN
                WRONG = WRONG + 1
```

```
                    GOSUB 900.2
                    IF WRONG > 2 THEN 450
                    ON (RND*5)+1 GOTO 400.071, 400.072, 400.073, 400.074, 400.075, 400.076
400.071   PRINT "SORRY, ";NAME$;" TRY AGAIN":GOTO 400
400.072   PRINT "THAT'S NOT IT - ONCE MORE, OK":GOTO 400
400.073   PRINT "TRY AGAIN, ";NAME$:GOTO 400
400.074   PRINT "YOU CAN GET IT, TRY AGAIN":GOTO 400
400.075   PRINT "NOPE, ";NAME$;" ONE MORE TIME":GOTO 400
400.076   PRINT "THINK ABOUT THAT AGAIN":GOTO 400
REM       GIVE ANSWER ROUTINE
REM               RETURNS TO GET NEXT PROBLEM
450       IF TYPE$="FLASH" THEN FLASH(I,J)=ANSW: RETURN
          GOSUB 900.2
          PRINT "I GUESS THIS ONE'S A LITTLE TOUGH FOR YOU"
          PRINT "I'LL TELL YOU THE ANSWER AND YOU TRY TO FIGURE IT OUT"
          PRINT "WHEN YOU'RE READY TO CONTINUE, TYPE RETURN "
          IF DIREC$="RL" THEN X1=X1-5
          GOSUB 900.5
          IF TYPE$="INTERMEDIATE" AND LEVEL=1 THEN PRINT USING "######"; ANSW1
          IF TYPE$="INTERMEDIATE" AND LEVEL=2 THEN PRINT USING "######"; ANSW2
          IF TYPE$<>"INTERMEDIATE"  THEN PRINT USING "######"; ANSW
          INPUT " "; LINE A$
          RETURN
REM       GET ANSWER ROUTINE
REM               CHECKS TIME LIMIT & PROMPTS IF EXCEEDED (EXCEPT FOR FLASH CARD)
REM       INPUTS ANSWER DIGIT BY DIGIT MAINTAINING CURSOR POSITIONING
REM               CHECKS FOR CR - TO INDICATE ENTRY COMPLETE
REM               CHECKS FOR DELETE CHAR
REM               CHECKS FOR NON-NUMERIC - DOES NOT PUT IT OUT, BUT RINGS BELL
REM       BUILDS ANSWER IN A WORK STRING
500       WORK$ = ""
          XSAVE=X1
500.01    GOSUB 900.3
          IF TIME = LIMIT AND TYPE$ = "FLASH" THEN 500.09
          IF TIME = LIMIT THEN \
                GOSUB 900.2:\
                PRINT "YOU SEEM TO BE HAVING TROUBLE":\
                PRINT "IF YOU DON'T UNDERSTAND IT, TYPE RETURN":\
                GOSUB 900.6:\
                GOTO 500.01
          DIGIT = (INP(2) AND 127)
          DIGIT$ = CHR$(DIGIT)
          IF DIGIT$=CHR$(CR) THEN PRINT CHR$(LF);CHR$(CR):GOTO 500.09
          IF DIGIT$=CHR$(DEL) OR DIGIT$="_" THEN GOSUB 900.4:GOTO 500.01
          NUMERIC=MATCH("#",DIGIT$,1)
          IF NUMERIC <> 1 THEN DIGIT=BELL:GOSUB 900.8:GOTO 500.01
          DIGIT=ASC(DIGIT$):GOSUB 900.8
          IF DIREC$="LR" THEN WORK$=WORK$+DIGIT$:X1=X1+1:GOTO 500.01
          DIGIT=BACKSP:GOSUB 900.8:WORK$=DIGIT$+WORK$:X1=X1-1
          GOTO 500.01
500.09    IF XSAVE <> X1 THEN X1=XSAVE:ANSWER$ = WORK$ ELSE ANSWER$=""
          RETURN
REM       GET RANDOM NUMBERS ROUTINE
REM               ALSO SETS CORRECT ANSWER VARIABLES
900.1     A=INT(RND*NUM1)+NUM3
          B=INT(RND*NUM2)+NUM3
          ANSW=A*B
          IF TYPE$="REG" OR TYPE$="FLASH" THEN RETURN
          B$=STR$(B)
          B1=VAL(RIGHT$(B$,1)):B2=VAL(LEFT$(B$,1))
          ANSW1=A*B1:ANSW2=A*B2
          RETURN
REM       BLANK BOTTOM OF SCREEN ROUTINE
900.2     IF TYPE$="FLASH" THEN ROW=52 ELSE ROW=48
          COL=32
          PRINT CHR$(ESC);"=";CHR$(ROW);CHR$(COL);
          PRINT BLANK$:PRINT BLANK$:PRINT BLANK$
          PRINT CHR$(ESC);"=";CHR$(ROW);CHR$(COL);
          RETURN
REM       TIMING ROUTINE
REM               CHECKS IF ANY DATA HAS BEEN ENTERED
REM               TIMING OUT WHEN LIMIT REACHED
900.3     TIME=0
900.31    IF (INP(3) AND 2) = 2 THEN RETURN
          TIME=TIME+1
          IF TIME < LIMIT THEN 900.31
          RETURN
REM       DELETE CHAR ROUTINE
REM               DIRECTION OF ANSWER ENTRY ACCOUNTED FOR
900.4     IF DIREC$ = "LR" THEN\
                DIGIT=BACKSP:GOSUB 900.8:\
                DIGIT=ASC(" "):GOSUB 900.8:\
                DIGIT=BACKSP:GOSUB 900.8:\
                WORK$=LEFT$(WORK$,LEN(WORK$)-1):\
                RETURN
          DIGIT=ASC(" "):GOSUB 900.8:GOSUB 900.8
          DIGIT=BACKSP:GOSUB 900.8
          WORK$=RIGHT$(WORK$,LEN(WORK$)-1)
          RETURN
REM       POSITION CURSOR ROUTINE - IF NEXT OUTPUT USES PRINT STATEMENT
REM               Y1 = # LINES DOWN : X1 = # CHARS OVER
900.5     POS$=CHR$(ESC)+"="+CHR$(Y1+31)+CHR$(X1+31)
          PRINT POS$;
          RETURN
REM       POSITION CURSOR ROUTINE - IF NEXT OUTPUT USES OUT CHAR ROUTINE
REM               Y1 = # LINES DOWN : X1 = # CHARS OVER
900.6     DIGIT=ESC:GOSUB 900.8
          DIGIT=ASC("="):GOSUB 900.8
          DIGIT=Y1+31:GOSUB 900.8
          DIGIT=X1+31:GOSUB 900.8
          RETURN
REM       WAIT ROUTINE
REM               PROVIDES DELAY LOOP BEFORE GOING ON TO NEXT PROBLEM
900.7     FOR WAIT = 1 TO 500
          NEXT WAIT
          RETURN
REM       OUT CHAR ROUTINE
REM               PUTS OUT 1 CHAR DIRECTLY TO TERMINAL OUTPUT PORT
900.8     IF (INP(3) AND 4) <> 4 THEN 900.8
          OUT 2,DIGIT
          RETURN
          END
```

# THEOREM-PROVING WITH EUCLID

## Artificially Intelligent CAI in Extended BASIC

Thomas J. Kelanic

EUCLID is a tool for individualizing instruction in high-school geometry courses. With EUCLID, students can work at their own rates, and along their own routes, in constructing proofs. This article gives a brief description of the operation of EUCLID and indicates directions to be taken in its continuing development.

EUCLID is the fourth version of a theorem-proving computer program written by the author since 1972. The first three versions were written under the auspices of Project Solo and Solo-Works at the University of Pittsburgh. The present version was written at Taylor Allderdice High School, beginning in May, 1977. EUCLID consists of two computer program segments called EUCLID and GEOMTREE1, sharing three initially empty random-access files called PROFILE, CONFILE, and PARTRUE. Together, EUCLID and GEOMTREE1 contain 678 lines of Extended BASIC occupying 14K of memory. An additional 9K of memory is reserved for all three files.

### Syntax

A "set" is a character string consisting of the first three letters of the corresponding term from geometry. SEG, ANG, and TRI are sets corresponding to segment, angle, and triangle, respectively. The symbol # denotes "the measure of," and is used to form the additional sets #SEG, #ANG, and #TRI, representing length, degrees, and area, respectively.

A "relation" is either CON for "is congruent to," or "=".

An "identifier" is a character string consisting of one, two, or three characters, located one blank space to the right of a set, and is used to identify a particular set. The triangle with vertices A, B, and C is identified as TRI ABC. All identifiers for segments and angles, but only the first triangle found in a string, are alphabetized by the program for unique identification. For example, SEG BA is changed to SEG AB, ANG BVA to ANG AVB, and TRI CBA TRI DEF is changed to TRI ABC TRI FED. Identifiers may contain any character except "&" and blank characters.

A "sentence" is a character string containing a set, identifier, relation, set, identifier sequence, with blank characters serving as delimiters. An example of a sentence is SEG A3 CON #TRI %C5. A sentence may contain a maximum of 31 characters.

A "statement" is a sentence which has been determined by the program to be either a given assumption input by the user; a conclusion derived by the program from any combination of previous statements; or a sentence possessing a reflexive property, whether input by the user or derived by the program. A statement must contain a consistent set, relation, set subsequence. The previous example of a sentence does not qualify as a statement, but SEG A3 CON SEG %C may.

A "conclusion" is a potential statement awaiting selection by the user or the program.

A "reason" is a character string which identifies the definition, property, or postulate of Euclidian Geometry used by the program to logically justify a statement. A reason may contain one to three three-character fields, followed by STA, followed by the statement numbers of the previous statements which imply the statement being

---

Thomas J. Kelanic, Taylor Allderdice High School, 2409 Shady Ave. Pittsburgh, PA 15217.

justified. An example of justification by the Transitive Property of Congruence of statement number seven, which is implied by statements one and two, is the following reason: 7. TRA PRO CON STA 1 2.

A "proof" is a sequence of statements, supported by reasons, which leads, by logical implication, to a desired concluding statement.

## Implication Functions

An "implication function" is a portion of a computer program which applies tests to one or two statements and may then synthesize a resultant conclusion and supporting reason. The Symmetric Implication Function, for example, can operate on the statement numbered three:

      3. SEG AB CON SEG XY

to produce the conclusion:

      SEG XY CON SEG AB

and the reason:

      SYM PRO CON STA 3

The Transitive Implication Function, for example, can operate on statements two and five:

      2. #SEG AB = #SEG CD
      5. #SEG CD = #SEG DE

to produce the conclusion:

      #SEG AB =SEG DE

and the reason:

      TRA PRO = STA 2 5

## Hint Functions

To prove that two triangles are congruent by the A.S.A. Postulate of Plane Geometry, three statements are required. Two of these three state congruence of pairs of corresponding angles, and one states congruence of the pair of corresponding included sides. Instead of using a three-input implication function, the program uses a "hint function" and a single-input implication function. Of the three input statements required for the conclusion, any two are independent, and one is dependent. In other words, any one of the three required inputs is uniquely determined by the other two! The uniquely determined third statement is called a "hint," and the function producing it is called a "hint function." Hints are so-named because they may be output to the user, or used by the program itself, as a means of guidance. The single-input implication function which operates on hints to produce the conclusion of three required statements is called "The Hint-Implication Function." For example, the S.A. Hint Function can operate on statements one and two:

      1. ANG ABC CON ANG DEF
      2. SEG AB CON SEG DE

to produce two hints, two tentative conclusions, and two tentative reasons:

      hint #1:    SEG BC CON SEG EF
tentative conclusion #1:    TRI ABC CON TRI DEF
   tentative reason #1:    SAS STA 1 2
      hint #2:    ANG BAC CON ANG EDF
tentative conclusion #2:    TRI ABC CON TRI DEF
   tentative reason #2:    ASA STA 1 2

Later on, The Hint-Implication Function can operate on a statement such as number five:

      5. ANG BAC CON ANG EDF

to produce

    conclusion  :   TRI ABC CON TRI DEF
       reason  :   ASA STA 1 2 5

or on statement number four:

      4. SEG BC CON SEG EF

to produce:

    conclusion:   TRI ABC CON TRI DEF
     reason:   SAS STA 1 2 4

The Hint-Implication Function does not permit redundancy. If statement number four does result in the above conclusion, then all six outputs from the S.A. Hint Function are erased from the file PARTRUE.

## Program Segment GEOMTREE1

Program segment GEOMTREE1 comprises 325 lines of Extended BASIC, and contains the following functions, among others:

1. Single-Input Implication Functions:
   A. The Hint-Implication Function
   B. The Definition of Congruent Triangles
   C. The Symmetric Properties of Congruence and Equality
2. Dual-Input Implication Functions:
   A. The Transitive Properties of Congruence and Equality
3. Dual-Input Hint Functions:
   A. The A.A. Hint Function (for A.S.A.)
   B. The S.A. Hint Function (for A.S.A. and S.A.S.)
   C. The S.S. Hint Function (for S.A.S. and S.S.S.)

```
END AT 2510
*RUN
INPUT MODE 1,2,3, OR 4 ? 1
INPUT "GIVEN" G1 ? ANG XAB CON ANG FAX
INPUT "GIVEN" G2 ? #ANG BXA = #ANG AXF
INPUT "GIVEN" G3 ? SEG BC CON SEG FE
INPUT "GIVEN" G4 ? ANG CBX CON ANG EFX
INPUT "GIVEN" G5 ? #SEG CD = #SEG ED
INPUT "GIVEN" G6 ? PROVE
INPUT "TO PROVE" ? #ANG CDX = #ANG XDE
STA 0 CON 0 HIN 0
STA 1 CON 1 HIN 0
STA 2 CON 2 HIN 1
STA 3 CON 3 HIN 0
STA 4 CON 7 HIN 0
STA 5 CON 8 HIN 0
STA 6 CON 9 HIN 2
STA 7 CON 10 HIN 0
STA 8 CON 14 HIN 0
STA 9 CON 15 HIN 2
STA 10 CON 15 HIN 4
STA 11 CON 15 HIN 4
STA 12 CON 15 HIN 4
STA 13 CON 15 HIN 8
STA 14 CON 16 HIN 8
STA 15 CON 20 HIN 8
```

```
GIVEN: G1: ANG BAX CON ANG FAX
       G2: #ANG AXB = #ANG AXF
       G3: SEG BC CON SEG EF
       G4: ANG CBX CON ANG EFX
       G5: #SEG CD = #SEG DE
PROVE:===> #ANG CDX = #ANG EDX   (MODE: .1)
```

| STATEMENTS: | REASONS: |
|---|---|
| 1. #ANG BAX = #ANG FAX | 1. GIV G1 |
| 2. #ANG AXB = #ANG AXF | 2. GIV G2 |
| 3. #SEG AX = #SEG AX | 3. REF PRO = STA 3 |
| 4. TRI ABX CON TRI AFX | 4. ASA STA 1 2 3 |
| 5. #SEG BC = #SEG EF | 5. GIV G3 |
| 6. #ANG CBX = #ANG EFX | 6. GIV G4 |
| 7. #SEG BX = #SEG FX | 7. DEF CON TRI STA 4 |
| 8. TRI BCX CON TRI FEX | 8. SAS STA 5 6 7 |
| 9. #SEG CD = #SEG DE | 9. GIV G5 |
| 10. #SEG AB = #SEG AF | 10. DEF CON TRI STA 4 |
| 11. #ANG ABX = #ANG AFX | 11. DEF CON TRI STA 4 |
| 12. #TRI ABX = #TRI AFX | 12. DEF CON TRI STA 4 |
| 13. #SEG CX = #SEG EX | 13. DEF CON TRI STA 8 |
| 14. #SEG DX = #SEG DX | 14. REF PRO = STA 14 |
| 15. TRI CDX CON TRI EDX | 15. SSS STA 9 13 14 |
| 16. #ANG CDX = #ANG EDX | 16. DEF CON TRI STA 15 |

```
Q.E.D.  EUCLID  (3/24/1978)
-----------------------------------------
61 CPU USED.

RUN
INPUT MODE 1,2,3, OR 4 ? 2
INPUT "GIVEN" G1 ? #SEG BC = #SEG ZY
INPUT "GIVEN" G2 ? #SEG BA = #SEG ZX
INPUT "GIVEN" G3 ? #ANG CBA = #ANT YZX
INPUT "GIVEN" G3 ? #ANG CBA = #ANG YZX
INPUT "GIVEN" G4 ? PROVE
INPUT "TO PROVE" ? SEG XY CON SEG AC

GIVEN: G1: #SEG BC = #SEG YZ
       G2: #SEG AB = #SEG XZ
       G3: #ANG ABC = #ANG XZY
PROVE:===> SEG XY CON SEG AC   (MODE: 2)
STA 0 CON 0 HIN 0
INPUT STATEMENT 1 ? G2
```

| 1. SEG AB CON SEG XZ | 1. GIV G2 |
|---|---|

```
NEW CONCLUSIONS FOUND:
C1: SEG XZ CON SEG AB
STA 1 CON 1 HIN 0
```

The over-all logic of GEOMTREE1 may be viewed under failure conditions. Failure of the examination of sets by 2A, above, leads to 3B Failure of the examination of identifiers by 2A leads to 3A or 3C. Hints, tentative conclusions, and tentative reasons are stored on file PARTRUE. Conclusions and reasons are stored on file CONFILE.

**Program Segment—EUCLID**

Program segment EUCLID contains the general logic of interactive proofs, a Turing Controller, and a User Command Interpreter. This program segment comprises 353 lines of Extended BASIC.

The general logic contained in EUCLID determines whether or not input from the user or the Turing Controller qualifies as a statement, as previously defined. Given assumptions, statements, and reasons are stored on the file PROFILE by EUCLID, for subsequent output as a finished proof.

The Turing Controller section of EUCLID provides for fully automatic theorem-proving by determining which step to take next. The Turing Controller admits defeat with the message EUCLID HAS FAILED, only after every valid possibility has been tried, but the final state, reaching the desired conclusion, has not been attained. The Turing Controller asks the following questions, where the word "match" is used to mean "identically equal to":

1. Does any Conclusion match the desired conclusion?
2. Has the Hint-Implication Function produced a conclusion?
3. Has a Dual-Input Implication Function produced a conclusion?
4. Does a hint possess a reflexive property?
5. Does a hint match a conclusion?
6. Have all given assumptions been used?
7. Have all conclusions been used?

The User Command Interpreter operates on all input strings having fewer than four characters. Commands beginning with a G or a C and immediately followed by one or two digits tell EUCLID which given assumption or conclusion has been selected for use as the next statement. For example, G5 or C10 are user commands indicating selection of the fifth given assumption or the tenth conclusion. Other user commands are used to output partial contents of the three files for review purposes. The command GIV results in output of the given assumptions stored, one per record, on the zeroth through n-1 records of PROFILE, where n is the number of given assumptions. The command STA results in output of a list of the statements made thus far in the proof, and stored on the nth, n+2, n+4, ... records of PROFILE. The command PRO results in output of the statement to be proved, located on record number three of CONFILE. The first six records of CONFILE are used for data storage and transmission from one program segment to the other. The command CON results in output of the remaining conclusions stored on the even-numbered records of CONFILE, beginning at record six. The command HIN results in output of the remaining hints stored on every third record of PARTRUE, beginning at record zero. To terminate program execution, the user may input END or hit the ESC key. Immediately prior to each program-prompt for user input of a new statement, the available contents of the three files PROFILE, CONFILE, and PARTRUE are displayed by the message STA X CON Y HIN Z, where X, Y, and Z are the appropriate numbers.

Program segment EUCLID offers the user one of four modes of operation. In all four modes, the user must input the given assumptions and the statement to be proved. The four modes are as follows:

1. *Demonstration Mode.* The Turing Controller directs EUCLID in an attempt to prove the theorem automatically. The program supplies all statements, reasons,

```
INPUT STATEMENT 2 ? G2
REDUNDANT SEE STA 1
INPUT STATEMENT 2 ? G1

2.  SEG BC CON SEG YZ            2.  GIV G1

NEW CONCLUSIONS FOUND:
C2:  SEG YZ CON SEG BC
STA 2 CON 2 HIN 2
INPUT STATEMENT 3 ? HIN

HINTS:
ANG ABC CON ANG XZY ?
SEG AC CON SEG XY ?
INPUT STATEMENT 3 ? G3

3.  ANG ABC CON ANG XZY          3.  GIV G3

NEW CONCLUSIONS FOUND:
C3:  TRI ABC CON TRI XZY
C4:  ANG XZY CON ANG ABC
STA 3 CON 4 HIN 0
INPUT STATEMENT 4 ? C3

4.  TRI ABC CON TRI XZY          4.  SAS STA 1 2 3

NEW CONCLUSIONS FOUND:
C5:  TRI XYZ CON TRI ACB
C6:  SEG AC CON SEG XY
C7:  ANG BAC CON ANG YXZ
C8:  ANG ACB CON ANG XYZ
C9:  #TRI ABC = #TRI XZY
STA 4 CON 8 HIN 0
INPUT STATEMENT 5 ? C3
INVALID CODE
INPUT STATEMENT 5 ? PRO
PROVE:===> SEG XY CON SEG AC  (MODE: 2)
INPUT STATEMENT 5 ? SEG AC CON SEG XY

5.  SEG AC CON SEG XY            5.  DEF CON TRI STA 4

NEW CONCLUSIONS FOUND:
C10: SEG XY CON SEG AC
STA 5 CON 8 HIN 0
INPUT STATEMENT 6 ? CON

CONCLUSIONS:
C1:  SEG XZ CON SEG AB
C2:  SEG YZ CON SEG BC
C4:  ANG XZY CON ANG ABC
C5:  TRI XYZ CON TRI ACB
C7:  ANG BAC CON ANG YXZ
C8:  ANG ACB CON ANG XYZ
C9:  #TRI ABC = #TRI XZY
C10: SEG XY CON SEG AC
INPUT STATEMENT 6 ? C10

6.  SEG XY CON SEG AC            6.  SYM PRO CON STA 5

INPUT YOUR FULL NAME ? THOMAS J. KELANIC
------------------------------------------
GIVEN:  G1:  #SEG BC = #SEG YZ
        G2:  #SEG AB = #SEG XZ
        G3:  #ANG ABC = #ANG XZY
PROVE:===> SEG XY CON SEG AC  (MODE: 2)
------------------------------------------
   STATEMENTS:                 REASONS:
------------------------------------------
1.  SEG AB CON SEG XZ         1.  GIV G2
2.  SEG BC CON SEG YZ         2.  GIV G1
3.  ANG ABC CON ANG XZY       3.  GIV G3
4.  TRI ABC CON TRI XZY       4.  SAS STA 1 2 3
5.  SEG AC CON SEG XY         5.  DEF CON TRI STA 4
6.  SEG XY CON SEG AC         6.  SYM PRO CON STA 5
------------------------------------------
Q.E.D.  THOMAS J. KELANIC  (3/24/1978)
------------------------------------------
53 CPU USED.

END AT 2510
*RUN
INPUT MODE 1,2,3, OR 4 ? 3
INPUT "GIVEN" G1 ? SEG RS CON SEG 21
INPUT "GIVEN" G2 ? SEG  ST CON SEG 32
INPUT "GIVEN" G2 ? SEG ST CON SEG 32
INPUT "GIVEN" G3 ? SEG 13 CON SEG RT
INPUT "GIVEN" G4 ? PROVE
INPUT "TO PROVE" ? TRI TSR CON TRI 321

GIVEN:  G1:  SEG RS CON SEG 12
        G2:  SEG ST CON SEG 23
        G3:  SEG 13 CON SEG RT
PROVE:===> TRI RST CON TRI 123  (MODE: 3)
STA 0 CON 0 HIN 0
INPUT STATEMENT 1 ? G1

1.  SEG RS CON SEG 12            1.  GIV G1 (O.K)

NEW CONCLUSIONS FOUND:
C1:  SEG 12 CON SEG RS
STA 1 CON 1 HIN 0
INPUT STATEMENT 2 ? G2

2.  SEG ST CON SEG 23            2.  GIV G2 (O.K)
```

conclusions, and hints to itself. The user sits back and watches the changing file status message.

2. *Practice Mode.* EUCLID supplies all reasons, all conclusions, and permits user access to hints and conclusions. The user must supply all statements.

3. *Quiz Mode.* EUCLID supplies all conclusions, checks user reasons, and permits user access to hints and conclusions. The user must supply all statements and all reasons.

4. *Test Mode.* EUCLID checks the reasons. User access to hints, conclusions, and file status is denied. The user must supply all statements, reasons, and conclusions.

Also contained in EUCLID is The Implicit Definition of Congruence, which automatically selects the proper subsequence of set, relation, and set for each statement of congruence or equality, based on the relation found in the statement to be proved. If the user wishes to conclude with a statement of equality, for example, all statements of congruence, except for congruent triangles, are automatically changed to statements of equality. No reason is given for implicit definitions, but it simplifies EUCLID's task.

**Future Development**

Both program segments can easily be modified to accept other sets such as SGM, NGL, and TRN, instead of SEG, ANG, and TRI, since only one line in each program segment contains sentence vocabulary. This means, of course, that EUCLID is potentially multi-lingual. Additional relations, such as +, -, *, /, >, and < will be used in additional GEOMTREES. Other sentence sequences are presently acceptable to the program, and will be used in additional GEOMTREES. The new GEOMTREES will contain additional definitions, properties, and postulates from Plane Geometry. Eventually, it will be possible to prove practically all of the approximately three-hundred or so theorems usually found in the traditional high-school geometry course. It is not known if indirect proofs can be included, but probably so. In the future, ALGTREES will be written and linked to EUCLID for use in equation-solving. A student might issue the command D5, for example, to see the result of dividing each term of an equation by five. Also in the future, TRIGTREES will be linked to EUCLID for use in proving trigonometric identities. Will there be a CALCTREE? A BOOLTREE? A TREETREE? Compatible graphics will be developed for use with EUCLID. The graphics display will show a blank screen, initially. As statements are made in a proof, labeled points and segments will be added to the display. The segments will be stretched, shrunken, rotated, and translated to produce a representation of the relationships established in the proof. The evolving diagram will reach a final state that depicts all relationships established by the completed proof, as well as those relationships not established.

Copies of the program listing are available from the author at the following address:

Thomas J. Kelanic
Taylor Allderdice High School
2409 Shady Ave.
Pittsburgh, PA 15217

Requests must be accompanied by a check in the amount of $2.00 and payable to "Taylor Allderdice High School" to cover costs of copying and first-class postage.

**Bibliography**

"Extended BASIC Users Manual," 093-000065-06, Data General Corp., rev. 06, February, 1975.

Kelanic, Thomas J. "Geometric Proofs," *Creative Computing*, Vol. 2, No. 2. Mar-Apr, 1976, 60-1.

Nevins, A.J. "Plane Geometry Theorem Proving Using Foreward Chaining," *Artificial Intelligence*. VI. No 1 Jan, 1975, 1-23. ∎

```
NEW CONCLUSIONS FOUND:
C2: SEG 23 CON SEG ST
STA 2 CON 2 HIN 2
INPUT STATEMENT 3 ? HIN

HINTS:
ANG RST CON ANG 123 ?
SEG RT CON SEG 13 ?
INPUT STATEMENT 3 ? SEG 13 CON SEG RT
INPUT REASON 3 ? G3

3. SEG 13 CON SEG RT              3. GIV G3 (ERR)
NEW CONCLUSIONS FOUND:
C3: SEG RT CON SEG 13
STA 3 CON 3 HIN 2
INPUT STATEMENT 4 ? C3
INPUT REASON 4 ? SYM PRO CON STA 3

4. SEG RT CON SEG 13              4. SYM PRO CON STA 3 (O.K)
NEW CONCLUSIONS FOUND:
C4: TRI RST CON TRI 123
STA 4 CON 3 HIN 0
INPUT STATEMENT 5 ? C4
INPUT REASON 5 ? SSS STA 1 2 4

5. TRI RST CON TRI 123            5. SSS STA 1 2 4 (O.K)

INPUT YOUR FULL NAME ? THOMAS J. KELANIC
----------------------------------------------------------
GIVEN: G1: SEG RS CON SEG 12
       G2: SEG ST CON SEG 23
       G3: SEG 13 CON SEG RT
PROVE:===> TRI RST CON TRI 123   (MODE: 3)
----------------------------------------------------------
    STATEMENTS:              REASONS:

1. SEG RS CON SEG 12       1. GIV G1 (O.K)
2. SEG ST CON SEG 23       2. GIV G2 (O.K)
3. SEG 13 CON SEG RT       3. GIV G3 (ERR)
4. SEG RT CON SEG 13       4. SYM PRO CON STA 3 (O.K)
5. TRI RST CON TRI 123     5. SSS STA 1 2 4 (O.K)
----------------------------------------------------------
Q.E.D.   THOMAS J. KELANIC   (3/24/1978)
----------------------------------------------------------
43 CPU USED.
```

```
*RUN
INPUT MODE 1,2,3, OR 4 ? 4
INPUT "GIVEN" G1 ? SEG AC CON SEG YX
INPUT "GIVEN" G2 ? #SEG BA = #SEG YZ
INPUT "GIVEN" G3 ? ANG BAC CON ANG ZYX
INPUT "GIVEN" G4 ? PROVE
INPUT "TO PROVE" ? #TRI CBA = #TRI XZY

GIVEN: G1: SEG AC CON SEG XY
       G2: #SEG AB = #SEG YZ
       G3: ANG BAC CON ANG XYZ
PROVE:===> #TRI ABC = #TRI YZX   (MODE: 4)
INPUT STATEMENT 1 ? G1

1. #SEG AC = #SEG XY              1. GIV G1 (O.K)
INPUT STATEMENT 2 ? G2

2. #SEG AB = #SEG YZ              2. GIV G2 (O.K)
INPUT STATEMENT 3 ? G1
REDUNDANT SEE STA 1
INPUT STATEMENT 3 ? G3

3. #ANG BAC = #ANG XYZ            3. GIV G3 (O.K)
INPUT STATEMENT 4 ? TRI ABC CON TRI XYZ
INVALID STATEMENT
INPUT STATEMENT 4 ? TRI ABC CON TRI XZY
INVALID STATEMENT
INPUT STATEMENT 4 ? TRI ABC CON TRI YZX
INPUT REASON 4 ? SAS STA 1 2 3

4. TRI ABC CON TRI YZX            4. SAS STA 1 2 3 (O.K)
INPUT STATEMENT 5 ? #TRI ABC = #TRI YZX
INPUT REASON 5 ?

5. #TRI ABC = #TRI YZX            5. DEF CON TRI STA 4 (ERR)

INPUT YOUR FULL NAME ? THOMAS J. KELANIC
----------------------------------------------------------
GIVEN: G1: SEG AC CON SEG XY
       G2: #SEG AB = #SEG YZ
       G3: ANG BAC CON ANG XYZ
PROVE:===> #TRI ABC = #TRI YZX   (MODE: 4)
----------------------------------------------------------
    STATEMENTS:              REASONS:

1. #SEG AC = #SEG XY       1. GIV G1 (O.K)
2. #SEG AB = #SEG YZ       2. GIV G2 (O.K)
3. #ANG BAC = #ANG XYZ     3. GIV G3 (O.K)
4. TRI ABC CON TRI YZX     4. SAS STA 1 2 3 (O.K)
5. #TRI ABC = #TRI YZX     5. DEF CON TRI STA 4 (ERR)
----------------------------------------------------------
Q.E.D.   THOMAS J. KELANIC   (3/24/1978)
----------------------------------------------------------
35 CPU USED.
```

# Programming Style

# How to Hide Your Basic Program

John M. Nevison

The heyday of the secretive programmer is over. Today he* is forced to fight a rearguard action. The machine language of the fifties gave way to Fortran. In the sixties, Fortran gave way to Basic and Cobol. The seventies has seen contorted code yielding to structured programming. Professional practices have made life harder and harder for the secretive programmer. No wonder he hurries home from the office each night to the limited memory and cramped code of the personal computer. Here, in private, he can continue polishing his ability to obscure code from the prying eye of the reader.

But even here the future threatens. A Basic program can be well-styled on minicomputers, and books have appeared that show how micro Basics can be styled to reveal the program's ideas to the reader. New disc-resident Basics are widening the opportunity to style Basic on micros. Memory that is

## Now anyone can mystify the reader with inscrutable code by following these four simple rules of style.

presently quite expensive will become quite cheap. One authority predicts that "a megabit storage chip will cost approximately $30 by 1985." Soon, perhaps within the year, bubble memory will make the secretive programmer's favorite excuse, limited memory, a thing of the past.

In addition to losing his technological excuses for writing hard to read code, the secretive programmer will be besieged with readers who, from time to time, will chance upon a well-styled program, read it, and demand that *all* programs be well-styled and

*While the masculine pronoun is used throughout this article, the person referred to may be of either sex.

John M. Nevison, 3 Spruce St., Boston, MA 02108.

easy to read. The inscrutable program may be doomed! In order to survive today's threats to his art, the secretive programmer must set some rules of style.

1. *Confuse Naked Code with a Well-Dressed Program*

Always call small fragments of working code "programs" and the reader won't know what he's missing. If he gets the idea that a program should be easy to read and understand, the program's mystery is seriously threatened. Do everything possible to suppress the notion that a finished program should, like an essay, have a title, a date, an author's name, and an opening statement of purpose.

**Example Before**

```
100 REM     SORT                 16 SEPTEMBER 1977    JOHN M. NEVISON
110
120 REM     SORTS A MIXED BATCH OF NUMBERS, B(), INTO ASCENDING
130 REM     ORDER.  ESPECIALLY GOOD FOR BATCHES OF LESS THAN 50.
140
142 REM     REFERENCE:  JOHN M. NEVISON, "THE LITTLE BOOK OF BASIC
144 REM                 STYLE:  HOW TO WRITE A PROGRAM YOU CAN READ,"
146 REM                 READING, MASS:  ADDISON-WESLEY PUBLISHING
147 REM                 COMPANY, 1978.
148
150 REM     VARIABLES:
160 REM         B()...THE BATCH OF NUMBERS
170 REM         I.....THE INDEX VARIABLE
180 REM         L.....THE LENGTH OF THE CURRENT LIST
190 REM         X.....THE EXCHANGE VARIABLE
200
210 REM     CONSTANT:
220         LET N9 = 38                      'NUMBER OF DATA
230
240 REM     DIMENSIONS:
250         DIM B(38)
260
270 REM     MAIN PROGRAM
280
290 REM     READ IN N9 RANDOM NUMBERS, SORT THEM,
300 REM     AND PRINT THEM OUT.
310
315         LET X = 0
320         FOR I = 1 TO N9
330             LET B(I) = INT(RND*25 +1)
340             PRINT B(I);
350         NEXT I
360         PRINT
366         PRINT
370
380         FOR L = N9 TO 2 STEP -1
390             FOR I = 1 TO L-1
400                 IF B(I) <= B(L) THEN 440
410                     LET X = B(I)
420                     LET B(I) = B(L)
430                     LET B(L) = X
440
450             NEXT I
460         NEXT L
470
480         FOR I = 1 TO N9
490             PRINT B(I);
500         NEXT I
510         PRINT
520
530         END
```

```
210 REM      CONSTANT:
220      LET N9 = 38                        'NUMBER OF DATA
230
240 REM      DIMENSIONS:
250      DIM B(38)
260
270 REM      MAIN PROGRAM
280
290 REM      READ IN N9 RANDOM NUMBERS, SORT THEM,
300 REM      AND PRINT THEM OUT.
310
315      LET X = 0
320      FOR I = 1 TO N9
330          LET B(I) = INT(RND*25 +1)
340          PRINT B(I);
350      NEXT I
360      PRINT
366      PRINT
370
380      FOR L = N9 TO 2 STEP -1
390          FOR I = 1 TO L-1
400              IF B(I) <= B(L) THEN 440
410              LET X = B(I)
420              LET B(I) = B(L)
430              LET B(L) = X
440
450          NEXT I
460      NEXT L
470
480      FOR I = 1 TO N9
490          PRINT B(I);
500      NEXT I
510      PRINT
520
530      END
```

Notice how the beheaded code is much more obscure. When the introduction is missing, the reader doesn't know whom to ask about the program. He doesn't know when it was written or why, or what the variables really mean. The odds are that he won't take the trouble to find out either. The program has a much better chance of passing by unexamined.

## 2. Never Comment Code

Even after the introduction has been stripped away, a program will frequently have scraps of comment dressing blocks of code. Expunge these notes mercilessly. Never give the reader any explanation beyond the code itself. Be careful to avoid any PRINT statements that might reveal what the code is doing.

## 3. Strain the Reader's Eye

What he can't see he can't understand. English has adopted many rules of spacing that the secretive programmer should avoid. The general practice of the secretive programmer should be donotuseaspaceifyoucanavoidit.

The first kind of space to avoid is the blank line.

After the last REM statement is removed, only the heartiest of readers would brave this code. To the true secretive programmer, REM means REMove.

Sequential units of the program blur together when the blank lines are removed. The reader can no longer see quickly where one part ends and the next begins. Many Basics currently help the secretive programmer here by not allowing a blank line, but future Basics will allow this dangerous line. Guard against its use.

The second space to avoid is indentation.

```
220 LET N9 = 38
250 DIM B(38)
315 LET X = 0
320 FOR I = 1 TO N9
330 LET B(I) = INT(RND*25 +1)
340 PRINT B(I);
350 NEXT I
360 PRINT
366 PRINT
380 FOR L = N9 TO 2 STEP -1
390 FOR I = 1 TO L-1
400 IF B(I) <= B(L) THEN 450
410 LET X = B(I)
420 LET B(I) = B(L)
430 LET B(L) = X
450 NEXT I
460 NEXT L
480 FOR I = 1 TO N9
490 PRINT B(I);
500 NEXT I
510 PRINT
530 END
```

Indentation can reveal the most difficult logical feature of most programs: the loop. Remove indentation, and loops regain their rightful mystery. The reader must now ferret them out one at a time. In fact, with both blank lines and indentation removed from the program, the logical structure is completely hidden from the reader. He must take the program one line at a time and slowly construct his own guess at what the structure of the program might be.

The third space to avoid is line spaces.

The reader must now read each line one character at a time. Almost no one

```
220      LET N9 = 38
230
250      DIM B(38)
260
315      LET X = 0
320      FOR I = 1 TO N9
330          LET B(I) = INT(RND*25 +1)
340          PRINT B(I);
350      NEXT I
360      PRINT
366      PRINT
370
380      FOR L = N9 TO 2 STEP -1
390          FOR I = 1 TO L-1
400              IF B(I) <= B(L) THEN 440
410              LET X = B(I)
420              LET B(I) = B(L)
430              LET B(L) = X
440
450          NEXT I
460      NEXT L
470
480      FOR I = 1 TO N9
490          PRINT B(I);
500      NEXT I
510      PRINT
520
530      END
```

```
220      LET N9 = 38
250      DIM B(38)
315      LET X = 0
320      FOR I = 1 TO N9
330          LET B(I) = INT(RND*25 +1)
340          PRINT B(I);
350      NEXT I
360      PRINT
366      PRINT
380      FOR L = N9 TO 2 STEP -1
390          FOR I = 1 TO L-1
400              IF B(I) <= B(L) THEN 450
410              LET X = B(I)
420              LET B(I) = B(L)
430              LET B(L) = X
450          NEXT I
460      NEXT L
480      FOR I = 1 TO N9
490          PRINT B(I);
500      NEXT I
510      PRINT
530      END
```

```
220LETN9=38
250DIMB(38)
315LETX=0
320FORI=1TON9
330LETB(I)=INT(RND*25+1)
340PRINTB(I);
350NEXTI
360PRINT
366PRINT
380FORL=N9TO2STEP-1
390FORI=1TOL-1
400IFB(I)<=B(L)THEN450
410LETX=B(I)
420LETB(I)=B(L)
430LETB(L)=X
450NEXTI
460NEXTL
480FORI=1TON9
490PRINTB(I);
500NEXTI
510PRINT
530END
```

# How To Hide Your Basic Program Round 2

**Robert S. Jaquiss, Sr.**

but the most die-hard fanatic will attempt to understand the program at this stage. The program's privacy is almost completely assured.

```
220LETN9=38
250DIMB(38)
315LETX=0
320FORI=1TON9
330LETB(I)=INT(RND*25+1)
340PRINTB(I);
350NEXTI
360PRINT
366PRINT
380LETL=N9
382IFL=2THEN530
390FORI=1TOL-1
400IFB(I)>=B(L)THEN450
410LETX=B(I)
420LETB(I)=B(L)
430LETB(L)=X
450NEXTI
455PRINTB(L);
460LETL=L-1
462GOTO382
530PRINTB(2);B(1)
535PRINT
540END
```

4. *Contort the Logic*

Structure is the secretive programmer's nemesis. By following the first three rules for obscure programs, the secretive programmer will frequently end up with contorted logic. Nonetheless, the code should be examined to be sure its logical flow is confusing. A little extra work can yield a lot of confusion.

Avoiding an easy-to-understand FOR-NEXT makes the program much more difficult to comprehend. The only thing this piece of code has in common with the original program is its output. Very few readers could verify that fact without running the code.

With these four simple rules of style, even the weakest secretive programmer can learn to hide his Basic program. The test of the truly obscure program is that it must be run on a computer to be understood. As a consequence, the secretive programmer, when confronted by an old piece of his own code, will be unable to guess why it was written or what it did. His confusion is his ultimate reassurance. For if he does not understand his own code, he can rest assured that no one else will.

**Author Note**

The author has been writing illegible Basic programs for thirteen years. A great deal of this time he was at Dartmouth College (where Basic was invented in 1964 by Thomas E. Kurtz and John G. Kemeny). Recently he has become a convert to writing well-styled programs and now refuses to read any of his own old programs. His articles have appeared in *Creative Computing* (Vol. 1, No: 1), *Science,* and the publications of the ACM. His new book, *The Little Book of Basic Style,* has just been published by Addison-Wesley. ∎

---

*This article is a follow-up to the article "How to Hide Your Basic Program" by John M. Nevison which appeared in the January 1979 issue of Creative Computing.*

I agree with many of the concepts that Nevison is trying to preach in this sarcastic article. Since Nevison has been programming for 13 years, he has undoubtedly observed the (recent) development of STYLE in regard to BASIC. He is also probably aware that a program written in the style of his first example uses much memory—more memory than I have had until recently. Until three years ago I was limited to about 1500 characters for a program. Programming in FOCAL, one of my students was able to write a program to solve 9 simultaneous linear equations; using little text and, of course, no remarks.

When my school got a computer that had a larger workspace, I also began to insist on well-documented programs from students (and myself). We sometimes have to un-learn our cherished primitive concepts such as using every conceivable trick to save memory locations. We do learn to use prompts and labels and remarks, etc., in my classes. Once in a while the phrase "structured programming" is heard.

Mr. Nevison's program line 130 REM, suggests that the program is 'especially good for batches of *less* than 50'. Good thing since B will later be dimensioned at 38. He might have suggested 38 or *fewer* numbers.

Line 180. L is also an index variable as I is in line 170.

Line 220 (LET N9=38) is a nice touch. This may make it easier to change the number of numbers to be sorted; or should that be the number of positive integers less than 26 to be sorted? I don't understand why N9 is chosen as a variable. Why not N7 or even just plain N? Also why 38? DIMension B(50).

Line 270 (REM MAIN PROGRAM) is hardly necessary since there appears to be no program other than the MAIN PROGRAM.

---

Robert S. Jaquiss, Sr. Instructor, Computer Science, North Salem High School, Salem, Oregon 97301 and Instructor, Computer Science, Willamette University.

---

Line 290 REM READ IN N9 RANDOM NUMBERS, SORT THEM,
Line 300 REM AND PRINT THEM OUT.

These REMarks are of less than no significance, they are of negative value. The whole program does these things. "SORT THEM" would be better placed just prior to line 380, whereafter the sorting is done.

"PRINT THEM OUT" would be more appropriate before line 480 where the results are about to be printed.

The REMark "READ IN N9 RANDOM NUMBERS" is confusing to me. "READ" is a reserved word in BASIC and means READ-DATA, or read a file (INPUT). Perhaps one could *generate* a list of RANDOM NUMBERS.

Line 315 (LET X=0) is not necessary at all, since a value will be assigned to X in line 410. In some cases, a variable should be initialized — at the beginning of the program — or just prior to where it is to be used (line 370). In this case, X should not be initialized any more than I, or L, which are assigned in the FOR-NEXT loops.

Lines 110, 140, 440: I have never used a BASIC that allowed a line number all by itself this way. Thus line 400 (IF-THEN 440) would not execute in any BASIC I know because there would not be a line 440 to go to.

(Mr. Nevison, If one of my students handed in this program I would return it for revision. But, I have some high standards.)

The downhill development of the remainder of the article make a number of good (or shall we say bad) points. Under point 3 (Strain the Reader's Eye) Mr. Nevison could have mentioned using no line number ending with zero.

There are a few more points I would like to add to the list of ways to do it all wrong.

*5th Downhill Concept.*

Another downhill step that Nevison missed is the vile practice of multiple statements per line. In keeping with the 'how not to do it' teaching philosophy, I have prepared a worse program for illustrative purposes (Program SRT100). It is easier to write badly than goodly anyway. I used the same line numbering as Nevison as far as possible. Since I wanted to display a RUN of the program, I did make a few changes, such as printing the numbers on one line. One line of 38 numbers is too long, so I asked the computer to print 5 numbers to a line. This addition paid unexpected dividends in helping to further obscure the program as I was able to insert several GOTO's.

Let us examine the program LISTing. We begin with a statement of program purpose: to sort numbers. No mention is made of what numbers, how many numbers, or ascending or descending order. Some FOR-NEXT loops have been deleted and others added. I really did try to eliminate all the spaces but the spacing habit is too strong.

I submit that the logic is contorted. The multiple statements per line make the program listing much more difficult to read. The addition of subtle changes such as RND(X)*10*N make for more confusion than just RND*25. Surely it is not clear that in line 340 the jump to 990 terminates a line of five integers.

It may seem peculiar that in line 360 L=N+1 and on the next line, line 380, L=L+1. Line 455 was previously included in line 450 (which I thought more confusing) but the line was considered too long for publication.

I do not have a book to sell. But let's hear it for Bob. Isn't my program worse than Nevison's?

*6th Make it worse concept:*
Document the program in a misleading manner.

My second program (SRT300) is the same as my first program with a few additions. First, the interactive user is told the program purpose (this program sorts numbers) in an ambiguous manner. The user may assume whatever s/he likes. The next printed output follows immediately, before the user has time to comtemplate, "HOW MANY NUMBERS TO SORT?".

If the user wants to sort more than 50 numbers the program will malfunction (DIM B(50)). So the user is told the computer is 'too tired to sort that many numbers.' If the user asks to sort fewer than 4 numbers the message "TRIVIAL. DO IT YOURSELF." is printed. This nasty was added because the computer will accept input numbers only until three fewer numbers than the specified number have been typed in (line 150).

In the example RUNs several numbers were typed in on one line to conserve space. Imagine the user who expects to type in 10 numbers. As soon as the seventh number is typed the computer starts printing out numbers. What numbers? Not the numbers the user typed in, but some numbers the computer made up. The specified number of numbers is displayed in random order, descending order, and ascending order. In addition, if the specified N is a multiple of 5, the program falls through a crack and starts again at the beginning.

```
200 REM PROGRAM TO SORT NUMBERS
220 N=38\DIMB(50)\I=0
320 I=I+1\B(I)=INT(RND(X)*10*N)+1
340 PRINTB(I);\IFINT(I/5)=I/5THEN990
350 IFI<NTHEN320
360 PRINT\PRINT\L=N+1
380 L=L-1\I=0\X=0
400 I=I+1\IFB(I)<=B(L)THEN 440
420 X=B(I)\B(I)=B(L)\B(L)=X
440 IFI<L-1THEN400
450 PRINTB(L);
455 IFINT((N-L+1)/5)=(N-L+1)/5THEN980
460 IFL>2THEN380
470 GOTO500
480 PRINT B(L);
500 L=L-1\IFL>0THEN480
590 PRINT
600 PRINT\FORI=1TONSTEP5
610 FORJ=0TO4
620 IFI+J>NTHEN999
630 PRINTB(I+J);
640 NEXTJ\PRINT\NEXTI
980 PRINT\GOTO460
990 PRINT\GOTO350
999 PRINT\PRINT\END

READY
```

```
RUN
138  127  241  134  255
206  323   10  206  356
 48  148  371  197  177
168  369  109  330   68
202   93  127  230  248
148  327  273  128  274
171  197  221   91   39
  3  148  360

371  369  360  356  330
327  323  274  273  255
248  241  230  221  206
206  202  197  197  177
171  168  148  148  148
138  134  128  127  127
109   93   91   68   48
 39   10    3

  3   10   39   48   68
 91   93  109  127  127
128  134  138  148  148
148  168  171  177  197
197  202  206  206  221
230  241  248  255  273
274  323  327  330  356
360  369  371
```

```
100 REM    SORT           16 SEPTEMBER 1977    JOHN M. NEVISON
110
120 REM    SORTS A MIXED BATCH OF NUMBERS, B(), INTO ASCENDING
130 REM    ORDER.  ESPECIALLY GOOD FOR BATCHES OF LESS THAN 50.
140
142 REM    REFERENCE:   JOHN M. NEVISON, "THE LITTLE BOOK OF BASIC
144 REM                 STYLE:  HOW TO WRITE A PROGRAM YOU CAN READ,
146 REM                 READING, MASS:  ADDISON-WESLEY PUBLISHING
147 REM                 COMPANY, 1978.
148
150 REM    VARIABLES:
160 REM         B()....THE BATCH OF NUMBERS
170 REM         I.....THE INDEX VARIABLE
180 REM         L.....THE LENGTH OF THE CURRENT LIST
190 REM         X.....THE EXCHANGE VARIABLE
200
210 REM    CONSTANT:
220        LET N9 = 38                NUMBER OF DATA
230
240 REM    DIMENSIONS:
250        DIM B(38)
260
270 REM    MAIN PROGRAM
280
290 REM    READ IN N9 RANDOM NUMBERS, SORT THEM,
300 REM    AND PRINT THEM OUT.
310
315        LET X = 0
320        FOR I = 1 TO N9
330            LET B(I) = INT(RND*25 +1)
340            PRINT B(I);
350        NEXT I
360        PRINT
366        PRINT
370
380        FOR L = N9 TO 2 STEP -1
390            FOR I = 1 TO L-1
400                IF B(I) <= B(L) THEN 440
410                    LET X = B(I)
420                    LET B(I) = B(L)
430                    LET B(L) = X
440
450            NEXT I
460        NEXT L
470
480        FOR I = 1 TO N9
490            PRINT B(I);
500        NEXT I
510        PRINT
520
530        END
```

**Original Nevison Program**

```
90DIMB(50)\RANDOMIZE                         THIS PROGRAM SORTS NUMBERS.
100PRINT"THIS PROGRAM SORTS NUMBERS.         HOW MANY NUMBERS TO SORT?15
110PRINT"HOW MANY NUMBERS TO SORT";          TYPE IN THE NUMBERS TO BE SORTED.
120INPUTN\IFN>50THEN800                      PRESS 'CARRIAGE RETURN'AFTER EACH NUMBER.
125IFN<4THEN850                              ?12,34,56,78,5,4
130PRINT"TYPE IN THE NUMBERS TO BE SORTED.   ?678,234
140PRINT"PRESS 'CARRIAGE RETURN'";           ?34567
145PRINT"AFTER EACH NUMBER."                 ?567
150FORI=1TON-3\INPUTB(I)\NEXTI\I=0           ?345
320I=I+1\B(I)=INT(RND(X)*10*N)+1             ?76543
340PRINTB(I);\IFINT(I/5)=I/5THEN990           44   45   102  116   67
350IFI<NTHEN320                               44  148   147  100  145
360PRINT\PRINT\L=N+1                         129   56   141   50   44
380L=L-1\I=0\X=0
400I=I+1\IFB(I)<=B(L)THEN 440                148  147   145  141  129
420X=B(I)\B(I)=B(L)\B(L)=X                   116  102   100   67   56
440IFI<L-1THEN400                             50   45    44   44   44
450PRINTB(L);
455IFINT((N-L+1)/5)=(N-L+1)/5THEN980          44   44    44   45   50
460IFL>2THEN380                               56   67   100  102  116
470GOTO500                                   129  141   145  147  148
480PRINTB(L);                                THIS PROGRAM SORTS NUMBERS.
500L=L-1\IFL>0THEN480                        HOW MANY NUMBERS TO SORT?13
590PRINT                                     TYPE IN THE NUMBERS TO BE SORTED.
600PRINT\FORI=1TONSTEP5                      PRESS 'CARRIAGE RETURN'AFTER EACH NUMBER.
610FORJ=0TO4                                 ?1,2,3,4,5,6,7,8,9
620IFI+J>NTHEN999                            ?78
630PRINTB(I+J);                              125   37    50   37   95
640NEXTJ\PRINT\NEXTI                          89   16   110   89   19
650GOTO100                                    71  126   108
800PRINT"I AM TOO TIRED TO SORT";N;"NUMBERS. 126  125   110  108   95
810GOTO110                                    89   89    71   50   37
850PRINT"TRIVIAL.       DOIT YOURSELF"        37   19    16
860GOTO810
980PRINT\GOTO460                              16   19    37   37   50
990PRINT\GOTO350                              71   89    89   95  108
999PRINT\PRINT\END                           110  125   126
```

7th and 8th make it worse concepts.

A program LIST illustration is not provided. The 7th concept is to rearrange the lines of the program flow so that many unnessary GOTOs are inserted. Thus the program flow flits from here to there in the program rather than following the line numbers sequentially.

The 8th concept is enabled by the 7th. Insert lines of code that are never executed. If anyone should try to figure out what you have done and you have made good use of concepts 7 and 8, the task will be immeasurably more complicated for that hapless individual.

Old time programmers who used machine language were very good in the implementation of concepts 7 and 8. Among other things, the (priesthood) rule of 7&8 made for good job protection and added mystery. The boss could not fire such a programmer because no one else would be able to maintain the programs.

*Concept 9.*

Concept 9 is really not a programming concept, but has to do with the dissemination of computer programs. Here are a number of ideas that can be used singly or in combination.

  a. If you use a teleprinter, make sure that the ribbon is old and the type head is out of alignment before you prepare a printout for publication.
  b. If you are the publisher, make liberal use of a copier that reduces print size. If the print size is reduced by half, four times, the print will be one-sixteenth the original size and more letters will fit on a page. The fact that (a) combined with (b) will surely make it impossible for anyone over 30 years of age to read the article is not important. People over 30 are not important anyway.
  c. If there is any possibility that (a) and (b) might not make an article completely unreadable, another add-on possibility is to print gray ink on a gray background, or try green ink on a blue background, or pink on red. Not only is the article now unreadable, it will also be difficult to make a copy of the article on a copier.
  d. Some publishers are very adept at overexposure of copy on the copy machine or the use of too much ink. Either of these tricks make the characters bleed out in unexpected and wildly unreadable ways.
  e. Don't publish the last page of the program LISTing.

On the other hand, there are many attributes of good programming style that should be mentioned early on in teaching programming in any computer language. It is much more difficult to write about good programming practice than to poke fun at another's serious attempt to produce a good program. Let us resolve to think positively about programming; to emphasize good programming practices.

The first few lines of a program should identify the program, by name, if it has a name on a storage device. The programmer should be identified. If the program is a student assignment, the problem should be identified by name, page, problem number, etc. If the program is not original, credit should be given where credit is due.

Next should be a brief explanation of the program purpose and directions to use the program. All variables should be identified at the beginning and itialized, dimensioned or declared as needed. Some persons take pride in boxing in these (and other) program segments with rows of stars.

Certainly some attention should be given to having a main program and subprograms, subroutines or procedures. Of course, the subprograms should be set off in printed format, identified, and pass-to, pass-back parameters clearly stated. Global variables used in subprograms must be identified. Indeed, in one style of writing, the main program is almost nothing but subroutine calls.

For example, many programs ask if the user would like instructions. A 'yes' answer should result in a subprogram call, which may in turn read the instructions from a disk file and display them to the user. There is much to be said on the subject of good programming in BASIC.

An excellent reference is the new book "Basic with Style," by Paul Nagin and Henry F. Ledgard, published by Hayden Book Company, Inc. 1978. ■

# Short Programs

# ...short programs...

## Convergence

Certain constants such as e and can be calculated as the sum of a number of elements in a series. However, if such a series is calculated by hand, or even with a small calculator, it is often difficult to carry it to many digits of accuracy. The computer can easily calculate such series, although to carry the calculations to more significant digits than your compiler or interpreter permits, you'll have to use a modification of the program presented here.

**Convergence on e by an infinite series.** The logarithmic constant e can be represented by the series:
1 + 1/1 + 1/2 + 1/6 + 1/24 + 1/120 + ... If you decide to extend your precision, the value of e to 15 places is 2.718281828459045.

```
LIST

10 REM: CONVERGE ON E
20 D=1
30 E=1
40 I=0
50 I=I+1
60 D=D*I
70 E=E+1/D
80 PRINT I,E
90 GOTO 50
100 END

RUN
 1        2
 2        2.5
 3        2.66667
 4        2.70833
 5        2.71667
 6        2.71806
 7        2.71825
 8        2.71828
 9        2.71828
10        2.71828
11        2.71828
12        2.71828
13        2.71828
14        2.71828
15        2.71828
16        2.71828
```

**Convergence on π by infinite series.** The series to converge on π is:
1 - 1/3 + 1/5 - 1/7 + 1/9 ...
Since the series converges extremely slowly, only every 500th value is printed out in the program.

```
LIST

10 P=0              RUN
20 S=1              3.1436
30 I=1              3.14059
40 FOR J=1 TO 499   3.14226
50 P=P+S/I          3.14109
60 I=I+2            3.142
70 S=-S             3.14126
80 NEXT J           3.14188
90 PRINT P*4        3.14135
100 GOTO 40         3.14182
110 END             3.1414
                    3.14178
                    3.14144
                    3.14176
                    3.14146
                    3.14174
                    3.14148
                    3.14172
                    3.14149
                    3.14171
                    3.1415
                    3.1417
                    3.14151
                    3.14169
```

**Convergence on π by polygons.** Does a square look like a circle? Not really. How about an octagon? Well, more so. A 100-sided polygon? That's getting close, but how close? One way of determining "how close?" is to calculate the perimeter of a polygon inscribed inside a circle and another circumscribed outside of a circle and when the two get very close, you've practically got a circle. This is also an interesting, if not very efficient way, to calculate π.

Consider a polygon circumscribed around a circle of radius 1.
Perimeter = length of side x no. of sides

Since the tangent of X = AB/BC, but BC = 1, then tan (X) = AB and the length of a side = 2x tan (X). Since the circumference = 2 π r and r = 1, then π is simply the circumference (or perimeter of a n-sided polygon) divided by 2.

Similar trigonometry leads to the perimeter of an inscribed polygon being equal to no. of sides x  sin (X) x cos (X).

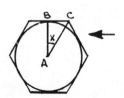

Unfortunately, there is one large fallacy in the program in that degrees must be converted into radians in statement 30. This means, of course, that you must already have the conversion factor, which is simply 2 π.

Does anyone want to guess what happened to this poor program after we got above a 768-sided polygon. What happened when we hit 50331700 sides?

```
LIST

10 N=6
20 N=2*N
30 X=6.2831853/N
40 PRINT N/2,N*SIN(X)*COS(X)/2,N*TAN(X)/2
50 GOTO 20
60 END
OK
```

| RUN | | |
|---|---|---|
| 6 | 2.59808 | 3.4641 |
| 12 | 3 | 3.21539 |
| 24 | 3.10583 | 3.15966 |
| 48 | 3.13263 | 3.14609 |
| 96 | 3.13935 | 3.14271 |
| 192 | 3.14104 | 3.14188 |
| 384 | 3.14144 | 3.14165 |
| 768 | 3.14158 | 3.14163 |
| 1536 | 3.14154 | 3.14155 | ← What's
| 3072 | 3.14169 | 3.14169 | happening?
| 6144 | 3.1414 | 3.1414 |
| 12288 | 3.14198 | 3.14198 |
| 24576 | 3.14083 | 3.14083 |
| 49152 | 3.14313 | 3.14313 |
| 98304 | 3.13853 | 3.13853 |
| 196608 | 3.14773 | 3.14773 |
| 393216 | 3.12932 | 3.12932 |
| 786432 | 3.16614 | 3.16614 |
| 1.57286E+06 | 3.09251 | 3.09251 |
| 3.14573E+06 | 3.23977 | 3.23977 |
| 6.29146E+06 | 2.94524 | 2.94524 |
| 1.25829E+07 | 3.53429 | 3.53429 |
| 2.51658E+07 | 2.35619 | 2.35619 |
| 5.03317E+07 | 0 | 0 | Gulp!
| 1.00663E+08 | 0 | 0 |
| 2.01327E+08 | 0 | 0 |
| 4.02653E+08 | 0 | 0 |
| 8.05306E+08 | 0 | 0 |
| 1.61061E+09 | 0 | 0 |
| 3.22123E+09 | 0 | 0 |
| 6.44245E+09 | 0 | 0 |
| 1.28849E+10 | 0 | 0 |
| 2.57698E+10 | 0 | 0 |
| 5.15396E+10 | 0 | 0 |
| 1.03079E+11 | 0 | 0 |
| 2.06158E+11 | 0 | 0 |
| 4.12317E+11 | 0 | 0 |
| 8.24634E+11 | 0 | 0 |
| 1.64927E+12 | 0 | 0 |
| 3.29854E+12 | 0 | 0 |
| 6.59707E+12 | 0 | 0 |
| 1.31941E+13 | 0 | 0 |

# ..short programs...

## Convergence Revisited

The "Convergence" Short Program on page 132 of the March-April 1978 issue of *Creative Computing* presented a method for calculating $\pi$ by inscribed polygons. I have a program which doesn't use $\pi$ to calculate $\pi$.

Since the circumference of a circle with radius R is $2\pi R$, we can approximate the circumference by inscribed polygons, dividing the perimeter by 2R to obtain an approximation to $\pi$. We can start with an inscribed square of side R*SQR(2) and double the number of sides for each calculation. If the old side length is S, then the length S' of a side of a new inscribed polygon with twice as many sides is obtained as follows:

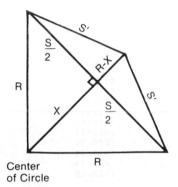

Two applications of Pythagorus' Theorem yield the two equations

$X^2 + (S/2)^2 = R^2$

$(R-X)^2 + (S/2)^2 = (S')^2$

Thus
$S' = \sqrt{(R - \sqrt{R^2 - (S/2)^2})^2 + (S/2)^2}$

It is easy to simplify this formula algebraically, but accuracy suffers if one does this. Also, one should use S*S instead of S**2, and SQR( ) instead of ( )**.5, although this may differ on various computers.

As the run shows, the calculation is accurate to 13 decimal places [compared with 4*ATN(1)], and actually differs only by $4*10^{-15}$. I ran it with the loop counter set at 30, but the value didn't change. I also experimented with different radii, and settled on R=10.

George W. Ball
Dept. of Mathematics
Alfred University
Alfred, NY 14802

```
LIST
100 REM PROGRAM TO CALCULATE PI BY INSCRIBED POLYGONS
110 REM GEORGE W. BALL
120 REM ALFRED UNIVERSITY
130 PRINT & PRINT
140 PRINT"THE MACHINE VALUE FOR PI IS";
150 PRINT PRC(1),4*ATN(1)
160 PRINT & PRINT
170 REM WE WILL USE A CIRCLE OF RADIUS 10
180 R=10
190 PRINT "SIDES","        PERIMETER"
200 PRINT
210 N=4, S=R*SQR(2)
220 FOR K=1 TO 28
230 PRINT PRC(1),N,(S*N)/(2*R)
240 Y=S*S/4
250 H=SQR(R*R-Y)
260 X=R-H
270 S=SQR(X*X+Y)
280 N=2*N
290 NEXT K
>RUN
11:09    MAR 09   RUNZBAA...

THE MACHINE VALUE FOR PI IS    3.141592653589793

    SIDES                PERIMETER

    4                    2.828427124746190
    8                    3.061467458920718
    16                   3.121445152258051
    32                   3.136548490545938
    64                   3.140331156954752
    128                  3.141277250932772
    256                  3.141513801144299
    512                  3.141572940367089
    1024                 3.141587725277158
    2048                 3.141591421511198
    4096                 3.141592345570116
    8192                 3.141592576584870
    16384                3.141592634338560
    32768                3.141592648776983
    65536                3.141592652386589
    131072               3.141592653288990
    262144               3.141592653514590
    524288               3.141592653570990
    1048576              3.141592653585090
    2097152              3.141592653588615
    4194304              3.141592653589496
    8388608              3.141592653589716
    16777216             3.141592653589771
    33554432             3.141592653589785
    67108864             3.141592653589788
    134217728            3.141592653589789
    268435456            3.141592653589789
    536870912            3.141592653589789

       290 HALT
```

# ...short programs...

## Systematic Savings Revisited

**Stu Denenberg**

Referring to "Systematic Savings" on page 132 of the Nov/Dec '77 *Creative Computing*, the fancy mathematics formula masks what is happening.

Why not just do the calculation as a person would do with a hand calculator? We could begin with the simpler problem of calculating compound interest and then slightly modify that procedure to do systematic investments. For example, the Basic program for compound interest is:

```
5 PRINT "AT END OF YEAR","BALANCE"
10 READ P,R,N
20 FOR I=1 TO N
30 P=P+P*R
40 PRINT I,,P
50 NEXT I
60 STOP
70 DATA 100,.1,10
80 END

RUN
AT END OF YEAR          BALANCE
1                       110
2                       121
3                       133.1
4                       146.41
5                       161.051
6                       177.156
7                       194.872
8                       214.359
9                       235.795
10                      259.374
```

Note especially that Line 30 is *not* P=P*(I + R); instead it stresses what we *actually* do when we calculate interest — namely multiply the principal by the interest rate and then add that back onto the principal to give the new principal.

Now the program to do systematic savings is exactly the same as the one for compound interest but instead of letting our 100 bucks lay around all lonely while it's compounding, we keep feeding in lumps of $100 at the end of each year so now the program looks like:

```
5 PRINT "AT END OF YEAR     AMOUNT INVESTED     TOTAL ACCUMULATED"
10 READ N,C,R
20 P=C
30 FOR I=1 TO N
40 P=P+P*R
50 PRINT TAB(5);I;TAB(25);I*C;TAB(45);P
60 P=P+C
70 NEXT I
80 STOP
90 DATA 10,100,.1
100 END

RUN
AT END OF YEAR      AMOUNT INVESTED     TOTAL ACCUMULATED
1                   100                 110
2                   200                 231
3                   300                 364.1
4                   400                 510.51
5                   500                 671.561
6                   600                 848.717
7                   700                 1043.59
8                   800                 1257.95
9                   900                 1493.74
10                  1000                1753.12
```

C is the constant amount we save each year. Line 60 is the only *real* difference between the two programs and it shows how we add in the constant savings to our principal each year.

Dr. Stuart Denenberg
Dept. of Computer Science
SUNY
Plattsburgh, NY 12901

## Compound Interest

If $1000 is deposited in a savings account paying 8% interest compounded n times a year, then this will accumulate to

$1000(1 + .08/n)$^n$

at the end of one year assuming that no deposits or withdrawals are made.

| n | 8% Compounded | Accumulation at end of one year. (Rounded to nearest cent) | |
|---|---|---|---|
| 1 | Yearly | $1000(1+.08/1)^1$ = | $1080.00 |
| 2 | Semiannually | $1000(1+.08/2)^2$ = | $1081.60 |
| 4 | Quarterly | $1000(1+.08/4)^4$ = | $1082.43 |
| 12 | Monthly | $1000(1+.08/12)^{12}$ = | $1083.00 |
| 365 | Daily | $1000(1+.08/365)^{365}$ = | $1083.28 |
| 8760 | Hourly | $1000(1+.08/8760)^{8760}$ = | $1083.29 |
| 525,600 | Every minute | $1000(1+.08/525600)^{525600}$ = | $1083.29 |
| 31,536,000 | Every second | $1000(1+.08/31536000)^{31536000}$= | $1083.29 |

Hardly worth quibbling over hours, minutes, and seconds.

# short programs.....

## Prime Factoring with Quasi-Primes

### Jay M. Jeffery

A program that yields the prime factors of integers comes in handy at various levels of mathematics education. It can be used to reduce fractions to lowest terms, to find greatest common factors, to find least common multiples, and to test if numbers are prime.

The following program in BASIC is operated very rapidly even with relatively large numbers. The feature that helps to increase the efficiency is the use of equations that generate all primes consecutively while eliminating many, though not all, composites. The set of numbers generated could be called quasi-primes. Using such a set as a source of divisors in a factoring algorithm reduces the number of unnecessary divisions considerably, compared to programs that employ consecutive integers or the set of odd integers,

Operation of the program is explained in the Remark statements.

```
100 PRINT "PRIME FACTORING PROGRAM"
110 PRINT "INPUT POSITIVE INTEGER TO BE FACTORED:"
120 INPUT X1
130 PRINT "THE PRIME FACTORS ARE AS FOLLOWS:"
140 REM:   X2 REPRESENTS THE DIVISORS USED TO TEST NUMBER.
150 X2=2
160 GOSUB 300
170 X2=3
180 GOSUB 300
190 REM:   LINES 170 THROUGH 230 GENERATE ALL PRIMES REQUIRED
200 REM:   AS DIVISORS. ALSO, SOME COMPOSITES ARE GENERATED,
210 REM:   BUT THE EQUATIONS REDUCE THE NUMBER CONSIDERABLY.
220 FOR N=0 TO SQR(X1)/6
230 X2=6*N+5
240 GOSUB 300
250 X2=6*N+7
260 GOSUB 300
270 NEXT N
280 GOTO 380
290 REM:   SUBROUTINE TESTS DIVISIBILITY.
300 X4=INT(X1/X2)*X2
310 X4=INT(X1/X2)*X2
320 IF X4#X1 THEN 370
330 PRINT X2;
340 X1=X1/X2
350 IF X1=1 THEN 390
360 GOTO 310
370 RETURN
380 PRINT X1
390 END
```

```
PRIME FACTORING PROGRAM
INPUT POSITIVE INTEGER TO BE FACTORED:
?56
THE PRIME FACTORS ARE AS FOLLOWS:
  2   2   2   7

PRIME FACTORING PROGRAM
INPUT POSITIVE INTEGER TO BE FACTORED:
?999991
THE PRIME FACTORS ARE AS FOLLOWS:
 17  59  997

PRIME FACTORING PROGRAM
INPUT POSITIVE INTEGER TO BE FACTORED:
?999919
THE PRIME FACTORS ARE AS FOLLOWS:
 991  1009

PRIME FACTORING PROGRAM
INPUT POSITIVE INTEGER TO BE FACTORED:
?8191
THE PRIME FACTORS ARE AS FOLLOWS:
 8191

PRIME FACTORING PROGRAM
INPUT POSITIVE INTEGER TO BE FACTORED:
?510510
THE PRIME FACTORS ARE AS FOLLOWS:
  2   3   5   7  11  13  17

PRIME FACTORING PROGRAM
INPUT POSITIVE INTEGER TO BE FACTORED:
?999199
THE PRIME FACTORS ARE AS FOLLOWS:
 999199.

PRIME FACTORING PROGRAM
INPUT POSITIVE INTEGER TO BE FACTORED:
?999919
THE PRIME FACTORS ARE AS FOLLOWS:
 991  1009

PRIME FACTORING PROGRAM
INPUT POSITIVE INTEGER TO BE FACTORED:
?9819
THE PRIME FACTORS ARE AS FOLLOWS:
  3   3  1091
```

Jay M. Jeffery
Mathematics Dept.
Hawken School
Box 249, County Line Road
Gates Mills, Ohio 44040

## Powers of 2

The Basic computer program below displays the numerical value of
$2^0, 2^1, 2^2, 2^3, \ldots, 2^{17}$
in the second column.

```
10 LET X=0
20 PRINT X,2↑X
30 LET X=X+1
40 IF X=18 THEN 60
50 GO TO 20
60 END

READY

RUNNK

0        1
1        2
2        4
3        8
4       16
5       32
6       64
7      128
8      256
9      512
10     1024
11     2048
12     4096
13     8192
14    16384
15    32768
16    65536
17   131072
```

—Want to play for 1¢ on the first hole, 2¢ on the second hole, 4¢ on the third hole, etc.?

Sorry. I can't afford $1,310.72 if I lose the last hole.

Could you put $2^{64}$ grains of wheat in this bag?

That would be 18,446,744,073,709,551,615 grains — Enough to cover the surface of the earth.

# short programs.....

## Double Subscripts Become Single

If you have a version of BASIC which does not allow double subscripts, have you looked wistfully at all the interesting games and programs that contain lines like:

100 A(I,J) = I + J/10...?

Well, the situation really is far from hopeless. As a simple example, suppose you wanted to create and print a table of numbers. You want the table to have five rows and seven columns, and each entry in the table is to be a decimal number with the row number on the left of the decimal point and the column number on the right. Here is one way to do it with double subscripts:

```
20 FOR I=1 TO 5
30 FOR J=1 TO 7
40 A(I,J)=I+J/10
50 PRINT A(I,J);
60 NEXT J
70 PRINT
80 NEXT I
90 END

run
1.1  1.2  1.3  1.4  1.5  1.6  1.7
2.1  2.2  2.3  2.4  2.5  2.6  2.7
3.1  3.2  3.3  3.4  3.5  3.6  3.7
4.1  4.2  4.3  4.4  4.5  4.6  4.7
5.1  5.2  5.3  5.4  5.5  5.6  5.7
```

Can we achieve the same result without the use of double subscripts? Happily, the answer is yes. Add the following line:

35 N = 7*(I-1) + J

and change lines 40 and 50 to:

40 A(N) = I + J/10
50 PRINT A(N);

The secret is at line 35. In general, to imitate the variable A(I,J), use A(C*(I-1) + J) where C is the number of **columns** in your array.

It's that simple.

With this technique, for instance, I have been able to translate Gregory Yob's "HUNT THE WUMPUS" into Radio Shack's Level I BASIC, which allows only single subscripts.

Well, what are you waiting for? Get out those back issues of **Creative Computing** that had all the games you **thought** you couldn't program.

Now you can.

James Garon
Math Dept.
Calif. State University
Fullerton, CA 92634

## Common Birthdays

```
10 PRINT "NUMBER OF      PROBABILITY THAT AT LEAST"
20 PRINT "PEOPLE         TWO HAVE SAME BIRTHDAY"
30 PRINT "--------       --------------------------"
40 Q=364/365
60 FOR N=2 TO 40
70 P=100*(1-Q)
80 PRINT "   ";N;TAB(20);INT(P*100+.5)/100;TAB(28);"%"
90 Q=Q*(365-N)/365
100 NEXT N
110 END
```

In a group of ten randomly selected people, there is about a 12% chance that two of them share a common birthday.

With 23 people, the probability is slightly greater than 50%.

With 40 people, the probability is about 89%.

Consider the set of all Presidents of the United States. Two of them, James Polk and Warren G. Harding were born on November 2.

It is interesting to note that John Adams, James Monroe, and Thomas Jefferson all died on July 4. Millard Fillmore and William Taft both died on March 8.

Sanderson M. Smith
Cate School
Carpinteria, CA

| RUN NUMBER OF PEOPLE | PROBABILITY THAT AT LEAST TWO HAVE SAME BIRTHDAY |
|---|---|
| 2  | .27   % |
| 3  | .82   % |
| 4  | 1.64  % |
| 5  | 2.71  % |
| 6  | 4.05  % |
| 7  | 5.62  % |
| 8  | 7.43  % |
| 9  | 9.46  % |
| 10 | 11.69 % |
| 11 | 14.11 % |
| 12 | 16.7  % |
| 13 | 19.44 % |
| 14 | 22.31 % |
| 15 | 25.29 % |
| 16 | 28.36 % |
| 17 | 31.5  % |
| 18 | 34.69 % |
| 19 | 37.91 % |
| 20 | 41.14 % |
| 21 | 44.37 % |
| 22 | 47.57 % |
| 23 | 50.73 % |
| 24 | 53.83 % |
| 25 | 56.87 % |
| 26 | 59.82 % |
| 27 | 62.69 % |
| 28 | 65.45 % |
| 29 | 68.1  % |
| 30 | 70.63 % |
| 31 | 73.05 % |
| 32 | 75.33 % |
| 33 | 77.5  % |
| 34 | 79.53 % |
| 35 | 81.44 % |
| 36 | 83.22 % |
| 37 | 84.87 % |
| 38 | 86.41 % |
| 39 | 87.82 % |
| 40 | 89.12 % |

```
10 DIM A(35)
20 FOR I=1 TO 5
30 FOR J=1 TO 7
35 N = 7*(I-1)+J
40 A(N) = I + J/10
50 PRINT A(N);
60 NEXT J
70 PRINT
80 NEXT I
90 END

run
1.1  1.2  1.3  1.4  1.5  1.6  1.7
2.1  2.2  2.3  2.4  2.5  2.6  2.7
3.1  3.2  3.3  3.4  3.5  3.6  3.7
4.1  4.2  4.3  4.4  4.5  4.6  4.7
5.1  5.2  5.3  5.4  5.5  5.6  5.7
```

# short programs.....

## Rorschach II

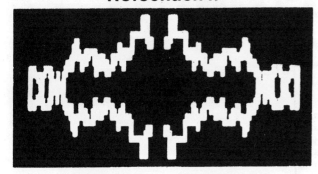

This program is written for the TRS-80, Level I. It simulates random Rorschach-like patterns. The program will continue to run with no user input. To stop, use the BREAK key.

```
100 CLS                          850 S.(X,47-Y)
150 X=1:Y=23                     870 S.(127-X,47-Y)
200 C=200                        900 N.B
300 F.B=1 TO C                   950 F.U=1 TO 1000:N.U:CLS
400 A(B)=RND(3)                  970 G.150
500 N.B                          1000 X=X+1
600 F.B=1 TO C                   1100 G.800
700 ON(A(B))G.1000,2000,3000     2000 Y=Y+1
800 IF(Y<1)+(Y>47)T.950          2100 G.800
820 S.(X,Y)                      3000 Y=Y-1
830 S.(127-X,Y)                  3100 G.800
```

Reprinted from Micronews, January 1979, published by the Cuatro Computer Club; Ave Los Pinos EDF, Airosa #5, La Florida—Caracas, VENEZUELA.

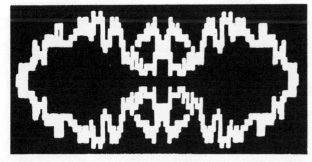

```
LIST
10 '        ==== INTEREST CALCULATION ====
20 '                    WITH
30 '        ====== DATE CONVERSION ======
40 '        ----------------------------
50 PRINT : PRINT : PRINT
60 PRINT "ENTER BEGINNING DATE (MM/DD/YY) : ";: INPUT A$
70 PRINT : PRINT "ENTER ENDING DATE (MM/DD/YY) : ";: INPUT B$
80 PRINT : INPUT "ENTER AMOUNT : ";A
90 PRINT : INPUT "ENTER INTEREST RATE : ";IT
100 PRINT: PRINT: PRINT: PRINT
110 IF IT>1 THEN IT=IT/100
120 M=VAL(LEFT$(B$,2))-VAL(LEFT$(A$,2))
130 D=VAL(MID$(B$,4,2))-VAL(MID$(A$,4,2))
140 Y=VAL(RIGHT$(B$,2))-VAL(RIGHT$(A$,2))
150 IF M<0 THEN M=12+M: Y=Y-1
160 IF D<0 THEN D=30+D: M=M-1: IF M<0 THEN Y=Y-1: M=M+12
170 PRINT TAB(20);"$";A;"   AT   ";(IT*100);"%"
180 PRINT TAB(10);"FROM :   ";A$;"   TO   ";B$
190 DAYS=D+(Y*360)+(M*30)
200 RATE=(IT*A)/360
210 AMT = DAYS * RATE
220 PRINT : PRINT TAB(20)"AMOUNT OF INTEREST"
230 PRINT TAB(22) USING "$$####.##";AMT
240 END
```

## Date Conversion

This short routine takes a beginning date and ending date, and calculates the elapsed time involved. The advantage of this routine is that it allows the dates to be entered in the standard MM/DD/YY format. Entered as strings, the numerical values are then extracted using string handling routines.

This routine can be very useful in a program where elapsed time is used as a factor in calculations. The interest calculation program shows an example of how this routine can be used in a program. This is only one example of how the routine can be used.

The calculations are based on 30 day months and 360 day years, as are most business oriented tasks such as the interest example.

Joe Ligori
2660 W. Ball Rd. #91
Anaheim, CA. 92804

```
LIST
10 '        ===== DATE CONVERSION =====
20 '        ---------------------------
30 PRINT : PRINT : PRINT
40 PRINT "ENTER BEGINNING DATE (MM/DD/YY) : ";: INPUT A$
50 PRINT: PRINT "ENTER ENDING DATE (MM/DD/YY) : ";: INPUT B$
60 PRINT: M=VAL(LEFT$(B$,2))-VAL(LEFT$(A$,2))
70 D=VAL(MID$(B$,4,2))-VAL(MID$(A$,4,2))
80 Y=VAL(RIGHT$(B$,2))-VAL(RIGHT$(A$,2))
90 IF M<0 THEN M=12+M: Y=Y-1
100 IF D<0 THEN D=30+D: M=M-1: IF M<0 THEN Y=Y-1: M=M+12
110 PRINT "MONTHS :";TAB(20);M
120 PRINT "DAYS   :";TAB(20);D
130 PRINT "YEARS  :";TAB(20);Y
140 END

ENTER BEGINNING DATE (MM/DD/YY) : ? 10/08/77

ENTER ENDING DATE (MM/DD/YY) : ? 03/04/78

MONTHS :          4
DAYS   :          26
YEARS  :          0

ENTER BEGINNING DATE (MM/DD/YY) : ? 08/28/76

ENTER ENDING DATE (MM/DD/YY) : ? 10/25/78

MONTHS :          1
DAYS   :          27
YEARS  :          2
```

```
RUN

ENTER BEGINNING DATE (MM/DD/YY) : ? 04/22/75

ENTER ENDING DATE (MM/DD/YY) : ? 10/24/78

ENTER AMOUNT : ? 1000

ENTER INTEREST RATE : ? 7


                $ 1000    AT    7 %
     FROM :   04/22/75    TO    10/24/78

            AMOUNT OF INTEREST
                  $245.39
```

# short programs...

## Indefinite Articles

When I was typing ANIMAL *(from BASIC Computer Games)* into my system, I noticed that the program did not provide for the proper article preceding a noun. This user defined function takes as its argument any word, and catenates "A" or "AN" as appropriate.

Here it is in two different BASICs:

DEF FNA(X$)="A"+LEFT$("N",INSTR(1,"aeiouAEIOU",LEFT(X$,1)))+" "+X$
DEF FNA(X$)="A"+SEG$("N",1,POS("aeiouAEIOU",SEG$(X$,1,1).1))+" "+X$

So the result of
PRINT FNA("HAWK"), FNA("OWL")
would be:

  A HAWK     AN OWL

Brian Hammerstein<br>
5700 Arlington Avenue<br>
Riverdale, NY 10471

# Puzzles, Problems and Programming Ideas

*A new learning activity from Creative Computing....*

# Reading, Writing, and Computing

Walter Koetke

When Alex was admitted to Hopeful Hospital he knew he was very ill. After a thorough examination Dr. Frank concluded that Alex had a severe case of sleepitis, and that the proper treatment was lots of activity and exercise. Because Hopeful Hospital was very modern and up-to-date, the medical facilities included a computer to assist in the diagnosis and treatment of unusual illnesses. Dr. Frank entered Alex's symptoms into the computer, and the computer concluded that Alex had a severe case of exhaustitis. The computer also indicated that the proper treatment for exhaustitis was no activity, no exercise and lots of sleep. Dr. Frank considered his original diagnosis and the computer diagnosis, then decided to go along with the computer. The doctor prescribed no activity, no exercise and lots of sleep. Alex died one week after beginning the prescribed treatment. After his death, a group of doctors re-examined Alex's records and demonstrated that Alex actually had sleepitis, and that the prescription of no activity, no exercise and lots of sleep was largely responsible for his death.

Who should be blamed for Alex's death?
   a) The computer — for making an incorrect diagnosis
   b) The programmer — because the computer's incorrect diagnosis was probably a programming error
   c) Dr. Frank — for making an incorrect diagnosis
   d) Alex — for getting sick in the first place
   e) No one — we all have to go sometime

If you answered a or b, or even gave serious consideration to answering a or b, then your computer literacy is indeed open to question. An elementary school education is inadequate if students leave without some idea of what computers can and cannot do. Society's use of computers is so extensive that even school dropouts are likely to use and certain to be affected by computers. Clearly then, the need for schools to carefully assess their efforts toward computer literacy of all students is essential.

If you need some support when you attempt to stir whatever is bogging down the implementation of computer literacy in your school, town or what-have-you, consider the following:

Dr. Arthur W. Luehrmann of Dartmouth College presented a paper titled "Should the Computer Teach the Student, or Vice Versa?" at the AFIPS 1972 Spring Joint Computer Conference. In this truly classic, penetrating paper Luehrmann raises questions such as "how much longer will a computer illiterate be considered educated? How long will he be considered employable and for what jobs..." Luehrmann's article inspired the title of this one — for developing skills in reading, writing and computing should now be the fundamental objective of education.

As you are probably aware, the NAEP (National Assessment of Educational Progress) conducted an extensive program in 1972-73 to determine the nature and effectiveness of mathematics education in the United States. In 1977-78, the NAEP will again assess the state of mathematics education — but something new will be added. At that time they will also separately assess the nature and effectiveness of computer literacy education. How well will the students of your school and your town reflect a basic knowledge of computer literacy?

Computer literacy can not only be profitable, but it also might record your name for posterity. Several students at the California Institute of Technology recently used a computer to help generate their entries for a sweepstakes type of contest being run by the McDonald's Corp. The students not only won first prize, a car — not a pile of hamburgers, but also won a large percentage of all the other prizes. While McDonald's was objecting and claiming that the students took unfair advantage of all other contestants, their rival, Burger King, was presenting CIT with a scholarship in the name of the student who originated the idea.

Since one of the primary purposes of this column is to suggest interesting problems, let's redirect our attention in that direction. Three of the following four problems are for those "getting started." Finding examples other than the very standard ones can be time consuming, particularly for teachers who are developing entire computer related units. Perhaps these will help.

1. Writing a program that generates or tests for prime numbers is a very good standard example for those learning to program. Usually, however, it is an end in itself. This problem is intended to provide an application for a prime testing algorithm.

Suppose that k people are standing in a circle. After choosing one person to begin, the people count-off in a clockwise direction around the circle. If a person counts-off with a prime number he must leave the circle. The winner is the last person who remains.

Consider one example, say k = 6.
Then the people appear as

The initial counting process would yield:

```
            1 STAY
              O
6 STAY  O              ⊗ 2 OUT

5 OUT ⊗                ⊗ 3 OUT
              O
            4 STAY
```

And continuing:

```
             1 STAY
             7 OUT ⊗
6 STAY
9 STAY ⊗               ⊗ 2 OUT
11 OUT

5 OUT ⊗                ⊗ 3 OUT
           O 4 STAY
             8 STAY
             10 STAY
```

Thus the fourth player is the winner.
The problem is to make a two-column table containing k and the player who would win for that value of k. Can you predict the player who wins for any given k?

2. The second problem can be rephrased in many different ways, all of which revolve about the question "Is each positive integer 1 through n a divisor of some integer that contains only the digits one and zero?" For example:

```
1  is a divisor of  1
2  is a divisor of  10
3  is a divisor of  111
4  is a divisor of  100
5  is a divisor of  10
6  is a divisor of  1110
   and so forth.
```

If you have an answer and can prove yourself right, send me your answer in care of Creative Computing. I haven't yet seen a valid proof of an answer. Writing a program to determine the dividend for a given divisor is apt to raise several ideas worth examining — and it's only a division problem.

3. One very standard, likely boring example for new programmers who are also studying algebra is often that of the quadratic formula. A very clever way of making the same points as well as a host of more interesting ones was first presented to the author at an NCTM (National Council of Teachers of Mathematics) workshop conducted by Helen Hughes. Given the quadratic equation $Ax^2 + Bx + C = 0$, where A, B and C are integers such that $1 \leqslant A \leqslant 10$, $0 \leqslant B \leqslant 10$, and $0 \leqslant C \leqslant 10$, write a program that will find the probability that
   a) the roots will be imaginary
   b) the roots will be rational and equal
   c) the roots will be rational but unequal
   d) the roots will be irrational.
How are these probabilities affected by varying the limiting values of A, B and C?

4. The final problem isn't computer related at all. In fact, it's not even going to help you get started at anything. Actually, it might finish age problems altogether. The next time you're asked to solve a problem about Dick being four years older than Fred was last Thursday, etc., pull this out of your pocket.

Ten years from now Tim will be twice as old as Jane was when Mary was nine times as old as Tim. Eight years ago, Mary was half as old as Jane will be when Jane is one year older than Tim will be at the time when Mary will be five times as old as Tim will be two years from now. When Tim was one year old, Mary was three years older than Tim will be when Jane is three times as old as Mary was six years before the time when Jane was half as old as Tim will be when Mary will be ten years older than Mary was when Jane was one-third as old as Tim will be when Mary will be three times as old as she was when Jane was born.

How old are they now?

I'm really not sure of the source of this timeless gem. A student gave it to me several years ago, and memory suggests he reported seeing it in an issue of the American Mathematical Monthly. At any rate, it might provide those who assign age problems a few moments to reconsider their usefulness.

**The problem of Mutab, Neba and Sogal** presented in the May - June issue has been resolved. The first correct solution was submitted by Charles Kluefil of Glen Oaks, New York, who reported that the chance of survival for Mutab is 30%, for Neda is 17-7/9%, and for Sogal is 52-2/9%. Unfortunately, the citizens of Aedi are still searching for a new leader. No solutions were received that qualified the sender as the new president.

"Today's topic of discussion will be, The Dehumanizing of Education"

# Programming Contest

Recently, I was fortunate to give the Awards Address at the University of Wisconsin-Parkside's **Computer Fair III**. An ongoing feature of this fair, so capably managed by Donald Piele, is a computer programming contest. At this year's fair, there were four divisions to the contest: A—Grades 10-12, B—Grades 7-9, Junior—Grades 4-6, and Pee Wee—Grades 3 and under.

Presented below are the rules for the programming contest and the problems for Divisions A and B.

— DHA

## Contest Rules

1. Division A: Grades 10-12. Complete both parts to each problem. Division B; Grades 7-9. Complete part A of each problem.
2. Team Size: You may have up to three members on your team.
3. Computer System: You will be provided a terminal to the HP-2000 BASIC timeshared system; optionally, you may use your own microcomputer system, provided that a hard copy of the programs and sample runs can be made. A printer is available for the Apple II, TRS-80, PET and other systems with RS232 interface capabilities. See us **before** you begin if you need to use such a printer.
4. Contest Problems: Your five team problems are attached to this sheet. You have 2 hours to solve them. An additional 15 minutes will be provided to produce hard-copy output if you bring your own computer.
5. Grading Procedure: Your solutions will be judged on following criteria:

   a. Does it run properly, using the test data requested in the problem?
   b. Is the program easy to read?
   c. Is the program logical, imaginative, creative?

   Ten points will be given for (a), and an additional 5 points will be given for both (b) and (c). No partial credit will be given for a program that does not run.

6. General: No outside help is allowed, including books, programs, or people outside your team. University personnel will be present to answer questions regarding the HP-2000 computer.

## 1. Patterns

**A.** Suppose you have a 4-by-4 checkerboard and four checkers. You are to place the four checkers on the checkerboard in such a way that each of the four rows, four columns, and two main diagonals contains exactly one checker. One such example is the following pattern:

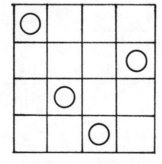

You are to write a program to print **all** the possible 4-by-4 checkerboard patterns which conform to the above rules. Each pattern should be displayed with a star (*) in the position of a checker, and a dot (.) in the position of a space. For example, the above pattern should be displayed as follows:

```
* . . .
. . . *
. * . .
. . * .
```

(Hint: To make the display more readable, you should print a blank between each of the characters on a line.)

**B.** More generally, your program should work for **any** n-by-n checkerboard, with the n checkers placed such that each of the rows, columns and two main diagonals contains exactly one checker. Your program should allow for input of the size n of the checkerboard. Each run should end by printing the total number of successful patterns found. The program should also request from the user whether the patterns themselves should be displayed; if not, **only** the **number** of successful patterns should be printed.

Test your program with $n=3$, $n=4$ and $n=5$ checkers. Include displays of the actual patterns for $n=4$, but do not include them for $n=3$ or $n=5$.

## 2. Letter Frequencies

**A.** In cryptography, it is often necessary to count the number of characters in a string. For example,

the string "ABAX1PX ?XX" has the following letter frequency table:

| letter | frequency |
|--------|-----------|
| A | 2 |
| B | 1 |
| X | 4 |
| 1 | 1 |
| P | 1 |
|   | 1 |
| ? | 1 |

You are to write a program to input a string of characters, and to output the string's letter frequency table as above. Test your program on the above example, and on the string 'BAA, BAA, BLACK SHEEP!"

**B.** Note that in displaying the letter frequency table, the letters should be displayed in **decreasing** order of frequency; that is, letters occurring most frequently should appear at the top of the list, and those occurring least frequently should appear at the bottom.

### 3. Pascal's Warped Triangle.

**A.** A variation on Pascal's Triangle is

```
row 1: ─────→ 1
row 2: ─────→ 1  1
              1 3 1
             1 5 5 1
            1 7 13 7 1
           1 9 25 25 9 1
```

Each new entry is obtained from the three entries above it by adding them together, as depicted above. The first and last entries in a row are always 1.

You are to write an algorithm which will produce the first eight rows of the above triangle. For simplicity, you should print each row beginning at the left hand margin.

**B.** You are to write an algorithm which will print the nth row of Pascal's Warped Triangle for any given input value n (where n < 100). Your program should use the least amount of storage possible.

Test your program by printing rows 2, 6, 8 and 13.

### 4. Sums of Digits.

**A.** Certain numbers are divisible by the sum of their decimal digits. For example, 18 is divisible by $1 + 8 = 9$.

You are to write a program which will find all three-digit numbers which are divisible by the sum of their digits.

Since there are over 200 of them, your program should merely produce the **number** of successes, rather than a complete list of them.

**B.** Run your program with n = 2, n = 3, and n = 4.

### 5. Rock, Scissors, Paper.

**A.** The children's game of Rock, Scissors, Paper is played by two opponents. At each turn, both players simultaneously signal their choices of either Rock, Scissors, or Paper. The winner of that turn is determined by the rule that Rock wins over Scissors (Rock breaks Scissors), Paper wins over Rock (Paper covers Rock), and Scissors wins over Paper (Scissors cuts Paper).

You are to write a program to play the game of Rock, Scissors, Paper between the computer and yourself. At each turn, the computer randomly chooses either Rock, Scissors or Paper (without letting you know). It

then prompts you for your choice. You enter R for Rock, S for Scissors, or P for Paper. The program then prints the winner of that turn. The game continues until you enter a Q to quit when it is your turn. The program should print a summary of the game's results.

Run your program twice, with about 20 turns on each run.

**B.** In addition, the computer should take into account your playing strategy. It should keep track of your plays, and should make its next choice based on what you have done on your previous turns. For example, if you have chosen P for paper on your last several turns, the computer should be more likely to choose S for scissors on its next turn. The choice of programming this strategy is up to you.

Run your program four times. You should play twice in such a way that the computer will show its ability to use its strategy to win. Then play twice to try to beat the computer, if you can. Use about 30 turns on each run.

# 40 Programming Ideas

by J. Cletheroe, Sandhurst School, England

Many of these ideas are not original, but they are gathered together here for convenience. They are presented in random order. [Condensed from a longer list which appeared in *Computer Education*.]

1. Evaluate 'pi'.
2. Provide an information retrieval system (for example information about characters in a book).
3. Provide an 'array arithmetic' package. Numbers are held with each digit in a separate cell of an array, and these numbers can be added, subtracted, etc.

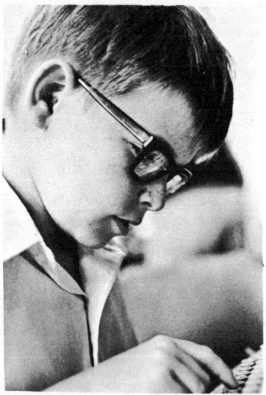

4. Round numbers (to so many decimal places or significant figures).
5. Compute the intersection and union of two numerical sets.
6. Test a number for being prime; print a list of prime numbers.
7. Find the HCF and/or LCM of two numbers.
8. Convert numbers from any base to base ten/base ten to any base/any base to any base.
9. Find the first few Perfect Numbers.
10. Print the Fibonacci Sequence.
11. Print the Fibonacci Sequence in a Modular Arithmetic — you get a set of 'rings'.
12. Work out square roots without using the square root function.
13. Simulate the action of the Absolute Value function.
14. Divide without using the / facility.
15. Print multiplication and division tables; repeat in a modular arithmetic.
16. Sort a two element/three element/n element set of numbers into order.
17. Print Pascal's Triangle.
18. Print Random Sentences, for example in the format (article) (noun) (verb) (article) (noun).
19. Solve problems like CROSS + ROADS = DANGER.
20. Convert from Arabic to Roman Numerals and vice-versa.
21. Matrix inversion.
22. Produce abstract (random) 'art'.
23. Game playing by heuristic (learning) methods.
24. Print squares and other shapes out of asterisks.
25. Print large letters out of asterisks (for posters etc.).
26. Simulate the action of the random number generator.
27. Print powers of 2 until the numbers have more than 100 digits.
28. Work out the best straight line through a set of points on a graph.
29. Output numbers in words (e.g. 512 gives FIVE HUNDRED AND TWELVE).
30. Text Analysis (frequency of letters, etc.).
31. Play the game of guessing a letter (is it a vowel?, has it any straight lines?, etc.).
32. Produce a plot of prime numbers (* for a prime, space for a non-prime). The user should select the number of elements per line of the plot. Are there any patterns?
33. Produce a table of n and the number of primes below n (call this m). Does m tend towards a function of n as n becomes large?
34. Simulate the growth of a colony of Amoebae (doubling in number every unit time, but allow for deaths due to the food supply running out and pollution building up). Can you achieve a stable state?
35. Produce a list of Primes using the 'Sieve of Eratosthenes' — strike out multiples of 2, find the next non-zero entry (which is the next prime), strike out multiples of this number, and so on.
36. Statistical work to test the randomness of the random number generator. For example, produce a table of the frequencies of ascending and descending runs of length n.

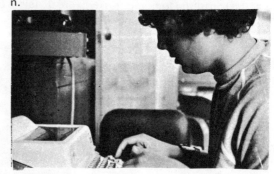

37. Print numbers with leading zeroes, to give

       1
      15
       3
    2357

    rather than

    1
    15
    3
    2357

    (which is what normally happens).
38. Scan a set of numbers and print the highest and the lowest.
39. Read in three numbers which are the lengths of the three sides of a triangle. Print (as appropriate) "ACUTE", "RIGHT ANGLED", "OBTUSE" or "NO TRIANGLE FORMED".
40. Store lists of variable length.

# Computer Conversations

**44**

$$A+B+C = A*B*C$$

Can you find three natural numbers that give the same result when added or multiplied?

**49**

A kangaroo is chasing a jackrabbit. The jackrabbit takes 3 jumps in the same length of time the kangaroo takes 2 jumps. But each jump the rabbit takes covers only half the distance of one of the kangaroo's jumps. The rabbit was 10 jumps ahead when the kangaroo first spotted it. How many jumps will the rabbit take before the kangaroo catches it?

**52**

Louise dropped a ball from her 3rd floor apartment window, which was 10 meters above the sidewalk. Her friend, Fletcher, counted the number of times it bounced. The ball rebounded half the distance on each bounce. How far had the ball traveled altogether when it hit the ground the 100th time?

**62**

A, B, C are each a single digit. What is the minimum value of ABC divided by A+B+C ? (The answer is not 1.)

**13**

K434KO

What value of K would make K434KO divisible by 36?

K = _____

"This problem is o.k. by me..."

**22**

Lot's of people have heard about Jack and his Beanstalk. But most of them don't know about the growth pattern of the beanstalk. On the first day it increased its height by $\frac{1}{2}$. On the second day it increased by $\frac{1}{3}$, on the third day by $\frac{1}{4}$, and so on. How long did it take to reach its maximum height (100 times its original height)?

IT TOOK _____ DAYS

**67**

A million people lived in the land of BASIC. Queen Terminalla was distributing her wealth to the people. She gave $1 to the first person. The next 2 people each got $2. The next 3 people each got $3, and so on. How much did the millionth person receive?

These activities are reprinted from "Computer Conversations" (a set of 41 colorful 14x21 cm cards) and "More Computer Conversations" (27 cards). "Computer Conversations" costs $3.95, teacher guide $2.95, "More Computer Conversations", $2.95, teacher guide $2.50. Postage on all orders $1.00. The Math Group, 5625 Girard Ave. So., Minneapolis, MN 55419.

© 1975-The Math Group, Inc.

# The Mechanical Mouse

## Can you flowchart his path through the four mazes?

John Maniotes
James S. Quasney

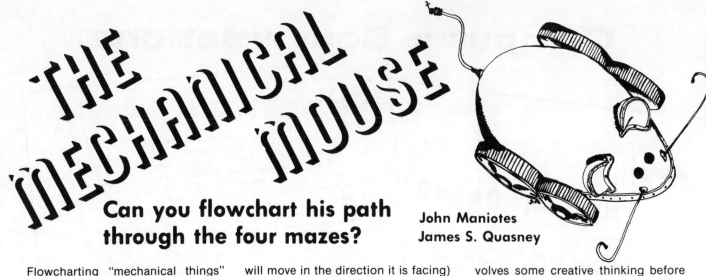

Flowcharting "mechanical things" has been around for quite a long time in beginning programming courses. A popular flowchart problem, which the senior author was exposed to in the late 1950's and which has since undergone many revisions, is *The Mechanical Mouse* problem. This is a fun-type flowchart problem that should delight the novice, intermediate, and professional programmer.

### The Problem

Draw *one* flowchart that will cause the Mechanical Mouse to go through any of the four mazes shown in the figure. At the beginning, an operator will place the mouse on the entry side of the maze, in front of the entry point, facing "up" towards the maze.

The instruction "Move to next cell" will put the mouse inside the maze. After that, the job is to move from cell to cell until the mouse emerges on the exit side.

If the mouse is instructed to "Move to next cell" when there is a wall in front of it, it will hit the wall and blow up. Obviously, the mouse must be instructed to test if it is "Facing a wall?" before any "Move."

The Mechanical Mouse's instruction set consists of the following:

A. Physical Movement
   (1) Move to next cell (the mouse will move in the direction it is facing)
   (2) Turn right
   (3) Turn left
   (4) Turn around (all turns are made in place, without moving to another cell)
   (5) Halt.

B. Logic
   (1) Facing a wall? (Through this test, the mouse determines whether there is a wall *immediately* in front of it; that is, on the border of the cell it is occupying and in the direction it is facing.)
   (2) Outside the maze?

If the mouse is outside the maze, the mouse can also make the following decisions:
   (3) On the entry side?
   (4) On the exit side?
   (5) Branch (unconditional to any part of the program).

### Types of Solutions

There is a variety of ways of attacking this problem and a variety of solutions.

Beginners seem to use two methods of *attack* to gain a solution. The first involves the "sledge-hammer" approach, where a flowchart is written to work for one maze and then additional logic is added in a piecemeal fashion to handle the remaining three mazes. Naturally a lot of trial and error is involved, and the flowchart solution is spread over several pages, making it difficult for one to comprehend the solution readily.

The second method of attack involves some creative thinking before the first flowchart symbol is ever drawn. The key centers around the definition of a *cell*. In this problem a cell is a "four-sided" figure with one or more sides missing. It is this symmetry that one wants the mouse to take advantage of so it can turn right or left or around accordingly.

The types of flowchart *solutions* generally fall into the "short" or "long" flowchart category with some solutions in between these two extremes. The short flowchart solutions have a few symbols (six to seven symbols, excluding Start and Halt) but subject the mouse to a lot of false and inefficient turns in each cell.

The long flowchart solutions have a lot of symbols (15 to 20) and subject the mouse to few false and inefficient turns in each cell. Other flowchart solutions are in between these two extremes and represent a compromise.

The short flowchart solutions have the advantage of using less "storage" than the long ones. However, the long flowchart solutions take less "execution time" for the mouse to carry out its objective. Hence, one has to weigh the amount of "storage" and "execution time" used to determine the "best" solution.

Note that one flowchart solution must work on all four mazes. The hardest maze for the beginner is usually maze 4. So don't be surprised if your flowchart works for the first three mazes but fails on the fourth maze.

As an extra-credit problem, enlarge each four-cell maze given to either a 9 or 16-cell maze and see if your existing solution still works for the new mazes as well as those shown in the figure.

For those who desire a solution to The Mechanical Mouse problem, please send the senior author a self-addressed stamped envelope (SASE) and enough postage for its return. For those who have other versions of this problem, we would be interested in corresponding with you. Either way, we hope you have fun with The Mechanical Mouse problem! ∎

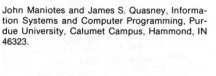

John Maniotes and James S. Quasney, Information Systems and Computer Programming, Purdue University, Calumet Campus, Hammond, IN 46323.

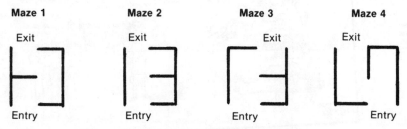

Four mazes where each maze has four cells.

# PINBALL MAZE

Start this maze at either of the two arrows at top and bounce from one "bumper" to another. Your object is to get a score of exactly 100 and then get out at one of the three bottom exits. You may not use any number more than once. No tilting please!

You might want to try to write a computer program to solve this maze but be warned: it is a challenging, non-trivial problem.

Reprinted from *Challenging Mazes* by Lee Daniel Quinn published by Dover Publications. Copyright 1975 by L.D. Quinn.

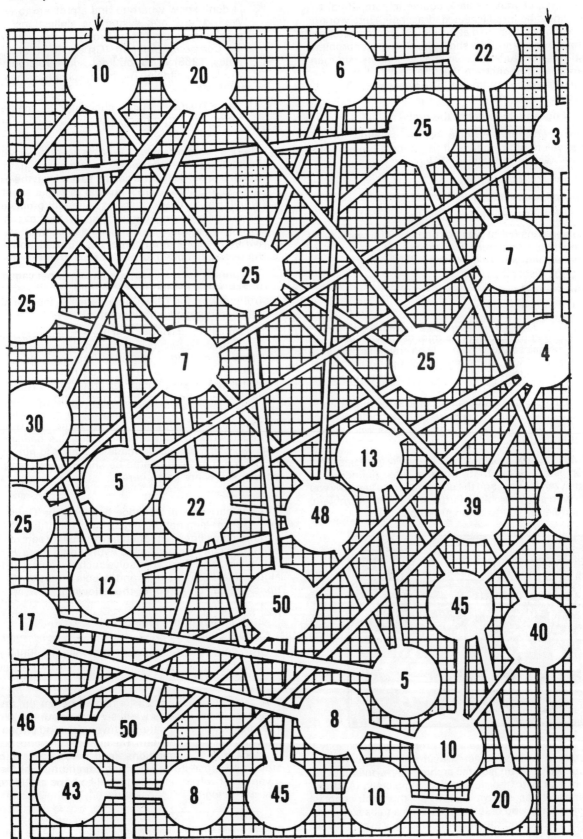

# Learning by Doing

by Fred Gruenberger

In a second, or intermediate, course in computing, the student should be lured (coerced, dragooned) into working on a term project, assigned at the start of the semester and due at final exam time. This should be a computer problem, preferably of the students' own choice, done in workmanlike manner to demonstrate that the student has learned something of the computing art. The final result should be packaged neatly, to include:

1. An English statement of the problem.
2. Flowcharts of the logic of the solution (or the equivalent of flowcharts, if the student prefers some other way of expressing his logic).
3. Listings of the program.
4. Results, clearly labelled.
5. A test procedure and test results.
6. Conclusions (what was learned from the work; what would be done differently if the project were to be repeated; limitations on the results; suggestions for further research, etc.).

The packaged term project should be saved for use in eventual job interviews.

Experience has shown that the student's chief problem is that of selecting the problem he intends to work on. In all likelihood, he has so far worked only on problems that were assigned (and hence clearly labelled computer problems), for which much of the analysis has already been done for him.

There is usually a fair amount of panic at the time this assignment is made, since it puts the student on his own for the first time. About a third of a typical class will suggest one of the following:

1. "May I use a problem I did in my Fortran class last semester?" No, that problem was defined and analyzed for you, and the object now is to see how you do with a new problem. Further, you *did* that problem; it's time for you to do another one.

2. "I've been assigned to a big problem at work; may I turn it in for this project?" No; that problem concerns your work; try an isolated problem here, one that you can deal with thoroughly and completely. (And experience has shown that when this restriction is relaxed, the end result always seems to be someone else's work, and the student can barely explain what went on.)

3. "I don't know where to find a problem to work on." Well, there are systematic collections of good problems; you might try browsing through *Problems for Computer Solution* (Gruenberger and Jaffray, Wiley, 1965) which outlines some 90 problems that would be suitable. The best problem is the one that interests you.

4. "I'm a business (music, chemistry, ..., mathematics) major; I'll do a problem in business (music, chemistry, ..., mathematics)."

There's the real problem. What is needed, early on in the semester, is a *proposal* by the student of just what he intends to do, so that he can be saved from the extremes of plunging into a problem that is either trivial or too grandiose. In the former case, he will produce something that requires little or no knowledge of computing; in the latter case, he will look sad at final exam time (when the computing center is saturated) and wail that he needs just one more run of his program.

The phrase "a business problem" is rather vague. A new attack on Bill of Materials scheduling? An inventory control program for 10,000 line items? A table of base pay times hours worked? Or what?

The trouble is, of course, that the whole idea is brand new to most students; he has never been placed in the awkward position of making a selection that is within his own capabilities (indeed, he has probably never been told what constitutes a decent computer problem). Further, he has never had to define a problem and outline an attack on it; this has been done for him for all of the 14 years he has been in school. It helps if he can see samples of what this is all about, so a collection of old term projects (both good ones and bad ones) should be made available to him. It would help even more if he could be shown some sample proposals. A few are given here.

I

A comparison will be made between the Gauss-Seidel and Gauss elimination algorithms for the solution of simultaneous linear equations. A number of sets of six such equations will be constructed with known roots. Programs will be written in Fortran to solve such systems with the two algorithms, both in single and double precision. The solutions will be compared for the following:

1. Computation time.
2. Accuracy of the results.
3. Compilation time.

One of the sets of equations will involve a singular matrix, to determine how this condition is handled by the programs.

II

A small inventory of 25 items will be set up, and daily changes to that inventory will serve as input to a program. The program is to update the inventory, and print a report showing, for each line item, the quantity on hand, items out of stock, reorder conditions, lead time to arrival of new stock, and those items requiring expediting. For the small inventory involved, all results will be hand calculated as a test of the logic.

---

Reprinted with permission from *Popular Computing*, Vol. 3, No. 11. Copyright 1975 *Popular Computing*, P.O. Box 272, Calabasas, CA 91302.

## III

The COBOL reference manual at our installation lists 58 error messages for errors that can occur at compile time. A program will be written that will trigger each of these error messages.

## IV

The melody of a song can be expressed as a series of numbers. The pitch of each note can be expressed by numbering the notes of the scale, and the duration of the note can be coded on a scale from 1 to 8. With a given melody so coded, an algorithm can be applied to it to translate it into a new melody. The simplest such algorithm would be to reflect each note around a middle value, to transform high notes into low notes and vice versa, leaving the duration of each note fixed.

Several algorithms will be devised, and applied to ten "standard" tunes. The resulting translated melodies will be played, and the most pleasing result will be recorded and submitted on a cassette as part of this project.

## V

The Los Angeles *Times* prints about 500 column-inches of text each day in its main news section. Some of these inches can be identified as politically oriented:

- A. Favorable to Republicans.
- B. Unfavorable to Republicans.
- C. Favorable to Democrats.
- D. Unfavorable to Democrats.
- E. Favorable to third-party candidates.
- F. Unfavorable to third-party candidates.
- G. Completely neutral.

In theory, material that reflects the political leanings of the newspaper's editors should be confined to the editorial pages, and the news pages should be unbiassed. In practice, the amount of text space allotted to a candidate or a party reflects the paper's views, however unconsciously.

The pages of each issue for the 8 weeks preceding the last general election will be examined, and a listing made of the column-inches in each of the first four categories given above, as objectively as possible. Ratios will be calculated of the following:

$$\frac{A}{B} \qquad \frac{C}{D} \qquad \frac{A}{C} \qquad \frac{A+D}{B+C}$$

for each day, and progressive totalled for the 56 days.

While this is not properly a computer problem (the small amount of arithmetic involved could easily be done by hand), the program will be useful for a much larger project jointly sponsored by the School of Journalism and the Department of Political Science. During the 12 weeks preceding the coming presidential election, the ratios will be calculated and plotted for each of 15 leading newspapers across the country.

## VI

Attached is a diagram of the maze in the gardens at Hampton Court Palace, constructed in the reign of William III. "The key to the centre is to go left on entering, then, on the first two occasions when there is an option, go right, but thereafter go left." The maze exists today, and for 2p a visitor may enter the maze, seek the centre, and retrace his way out.

Given a new maze, described in terms of the coordinates of the branch points, a program is to be written to explore the maze and output the directions for the choices to be made to proceed from the entry point either to the center or to a specified exit. The choices will be limited to two at each branch point.

One test of the program will be the reproduction of the directions quoted above for the Hampton Court maze. The computer solution for the other mazes will be checked by hand.

Plan of the maze at Hampton Court Palace, England. The path to be followed is the black line.

## VII

If the numbers from 1 to 20 are permuted, what is the distribution of runs up and runs down of the numbers? Consider the following possible permutations:

| | | | | | | | | | | | | | | | | | | | | |
|---|---|---|---|---|---|---|---|---|---|---|---|---|---|---|---|---|---|---|---|---|
| A | 1 | 2 | 3 | 4 | 5 | 6 | 7 | 8 | 9 | 10 | 11 | 12 | 13 | 14 | 15 | 16 | 17 | 18 | 19 | 20 |
| B | 1 | 20 | 2 | 19 | 3 | 18 | 4 | 17 | 5 | 16 | 6 | 15 | 7 | 14 | 8 | 13 | 9 | 12 | 10 | 11 |
| C | 20 | 18 | 16 | 14 | 1 | 3 | 5 | 7 | 12 | 10 | 8 | 6 | 4 | 2 | 9 | 11 | 19 | 17 | 15 | 13 |
| D | 2 | 12 | 17 | 4 | 13 | 10 | 1 | 5 | 15 | 19 | 20 | 7 | 16 | 14 | 6 | 18 | 3 | 8 | 9 | 11 |
| E | 11 | 19 | 2 | 7 | 12 | 1 | 8 | 6 | 18 | 13 | 15 | 20 | 17 | 10 | 16 | 14 | 3 | 4 | 5 | 9 |
| F | 1 | 15 | 14 | 2 | 16 | 13 | 3 | 17 | 12 | 4 | 18 | 11 | 5 | 19 | 10 | 6 | 20 | 9 | 7 | 8 |

For A, there is just one run up, and none down, but for permutations taken at random, the chance of arrangement A is just 1 in 20!, or about 1 in 2.4 times 10 to the 18th power.

For B, there are 10 runs up and 9 down, but this, too, is an unlikely arrangement.

For C, there are two runs up and three down.

For D, there are six runs up and five down.

For E, there are seven runs up and six down.

For F, there are seven runs up and six down.

The complete distribution of all possibilities for all 20! permutations could be determined by theory. I propose to approximate the distribution by sampling random permutations and counting the runs.

In order to do this, I will need a random number generator and an algorithm for forming random permutations. For the former, I will use the generator described on page 8 of issue 21 of *Popular Computing* (my work will be done in Fortran). For the latter, one of the following schemes will be used:

1. Generate an array of 40 numbers. In the even positions put the numbers from 1 to 20. In the odd positions, generate 20 random numbers. Then sort the 20 pairs, using the random numbers as the sort key. Although it is inefficient, I will bubble sort the 20 pairs. After sorting, the right hand number in each pair of numbers will be a random permutation of the numbers from 1 to 20.

2. Use the random number generator to generate integers in the range from 1 to 20, and let these integers be the subscripts for entries in an array of dimension 20. First zero out this array. Select elements in the array by the choice of subscript. If the chosen element is zero, fill it with the next consecutive integer, starting with 1. If the element is not zero (i.e., it has already been selected), proceed to another element. When all the elements are filled, the array contains the desired random permutation.

3. In the above scheme, the first 16 or so elements will be filled rapidly (that is, the condition of duplicates will not occur too often), but the last 4 or so may take an undue amount of time. A variation might be tried. Apply the scheme described until 16 elements are filled. Then insert the remaining four numbers into the four blank slots, picking one of 24 arrangements of those four positions at random, again using the random number generator.

At least 1000 random permutations will be generated, and a distribution made of the lengths of the runs. This distribution will be compared to the theoretical (if I can obtain that). The program will be generalized so that it can operate on dimensions other than 20. If time permits, runs will be made on permutations of 10 and 30 items.

## VIII

I would like to try to calculate the number (15000!). (I understand that the current record for factorials is 10000!, and that all the factorials by thousands are on file.) Even if I don't succeed in obtaining the desired result, I believe that working on the project will be worthwhile, and I will be able to at least provide hints and suggestions for the next person who tries to break the record.

Since this calculation will consume considerable amounts of machine time, I propose to calculate the following test data first:

1. The exact number of expected digits.
2. The number of low order zeros.
3. Some of the high order digits, and some of the low order non-zero digits.

In addition, I will use my program to calculate (and check with known results) 500!, 1000!, and 10000! before requesting a commitment of machine time for the long run.

I believe that I can hold intermediate results in storage by packing six decimal digits per machine word. I will need packing and unpacking subroutines, and a subroutine for decimal multiplication. I will test these subroutines before making any long runs.

## IX

Problem H5 in *Problems for Computer Solution* calls for the creation of abstract paintings by a computer program. The program is to select the size and shape of various geometric figures and their position on a canvas. I wish to explore this notion extensively with the aid of the plotter now available, which should aid greatly in the mechanical chore of examining the program's results. The plotter will allow up to three primary colors for the figures, and the choice of color will also be made by the program. The plotter routines also permit the figures to appear in outline form, or solid (filled in), as well as various degrees of cross-hatching.

It is stated in Problem H5 that an important aspect of the problem is the determination of when to stop. I propose to put this decision into the program in terms of the area covered by the random figures. This will be a bit tricky, since the figures overlap. However, if the total area covered by the figures is limited to some fraction of the available area, with or without overlap, and this limit is a parameter of the program, then the program will be able to output finished art without intervention in any quantity.

The ratio of successive terms of the Fibonacci sequence approximates the golden mean:

$$\frac{1 + \sqrt{5}}{2} = 1.6180339887498948482045\ldots$$

Thus, we see

$144/89 = \underline{1.6179775}$

$1597/987 = \underline{1.6180344}$

$121393/75025 = \underline{1.6180339886}$

Using the EXACMATH package, I propose to write a program to generate successive terms of the Fibonacci sequence, form the ratio, and determine to how many digits the ratio agrees with the golden mean. The limits of the EXACMATH package will let me carry these calculations to at least the 1000th term of the sequence. My output will be a table of values (term number against the number of digits of agreement) and a graph showing the rate of growth of the function being explored.

## X

The Raindrop Problem, which appeared in issue 6 of *Popular Computing*, called for selecting a point at random in the unit square as the center of a circle whose radius is taken at random between zero and 1/2 unit. The problem asked for the number of such circles needed to completely cover the unit square.

A crude solution will be attempted by subdividing the square into 400 smaller squares. A mathematical test can be devised to determine whether or not each of these smaller squares is covered by one of the circles. The results of 100 trials will be plotted, to obtain an approximation to the desired distribution. For a few of these trials, the method will be repeated with the large square subdivided into 900 smaller squares, to determine if the method could lead to a correct solution.

# Puzzles and Problems for Computers and Humans

▶ A fly and a mosquito start together to circle a building, but the fly circles the building in six minutes, the mosquito in four. How many minutes will elapse before the mosquito passes the fly?

▶ Write down five odd numbers so they total 14.

▶ A man, anticipating the rationing of gasoline, hoarded 100 gallons in a large tank behind his garage. One night a neighbor decided to siphon off ten gallons for his snowmobile. As the gasoline was run off, the villainous neighbor poured water in the tank at the same rate as he was filling his gas can, keeping the contents thoroughly mixed throughout.

If the whole caper took ten minutes, how many gallons of gasoline and how much water did the neighbor end up with in his ten-gallon can?

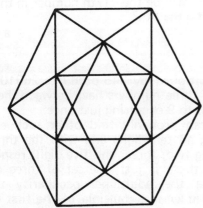

▶ How many triangles are there in this figure?

▶ Write a program to convert any number from 1 to 3000 to its equivalent Roman numeral. The seven Roman symbols are:

| | |
|---|---|
| M | 1000 |
| D | 500 |
| C | 100 |
| L | 50 |
| X | 10 |
| V | 5 |
| I | 1 |

The rules for forming Roman numerals are:
1. If a symbol precedes one of smaller value, its value is added.
2. If a symbol precedes one of larger value, its value is subtracted; then the difference is added to the rest of the number.
3. Numbers are written as simply as possible using only C, X, and I as subtrahends. Some examples are MCMLXIV, 1964; DXLIX, 549.

Your program should accept as input the decimal number and output the Roman numeral. Convert the following numbers in your contest entry: 1, 14, 400, 549, 999, 1964, 1975, 2500, 2994, 3000.

▶ In the octonary (modulus 7) system, write a program to find the seven 7-digit squares which contain no duplicate digits. Here is one: $1242^2$ = 1567204.

▶ We are told that the purchase price of Manhattan Island in 1626 or 1627 was $24. Assume the Indians invested their money on January 1, 1627 at 6% compounded annually. At that rate, the money would be worth a great deal by January 1, 1975. We know that if P dollars are invested at an interest rate of r (expressed as a decimal), the amount A the following year is given by:
$$A = P(1 + r) \quad (1)$$
and the total amount after n years is given by:
$$A = P(1 + r)^n \quad (2)$$

Write a program to calculate the value of the investment on January 1, 1975 to the nearest penny using two methods: (1) compute the amount year by year using Formula 1 and accumulate it, (2) compute the amount by Formula 2. In Method 1, round the result each year to the nearest cent. Your challenge, of course, is to maintain more significant digits than most computers are capable of handling.

You'll be interested to compare your answers with the 1974 assessed value of the land of some $6.4 billion.

▶ Arrange two of each of the digits 0 to 9 so as to form a 20-digit number. Your number may not begin with a zero. Then score your number as follows:

For every two consecutive digits that form a perfect square, score two points. For every three consecutive digits that form a perfect square, score three points. A four-digit square scores four points, and so on.

For example, if your number was 587382190249719503664, you would score two points for 49, two points for 36, two points for 64, and six points for 219024 — for a total of 12 points. You may *not* count 036 as a three-digit square.

What is the maximum number of points you can score?

---

Lynn D. Yarbrough comments that he has no idea of a reasonable way to attack the problem with a computer, but using a hand-held scientific calculator and some ingenuity seemed to yield a good solution. The number Lynn found, which scored 100 points, is:

| 49134681827562500379 | Square Root | Points |
|---|---|---|
| 49 | 7 | 2 |
| 134681827562500 | 11605250 | 15 |
| 1346818275625 | 1160525 | 13 |
| 81 | 9 | 2 |
| 1827562500 | 42750 | 10 |
| 18275625 | 4275 | 8 |
| 27562500 | 5250 | 8 |
| 275625 | 525 | 6 |
| 7562500 | 2750 | 7 |
| 75625 | 275 | 5 |
| 562500 | 750 | 6 |
| 5625 | 75 | 4 |
| 62500 | 250 | 5 |
| 625 | 25 | 3 |
| 2500 | 50 | 4 |
| 25 | 5 | 2 |
|  |  | 100 |

Lynn comments further, "the longest single square I have found with no tripled digits is 9987338075625 = $31602755^2$, which can be neatly combined with 25210441 = $5021^2$ into the sequence 99873380756252104416, but its score is not nearly so high (40)".

For those readers wishing to continue with Squaresville, there is a number with a score of 106 points. I leave it as a challenge for you to find it. — DHA

---

▶ Assume a life-span of 80 years. In what year of the 20th century (1900-1999) would you have to be born in order to have the maximum number of prime birthdays occurring in prime years? The minimum number?

Marsha Lilly
Sudbury, Mass.

▶ Arrange two of each of the digits 0 to 9 to form a 20-digit number. Your number may not begin with a zero. Score your number as follows:

For every two consecutive digits that form a perfect cube, score two points; for every three consecutive digits that form a perfect cube, score three points; and so on.

For example, if your number was 45864096859013312727, you would score two points for 64, four points for each for 4096, 6859 and 1331, and two points for each of the 27's ... which would give you a total of 18 points. You may not count 01331 as a five-digit cube.

Games & Puzzles

▶ A regular dodecahedron has twelve pentagonal sides and twenty vertices. Assuming that one face is in the X—Y plane with an edge along (0,0,0) to (0,1,0), what are the coordinates of the remaining 18 vertices?

Lynn Yarbrough
Lexington, Mass.

▶ There are 720 ways to arrange the digits 1 through 6 as six-digit numbers:
1 2 3 4 5 6
1 2 3 4 6 5
1 2 3 5 4 6
etc.

If you continue this sequence, in numerical order, what will be the 417th number in the series? What will be the *n*th?

Bill Morrison
Sudbury, Mass.

▶ 78 multiplied by 345 produces 26910. Notice that these three numbers have between them all of the digits 0 to 9 occurring just once.

You can probably find other such examples containing all ten digits and with the three numbers having two, three, and five digits respectively. However, there is just one set of three numbers which has the additional peculiarity that the second number is a multiple of the first number. Can you write a computer program to find this combination?

# GRID ADDRESSING

### by Gerard Akkerhuis

Grid addressing is a technique which has many applications in contemporary technology. An understanding of this method erases part of the mystery which surrounds the computer's instantaneous manipulation of both instructions and data through only data.

A fictitious kitchen grid will be used to illustrate the concept of grid addressing. A bachelor enjoyed waking to a warm room filled with soft music. He would rise, shave, have eggs, coffee, toast, and run the dishwasher. Then he would read the paper and watch the TV Morning Show. When he left the house for work, his answering service monitored his telephone calls.

After work the bachelor checked his answering service and listened to music while he shaved. Then he ate, ran the dishwasher, read the evening paper, watched the late show, had toast and milk, and went to bed with the heat turned down.

If the bachelor's appliance grid is that shown in the figure and if half of the electricity required to operate an appliance went up the vertical line and half across the horizontal line, what numbers would he give his grid for his morning program and what numbers would he give his grid for his evening program? We've filled in the first four morning operations in the grid (memory). Can you fill in the rest?

Gerald Akkerhuis, US Dependent Schools, European Area, APO, NY 09175.

*When I was in Europe at the 2nd World Computers in Education Conference in September 1975, Sam Calvin and Gerard Akkerhuis were kind enough to give Sandy and I lodging and "guide service" in the Frankfurt area. Gerard also gave me a wealth of outstanding material that he wrote for use in the USDESEA schools. Unfortunately, much of it is geared to the Interdata 7/16, a not-very-common computer in schools and homes. However, as we get the material generalized, it will appear on these pages (eventually). — DHA

## J, J, J & T!

(1) Joe, Jack, John and Tom live one on each floor of a four-story apartment.
(2) Their ages are 10, 9, 8, 5, but not necessarily in that order.
(3) Joe lives directly above the 9-year-old and directly below the 8-year-old.
(4) Jack has to pass by the 5-year-old to leave the building from his apartment.
(5) Jack is more than one floor away from Tom, who is more than one year younger than Jack.

**Find the ages and on which floor each of the boys lives.**

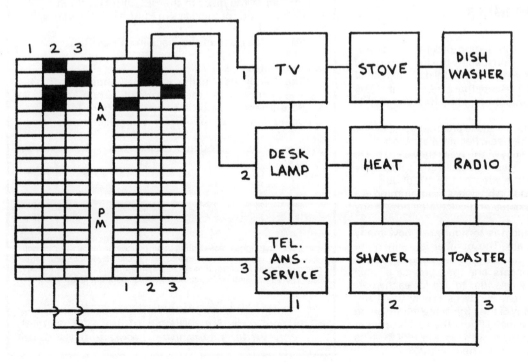

# Puzzles and Problems For Fun

### THE VICAR TOLD THE SEXTON...

The sexton at the local church was ill and did not attend the Sunday service. The vicar visited him after the service, and the following conversation took place.

Vicar: "There were only 3 people in the congregation — excluding myself — and the product of their ages was 2450. The sum of their ages was twice your own age. Can you tell me the ages of the members of the congregation?"
Sexton: "You haven't given me enough information."
Vicar: "It is sufficient for you to know that I was the oldest person there!"

Now there is no doubt that the vicar was a very mathematically minded man; but how old was he?

### ...AND THE SEXTON TOLLED THE BELL

By the following Sunday the sexton had recovered well enough to be able to ring the bells at the regular service. There were just two bells in the church belfry; the first can best be described as a 'ding' — the second, not unnaturally, as a 'dong'. Now an ancient by-law in the district proclaimed that no 'ding' could be rung exactly two chimes after another 'ding'; and that no 'dong' could be rung exactly three chimes after another 'dong'. So what was the longest sequence of chimes that the poor sexton was permitted to ring?

*Games & Puzzles*

### FOLDED

A piece of paper may be folded so that the bottom right corner just touches the left edge. Your problem is to determine, for any rectangular piece of paper, a method to produce the shortest possible crease. In the diagrams, crease CD is longer than AB but still not the shortest. What is? If you rotate the piece of paper 90 degrees (longest side horizontal) and repeat the exercise, what is your method then?

*Games & Puzzles*

# COMPUTER RECREATIONS

### by D. Van Tassel

## Calendars

Calendars provide some interesting programs. To program calendar programs you must know the following rule: A year is a leap year (that is, February has 29 days instead of 28 days) if it is a multiple of 4, except that a multiple of 100 is a leap year only if it is also a multiple of 400. January 1, 1800 was a Wednesday.

A common program on terminals is one that accepts any date and prints the day of the week. People like to input their birthdate and find out what day of the week they were born. This program can also be used to locate future three day weekends caused by holidays such as the 4th of July.

A common program needed in business programming is a program that accepts two dates and calculates how many days have passed.

A more interesting problem is to figure out how many different calendars there are. The number is really quite small. It is possible to print a perpetual calendar. That is print all the possible calendars and then provide a table showing the years and the calendar to use for each year.

The final calendar problem concerns Friday the 13th. First calculate the probability of the 13th of the month being on a Friday. Next count how many Friday the 13th's occur in this century. Is the 13th of the month more or less likely to be a Friday than any other day of the week? Why?

# Thinkers' Corner

### by Layman E. Allen © 1975

### MATHEMATICS PUZZLES

How many of the problems (a) through (f) below can be solved by forming an expression equal to the GOAL? (Suppose that each symbol below is imprinted on a disc.) The expression must use:
(1) only single digits combined with operators,
(2) all of the discs in the REQUIRED column,
(3) as many of the discs in PERMITTED as you wish, and
(4) exactly one of the discs in RESOURCES.

Special Rules:
The '*' indicates "to the power of." Thus $3*2 = 3^2 = 9$.
The '√' indicates "the nth root of." Thus $^3\sqrt{8} = 2$.
Parentheses can be inserted anywhere to indicate grouping, but never to indicate multiplication.

| Problem | GOAL | REQUIRED | PERMITTED | RESOURCES |
|---|---|---|---|---|
| (a) | 11 | 67 − | 13+ | +− ÷ √246 |
| (b) | 13 | 1+ | 34− | ×÷ √1359 |
| (c) | 1 | 18 ÷ | 4+ | +− ÷ 1279 |
| (d) | 7 | 3− | 248 | − × √678 |
| (e) | 2 | 3 ÷ | 37− ÷ | +− × 0135 |
| (f) | 1 | 9√ | 22+ | − × ÷ √− 28 |

If you enjoy this kind of puzzle, you might like playing EQUATIONS: The Game of Creative Mathematics. Free information about this and other instructional games is available upon request from The Foundation for the Enhancement of Human Intelligence, 1900-E Packard Road, Ann Arbor, MI 48104.

*Some Suggested Answers (frequently there are others):*
(a) 7 + 6 − 2
(b) (4 × 3) + 1
(c) (8 ÷ 4) − 1
(d) 2 − (3 − 8)
(e) (7 ÷ 3) − (1 ÷ 3)
(f) 2 √9 − 2

# Puzzles and Problems for Fun

## THE KEYRING PROBLEM

Consider a keyring with 5 keys. Because of the structure of a keyring, each key is adjacent to 2 other keys. The problem is to engrave a number on each key so that the keying possesses the following property:

For any number, n, between 1 and 21 inclusive, there exists an *adjacent* group of keys whose engraved numbers sum to n. For example: If 1-2-4-x-x is part of the keyring, here are the possibilities:

| n | keyring |
|---|---------|
| 1 = | 1 |
| 2 = | 2 |
| 3 = | 2+1 |
| 4 = | 4 |
| 5 = | ? |
| 6 = | 4+2 |
| 7 = | 4+2+1 |

Clearly, it is not possible to form a 5 with adjacent keys. Can you write a program to solve this problem in a *reasonable* amount of time?

*Rob Kobstad and Mike Lucey*
*Notre Dame, IN 46556*

P.S. A Fortran solution on a GA 18/30 required less than 1 minute to compile/solve the problem.

## THE FRIENDLY SKIES

Every hour on the hour a jet plane leaves New York for Los Angeles and, at the same instant, one leaves Los Angeles for New York. If each trip lasts exactly five hours, how many planes from L.A. will each plane from N.Y. see (assuming good visibility, of course)?

# COMPUTER RECREATIONS

*by D. Van Tassel*

## Syntax Messages

All programmers get tired of getting syntax error messages, but there is an interesting program to write where the goal is to get syntax error messages. Now any fool can get a lot of syntax messages (just forget to declare an array) but try to get the maximum number of *different* syntax error messages. To find out how many error messages are possible check your manuals for a complete listing of syntax error messages.

In order to not make it too easy let's try to get as many different syntax messages with as few statements as possible. We can set up a ratio as follows:

$$\text{ratio} = \frac{\text{\# of statements}}{\text{\# of different error messages}}$$

If you get a good solution send me the listing (but you must do the counting). If I get some real good solutions I will publish them in a later column. This problem is language dependent so I will try to publish solutions by language. (Send solutions to D. Van Tassel, Computer Center, Univ. of California, Santa Cruz, CA 95064).

# Thinkers' Corner

by Layman E. Allen © 1975

### SET THEORY PUZZLES

How many of the problems (a) through (f) below can you solve by forming an expression that will name the number of cards in the universe that is listed as the GOAL? (Suppose that each letter and symbol below is imprinted on a disc.)

The expression must use:
(1) all of the discs in the REQUIRED column
(2) as many of the discs in PERMITTED as you wish, and
(3) exactly one of the discs in RESOURCES

Universe of Cards:
1: B, D
2: A, B, D
3: B, D
4: B, C, D
5: A, C
6: A, C

Examples:
The expression A names 2 cards (2,6).
The expression A' (complement) names 4 cards (1,3,4,5).
The expression B∩D (intersection) names 2 cards (2,4).
The expression B∪D (union) names 5 cards (1,2,3,4,5).
The expression B-D (difference) names 1 card (1).

| Problem | GOAL | REQUIRED | PERMITTED | RESOURCES |
|---------|------|----------|-----------|-----------|
| (a) | 4 | U | ACDU | ABCU - ' |
| (b) | 3 | B | CDU∩ | BDU∩ - ' |
| (c) | 5 | CU | ABD∩ | BCU∩ - ' |
| (d) | 5 | C - | BCU∩ | CDDU∩ ' |
| (e) | 3 | AU' | BCU∩ | ABCD∩ - |
| (f) | 4 | BD - | ADU - | BCDU∩ ' |

*Some Suggested Answers (frequently there are others):*
(a) A∪B (b) B∩(C') (c) A∪C∪B
(d) (B-C)∪D (e) A'∪D (f) (A∪D)-(B∪D)'

If you enjoy this kind of puzzle, you might like playing ON-SETS: The Game of Set Theory. Free information about this and other instructional games is available upon request from The Foundation for the Enhancement of Human Intelligence, 1900-S Packard Rd., Ann Arbor, MI 48104.

# COMPUTER RECREATIONS

Dennie Van Tassel

## Bust Your Compiler

Many commands within your compiler have limitations. These limits are usually so large that you will seldom encounter them. For example, one popular compiler will allow about 400 parentheses in *one* statement before objecting. An interesting exercise is to find other limitations on your favorite compiler. Here are some suggestions:

(a) Maximum number of parentheses in one statement.
(b) Maximum size of a 1-dimensional array. Maximum dimensions of an array.
(c) Maximum length of literal or bit constant.
(d) Maximum length of a single statement.
(e) Maximum length comment or maximum number of consecutive comments.
(f) Maximum number of nested DO or FOR loops (or blocks or IF THEN ELSE).
(g) Maximum number of subroutines (or nested calls to subroutines).
(h) Maximum number of arguments in a subroutine.
(i) Maximum number of recursive calls.

Can you think of any other restrictions of this type? Hint: Try examining the list of error messages for your language compiler.

## Self-Reproducing Program Revisited

I received many solutions for this problem. For those of you that may have missed it in the Sep/Oct 1976 issue, here is the problem: Write a program that prints an exact copy of itself. No input statements are allowed.

Several people sent in solutions where they used the file the program was in or they created a file before hand, and then read the file. But this violated the rule that no input statements were allowed. Also there were several solutions sent in that required over a page of code.

Here are three good solutions, one in BASIC and two in FORTRAN. No COBOL solution was sent in, even though it is fairly easy in COBOL. It seems it should be possible to write a shorter BASIC version, but the solution is pretty good.

Basic solution by Donald Bell, a student at California State University at Fullerton.

```
10 DATA "B$= 'DATA '+CHR$(34)
20 DATA "FØR J=10 TØ 180 STEP 10
30 DATA "READ A$
40 DATA "PRINT J;B$;A$
50 DATA "IF J<>90 THEN 170
60 DATA "RESTØRE
70 DATA "B$= '
80 DATA "NEXT J
90 DATA "END
100 B$= 'DATA '+CHR$(34)
110 FØR J=10 TØ 180 STEP 10
120 READ A$
130 PRINT J;B$;A$
140 IF J<>90 THEN 170
150 RESTØRE
160 B$= '
170 NEXT J
180 END
```

## Run Times — The Most Important Variable is the Human Factor

78 multiplied by 345 equals 26910. Notice that these three numbers have between them all of the digits 0 to 9 occurring just once. Can you write a computer program to find all such combinations?

In the Jan-Feb 1975 issue of *Creative Computing,* we posed a problem to find all of the combinations of a 2-digit number multiplied by a 3-digit number equaling a 5-digit number which used all ten integers 0 to 9. (There are nine solutions.)

Geoffrey Chase, OSB, of the Portsmouth Abbey School in Rhode Island wrote five different programs to solve the problem on the same computer (PDP 8/e) and did an exhaustive analysis of the differences. Space does not permit us to print his entire discussion or the programs; however, the following is a brief summary.

| Language | Timing |
|---|---|
| FORTRAN/SABR Coding, using EAE subroutines | 0.9 sec |
| FORTRAN, no machine language patches | 3.2 |
| Compiled BASIC, using EAE | 15.5 |
| FOCAL, with some EAE floating point patches | 61.5 |
| Multi-user BASIC, no EAE | 108.0 |

We see that there is over a 100-to-1 spread with the easy-to-program languages taking considerably longer to run. However, one must ask the question whether the ultimate goal in a particular program should be efficiency in running or efficiency in coding. To give you some grist for thought, why not try to come up with an estimate of the following ratio for your computer installation.

$$\frac{\text{Cost to Program One Line of Code}}{\text{Cost to Execute One Line of Code}} = x$$

IBM estimates that the value of x for 360 and 370 series computer installations is approximately 100 million to 1. Obviously the ratio is different for a hobbyist or student programming a dedicated micro or mini. Nevertheless, the point is that the human factor is incredibly important.

That is not to say that the computer doesn't play an important role. Next issue we'll be publishing a set of timing comparison programs in Basic and Fortran along with timings on popular minis, micros and timesharing systems so you can compare your machine to others.

Fortran solution by Mark Barnett at Stanford University.

```
      REAL*8F(6)/48H(7X'REAL*8F(6)/48H'6A8,1H//7X'PRINTF,F'/7X'END')/
      PRINTF,F
      END
```

Fortran solution by Armond O. Friend of Brookline, Mass., a Freshman at MIT.

```
      WRITE(6,100)
      CALL EXIT
 100  FØRMAT(T7,12HWRITE(6,100)/T7,9HCALL EXIT/
     1 12(48H 100  FØRMAT(T7,12HWRITE(6,100)/T7,9HCALL EXIT/
     1/T6,6H12(48H),T69,2H)/,T7,2(31H/T6,6H12(48H),T69,2H)/,T7,2(31H)/
     1 T62,11H)/T7,3HEND),T6,2(28H1T62,11H)/T7,3HEND),T6,2(28H)/T7,3HEND)
      END
```

# puzzles & problems

## Number Game

Write a BASIC program to simulate this Number Game that appeared in the latest issue of *Zephrus: De-Schooler Primer*.

Equipment: 2 dice and a score sheet. Each die numbered from 1 to 6.

Score sheet numbered from 1 to 100 with a blank beside each number.

Rules: Each player (you vs. the Computer?) rolls the dice. The object is to fill the blanks on the score sheet using the numbers on each die in any arithmetic operation. Only one operation may be used.

Strategy: With a roll of 2 and 6 the player could elect to fill in either the 2, the 6, the 4 (6-2), the 8 (6+2), the 12 (6×2), the 3 (6/2), the 36 ($6^2$), or the 64 ($2^6$).

The winner? The player that fills the most blanks!

## False Cancellation

The equation 16/64 = 1/4 is a result obtained by the cancellation of the 6 in the numerator and denominator. Find all the cases in which AB/BC = A/C for A, B, and C integers between 1 and 9 inclusive. Do not consider obvious special cases such as 22/22, 33/33, etc.

## Squared Sums

The four digit number 3025 has the following property: if the number formed by considering only the first two digits (30) is added to the number formed by considering only the last two digits (25) (the total will be 55), and if this number (55) is squared, the result will be the original number:

$$(55)^2 = 3025$$

Find all 4 digit numbers having this property. Do *not* check numbers beyond 9900 since 9901 would be arranged as

$$99 + 01 = 100 \text{ and } (100)^2 = 10000$$

which is a *5* digit number.

## Sequential

What is the next number in the sequence:
9, 7, 7, 9, 13, 10, 9, ?

## Drop It

Cut out a cardboard rectangle 1-1/2" x 3". Drop it from a height of about 2 feet onto an 8-1/2" x 11" sheet of paper laid flat on a table. Outline the rectangle where it falls with a "Flair" type pen. Repeat 15 times.

Write a program to determine the probability of one rectangle touching one other, two others, and so on. How do the results from your program compare to your drawing? Try it for 30 drops. Any improvement?

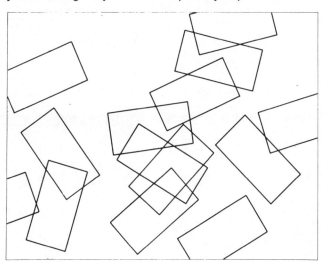

## Too Many Coconuts

There are 3 pirates and a monkey on a desert island who have gathered a pile of coconuts to be divided the next day. During the night one pirate arises, divides the pile into 3 equal parts and finds one coconut left over, which he gives to the monkey. He then hides his share away from the pile. Later during the same night, each of the other two pirates, in turn, arise and repeat the performance of the first pirate. In the morning all 3 pirates arise, divide the pile into 3 equal shares and find one left over which is given to the monkey. How many coconuts were in the original pile? Since the result is not unique, find all values from 1 to 1000 that satisfy the conditions.

# puzzles & problems

## New Life for Nim!

by B. M. Rothbart, London

Nim is perhaps the most popular game as a programming exercise in beginning computer science courses. However, as far as playing the game for the initiated player, it is no more than a test of adding binary numbers in one's head.

However, a small change produces a two-dimensional Nim which resists attempts at a definite analysis. Objects are placed in small groups forming a two-dimensional array as follows:

```
2 7 6 3
1 2 0 5
4 3 9 1
```

The two players then take turns to remove matches with the sole limitation that the matches removed on any turn must be from the same column or row. As in conventional Nim, the player who takes the last match can be either the winner or loser depending on the agreement reached before the game started.

Some trivial wins and losses are readily spotted, but the charm lies in the way in which even experienced players are unable to play by rote. Is there a foolproof method for playing? We leave that for readers to find out.

## Different Numbers

by
Eve R. Wirth

In this puzzle; first you must supply the 12 missing numbers and then add them up. This will give you a year in which one of the underlying principles of the computer was invented. (Hint: it was a mathematical principle.) What was the year, the principle, and who was the person?

1. _____ Nights" ak/a The Arabian Nights
2. _____ degrees in a circle
3. _____ original colonies
4. _____ signs in the Zodiac
5. _____ square inches = 1 square foot
6. _____ th Amendment (Women's Suffrage)
7. _____ Years War (Anglo-French wars)
8. _____ Heinz Varieties
9. _____ Degrees (boiling point of water Fahrenheit thermometer)
10. _____ feet = 1 fathom
11. _____ R's (basics of education)
12. _____ good turn deserves another.
_____ Total

**Answer to "Different Numbers"**

| | | | |
|---|---|---|---|
| 1. 1001 | 5. 144 | 9. 212 | |
| 2. 360 | 6. 19 | 10. 6 | |
| 3. 13 | 7. 100 | 11. 3 | |
| 4. 12 | 8. 57 | 12. 1 | |

The total is 1928 and if you have a good mathematics history book you will discover that Vannevar Bush devised differential analysis in that year. You will be forgiven if you could not find Bush (he is not widely recognized), however, you might have found that Fleming discovered Penicillin that year and Pressey built the first teaching machine.

## Simple (Crypt) Arithmetic

```
  N I N E
- F O U R
  F I V E
```

```
  O N E
  T W O
+ F I V E
  E I G H T
```

```
  O N E
  T W O
+ F O U R
  S E V E N
```

```
  S E V E N
  S E V E N
+   S I X
  T W E N T Y
```

```
  F I V E
- F O U R
    O N E
+   O N E
    T W O
```

```
  F O R T Y
    T E N
+   T E N
  S I X T Y
```

MULTIPLY

```
    G E T
    O N
    R O N
    G E T
  G R A N
```

DIVISION

```
         M A N
  B E )A B L E
       M N
        L L
        A T
         H E
         H E
```

# puzzles & problems

### Debugging the Raise

New program debuggers are paid $10,000 per year at Microbug, Inc. Every six months there is a salary review and good debuggers get a $400 annual raise. Now management has proposed that the salary review occur annually, and that good debuggers get an annual increase of $850. Good for the employees or the company? Why?
*DHA*

### Cubiquiz

Here is a number: 94217. (a) Drop one digit and rearrange the others to produce a perfect cube. (b) Drop another and rearrange to give another perfect cube. (d) Do the same again.
*Games & Puzzles*

### Perfect Rectangle

The reproduction is of a painting titled "Perfect Rectangle" by Mary Russell. The underlying idea is mathematical. A not-quite-square rectangle has been divided into the minimum number of squares that will fill its entire area, making it a "perfect" rectangle.

Write a program to determine for any rectangle with dimension L and W the minimum number of squares that will fill its area.
CAUTION! This rather innocent problem is not that easy to solve!

# puzzles & problems

## Bionic Toads

We have been training six bionic toads to do a new trick. When placed on seven glass tumblers, as shown below, they change sides so the three black ones are to the left and the white ones to the right, with the unoccupied tumbler at the opposite end—#7. They can jump to the next tumbler (if unoccupied), or over one or two toads to an unoccupied tumbler. The jumps can be made in either direction, and a toad may jump over his own or the opposite color, or both colors. Four successive sample jumps will make everything plain: 4 to 1, 5 to 4, 3 to 5, 6 to 3. Can you show how they do it in 10 jumps? Can you write a program to solve the problem?

---

*A man rarely succeeds at anything unless he has fun doing it.*

---

## Programming Problem...
## SATURDAY NIGHT TENNIS SCHEDULE

Given:
  6 tennis courts
  6 time periods (45 minutes each)
  18 couples

Objectives:
  • each couple plays together once (1st time period)
  • each person plays four times, sits out twice
  • each person plays one men's (women's) doubles match once; 3 mixed doubles matches.
  • each person should not play with or against the same person more than twice.

The task:
  Develop a schedule that meets the objectives, and also optimizes each individual's personal schedule. (Optimum means no more than two plays back to back) It's also nice to have both members of a couple arrive and leave together — schedules permitting!

The task minus one:
  What do you do if one couple doesn't show up?

Furthermore:
  How about a general program for N courts, M time periods and C couples.

# The Mechanical Man

## Ken Lebeiko

The mechanical man is designed as a beginning exercise in flowcharting. If you're learning programming on your own, try the mechanical man flowchart; even experienced people will have fun with it. If you're teaching flowcharting or programming, follow the steps below and it will be an exciting experience for both you and your class.

1. Copy the handout on the facing page, give it to your class and explain the problem. Keep it simple; state there is a clear path to the wall. An example may help; simply put a chair about six feet from the wall in the classroom.
2. Take the first few flowcharts done (make sure they aren't perfect) and ask the students to place them on the board. This will allow the slower students to get a general idea of what is to be done.
3. Work through the flowcharts without counters first. Before starting those flowcharts with counters, draw a counter and put a number in it (87). This is to show the effect of not clearing a counter in a program. Don't make a big thing of this; just say it was left from the last program. Everyone will see the results later.
4. Now have the student whose flowchart you are working with become the mechanical man. Use a pointer and instruct the student, following every detail of his flowchart. As the student or class see errors, don't let them change the flowchart.
   Ham it up; play it for all its worth. The class will have a good time and really learn a lot.
   Common things to happen are: people forget to turn twice and walk off somewhere else, or they forget to turn before sitting and end on the floor, and lots more—the list is long.
5. Before the students get tired of the problem, tell them to work in groups and test each others flowcharts for the perfect one.
6. Toward the end of class ask who has the "perfect" flowchart and have him put it on the board.
7. Have a member of that group sit in the chair, but move the chair right against the wall, leaving only enough room for the student's legs.
   This is called a special case. Discuss whether or not the flowchart will still work and what could be changed to make it work. This case could be used for homework. You can tell the students to get their families involved by having a parent become the mechanical man.
8. Another special case is placing the chair one step from the wall. Give this case as homework too. It demonstrates that it is very difficult to get an all-purpose program and that some special cases must be allowed for separately.

The author would like to invite any teacher with other ideas for teaching programming or flowcharting to share them with him (and us at *Creative* too!). Write Ken Lebeiko, Lockport Central HS, Lockport, IL 60441.

# The Mechanical Man

A mechanical man is sitting on a chair, facing a wall a short distance away. Draw the flowchart of the procedure to have the mechanical man walk to the wall and return to his initial position.

Permissible flowcharting symbols are:

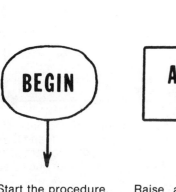
Start the procedure.

Raise arms straight out.

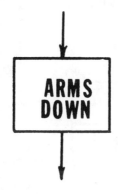
Lower arms.

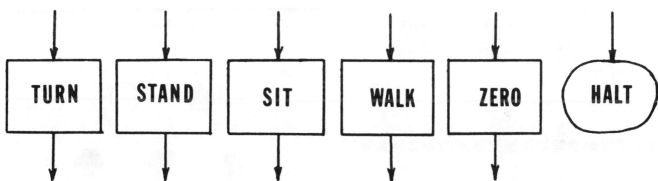

Turn 90° to the right. | Stand up. | Sit down. | Take one step. | Set the counter to zero. | Stop the procedure.

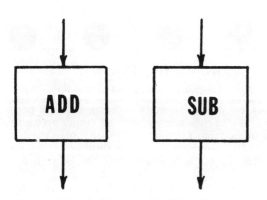
Add 1 to the counter. | Subtract 1 from the counter.

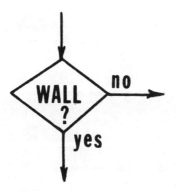
Test for touching wall. Arms must be straight forward to activate the finger-contact units.

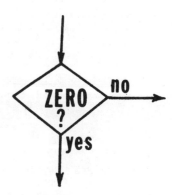
Test if counter is zero.

# puzzles & problems

## Crossnumber Puzzle

You shouldn't have too much trouble finding numbers which, when inserted in the blank spaces, complete all the equations. However, the big question is: how many solutions are there? Can you find them? Can you write a program on your computer to find them? (It's only eight trivial sumultaneous equations but...)

DHA

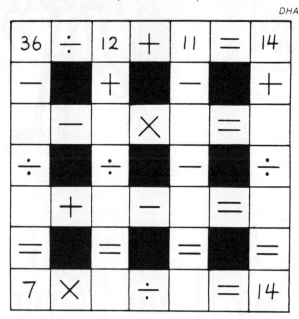

## The Classic Love-Bugs Problem

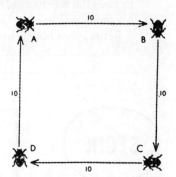

Four bugs, A, B, C, and D, occupy the corners of a square 10 inches on a side. A and C are male, B and D are female. Simultaneously A crawls directly toward B, B toward C, C toward D, and D toward A. If all four bugs crawl at the same constant rate, they will describe four congruent logarithmic spirals which meet at the center of the square.

How far does each bug travel before they meet? (The problem can be solved without calculus.)

Martin Gardner in
*Mathematical Puzzles & Diversions*

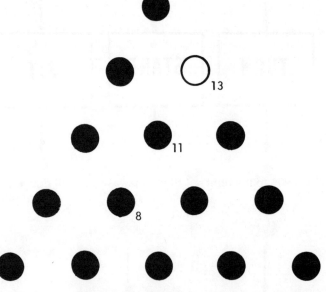

## Remove the Pegs

In the pegboard above, all 15 pegs are in at the beginning of the game. To start, remove any one peg. Then jump one peg over another into an empty hole and remove the jumped peg. For example, Peg 8 moves to Hole 13 and Peg 11 is removed. Continue until you have no jumps left. The object is to leave only one peg on the board.

Our question is no whether you can leave just one peg, but first *how many total ways* are there to leave only one peg. Second, *how many unique ways* are there to leave one peg eliminating solutions that are congruent by rotation or reversal.

*Institute for Advanced Computation Newsletter*

## Missing Digits

The famous computer scientist, Professor Abort Easycode, is engaged in testing his new computer by trying the $81 \cdot 10^5$ possible solutions to the problem of reconstructing the following exact long division in which all the digits, except one in the quotient, have been replaced by a star:

```
            **8**
      _____
*** / ********
      ***
      ____
       ****
       ***
       ____
        ****
        ****
```

(a) Each * denotes a digit between 0 and 9 and all leading digits are nonzero. Find a solution to the above.
(b) How many actual solutions are there?
(c) If you get a solution, send me the answer.

(Send solutions to D. Van Tassel, Computer Center, Univ of California, Santa Cruz, CA 95064).

# puzzles & problems

## Want To Be a Millionaire?
### David H. Ahl

I recently received the letter shown to the right giving me a unique opportunity to become a millionaire virtually overnight. I chose not to follow up on this fantastic opportunity not because it is in violation of the U.S. Postal Law (it is), but because a chain letter relies upon a geometric progression for its success. What's wrong with that? Simply that you very quickly use up the entire population of the U.S. and, indeed, the world before you "get rich."

Consider this letter which asks that each recipient send it to 500 other people (most chain letters settle for 10 other people).

| Position on List | Total People in Chain |
|---|---|
| 1 | 500 |
| 2 | 250,000 |
| 3 | 125,000,000 |
| 4 | 62,500,000,000 |
| 5 | 31,250,000,000,000 |
| 6 | 15,625,000,000,000,000 |

To continue would be ridiculous; however, bear in mind that to make any money by starting as the fourth name down (with a list of three names), your payoff comes from people at the sixth level. They would have to number some four million times the total population of the world for you to make your million.

Chain letters that ask you to send copies on to only ten people somehow sound more plausible, but are they? Work it out for yourself and see whether you can reasonably expect a payoff. Hint: there are about 70 million households in the USA.

---

$$$$$$$$$$$$ MILLIONAIRE'S NEWSLETTER $$$$$$$$$$$$

DEAR MILLIONAIRE: $$$$$$$$$$$$$$$$$$$$$$$$$$$$$$$$$$$$$$$$$$$$$$$$$$$$$$$$$$$

We know this is an unusual way to open this issue, ie, calling those of you who may feel far from it——MILLIONAIRE'S NEWSLETTER is going to be an unusual edition of the MILLIONAIRE'S NEWSLETTER because one subsidiary in Bessemer, Alabama, has hit upon something BIG he is sharing with the rest of us that took him from his meager position of clerk in a small hole-in-the-wall store of a dinky little town. His small take home pay was $90.00 a week.

BUT at this time Steve Gilbert of that small town is reported to now be sitting of $737,988.00 all gain in less than three months by using this letter.

We have been experimenting with this letter ourselves and in just five weeks after sending out a mailing of only 500 pieces we have our own success to relate to you.

After just five weeks we are $17,898.00 stronger in revenue procured from only 500 pieces mailed. We wish now we had mailed 1,000 instead of 500. Our next mailing will be 5,000 and we may retire at the end of the year.

You have been HAND-PICKED for this OPPORTUNITY and we URGE you to read this letter carefully. Because this is SOLID and we so firmly know it and have our own proof of it that we're doing something we've never done before and that is stake our reputation on it. Steve Gilbert says he started with a mailing list of 500 and then quickly——followed it with 1000 quantity mailings as the MONEY began to pour in faster and faster.

This is the most UNIQUE and original letter ever written and released to the public, so if you follow the forecoming procedures CAREFULLY you will soon be receiving CASH from 25,000 people who, like you, believe in helping others before they can expect others to help them.

Here is how you do it, Type the name list on a 3"x5" slip of paper OMITING the name in number one position and place your name and address in the THIRD position. MAIL $5.00 to the name in number one position if you don't, do not expect others to mail you your $5.00. Mail $5.00 as soon as you get this to #1 name. Paste the new list over the present list then photocopy it and take the new letter to your nearest printer who will copy them for 2 or 3 cents each. When your name reaches the #1 position, it will appear on enormously multiplied number of copies. At that point, you'll start receiving the $5.00 gifts. You need not worry about paying it back because your $5.00 sent to the number one name entitled you to MEMBERSHIP. Do not ALTER this LETTER. IT really is working for us and it will for you, ONLY if you use it right NOW. $$$$$$$$$$$$$$$$$$$$$$$$$$$$$$$$$$$$$$$$$$$$$$$$$$$$$$$$$$$$$$$$$$

If you do not have names and addresses of people to whom your letters can be mailed, you can send for Mailing List, Cost is about $38.50 per 1,000 names. Call or write: Addresses Unlimited, (213) 873-4114. Shipping including — 14760 Oxnard St., Van Nuys, Cal.

.................................................

NO. ONE-EDDIE DILL'S, Rt. No 1, Box 115-B, Blue Ridge, Ga. 30513
NO. TWO-G.L. GILLHAM, Rt. 1, Lot, 150, Franklin, Ga. 30217
NO. THREE* J. GOUGHNOUR, Box 389A, Mineral Point, Pa. 15942

---

## Train and Bee

Two trains are converging on the same track. One is going 11 km./hr and the other is going 22 km./hr. A bee is flying from one to the other at a constant rate of 33 km./hr., back and forth, back and forth. How far will the bee fly in the last hour before the trains collide?

— Sanderson M. Smith

## Goops, Gorps, and Gorgs

Every GOOP is a GORP.
Half of all GORGS are GORPS.
Half of all GORPS are GOOPS.
There are 40 GORGS and 30 GOOPS.
No GORG is a GOOP.
How many GORPS are neither GOOPS nor GORGS?

— Sanderson M. Smith

## Square Root, Root, Root, Root

What is the value of:

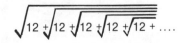 

$$\sqrt{12 + \sqrt{12 + \sqrt{12 + \sqrt{12 + \sqrt{12 + \ldots}}}}}\ ?$$

Sanderson M. Smith
Chairman, Math Dept.
Cate School
Carpinteria, CA

# Puzzles & Problems

## Superprimes

Definition: A **SUPERPRIME** is an integer such that it is prime and every integer obtained by deleting a digit from the right is a prime. E.g., 7331 is prime, 733 is prime, 73 is prime, and 7 is prime. Thus 7331 is a superprime.

Problems:

a. How many 2- and 3-digit superprimes exist and what are they?

b. Which digits of a superprime can be a 1, 2, 4, 5, 6, 8, and 0?

c. Are there any superprimes with more than 3 digits (other than 7331)?

*Macug Newsletter*

## Costume Party

At a party there are: 14 girls, 11 adults without costumes, 14 women, 10 girls with costumes, 24 people without costumes, 8 women with costumes, and 10 males with costumes. How many people are at the party?

## Yea, Team!

Of the members of three athletic teams in a certain school, 21 are on the basketball team, 26 are on the baseball team, and 29 are on the football team. 14 play baseball and basketball; 15 play baseball and football; and 12 play football and basketball. Eight are on all three teams. How many members are there altogether?

*The Mathematics Student*

## Emily Lime

Emily Lime
Has a marvelous time
Giving her friends palpitations:
Arranged in this rhyme
As  EMIL
      Y
    ----
    LIME

She's a problem in multiplication. (Each letter stands for a different digit.)

## Run Jeff, Run

If Matthew can beat Jeff by one-tenth of a mile in a two-mile race and Jeff can beat Steven by one-fifth of a mile in a two mile race, by what distance could Matthew beat Steven in a two-mile race?

*The Mathematics Student*

## Thinkers' Corner

© **Layman E. Allen**

### WORD PUZZLES

How many of the problems (a) through (f) below can you solve by forming a network of words that have exactly as many letters as the number listed as the GOAL? (Suppose that each symbol below is imprinted on a disc.)

To qualify as a network

(1) all sequences of discs across and down must be words,

(2) the words must have two or more letters and not be proper names,

(3) all of the discs in the REQUIRED column must be used,

(4) as many of the discs in PERMITTED as you wish may be used, and

(5) at most one of the discs in RESOURCES may be used.

Example:  The number of letters in the words of the network

CAT   is 7:      CAT=3, TO=2, ON=2
ON                 3 + 2 + 2 = 7

The number in the network CAT is 3.

| PROB. | GOAL | REQUIRED | PERMITTED | RESOURCES |
|---|---|---|---|---|
| [a] | 7  | S M     | E H T   | E F L P R       |
| [b] | 10 | A E O   | M S P Z | G I K O R T U   |
| [c] | 12 | F O O   | F O O   | M N O P Q R S T |
| [d] | 14 | C V X E | E H O N | B D J K L M W Y |
| [e] | 18 | A A C P | E O R T | A B D E F H I R |
| [f] | 22 | E L R M T | A E H O O V | B E J L N P S U Y Z |

If you enjoy this kind of puzzle, you may like playing ON-WORDS: The Game of Word Structures. Free information about this and other instructional games is available upon request from The Foundation for the Enhancement of Human Intelligence, 1900-W Packard Road, Ann Arbor, MI 48104.

*Some Suggested Answers (frequently there are others):*

[a]  ME / SHE
[b]  MAT / SO / TOO
[c]  TOO / OFF
[d]  HEX / COME / AV
[e]  COT / ARE / PEA
[f]  LOVE / HARM / TOE / Z

# Puzzles & Problems

## Monkey and Banana

A rope over the top of a fence has the same length on each side. Weighs 1/3 lb. per ft. On one end hangs a monkey holding a banana, and on the other end a weight equal to the weight of the monkey. The banana weighs 2 oz. per inch. The rope is as long as the age of the monkey, and the weight of the monkey (in ounces) is as much as the age of the monkey's mother. The combined ages of monkey and mother are 30 years. 1/2 the weight of the monkey, plus the weight of the banana, is 1/4 as much as the weight of the weight and the weight of the rope. The monkey's mother is 1/2 as old as the monkey will be when it is 3 times as old as its mother was when she was 1/2 as old as the monkey will be when it is as old as its mother will be when she is 4 times as old as the monkey was when it was twice as old as its mother was when she was 1/3 as old as the monkey was when it was as old as its mother was when she was 3 times as old as the monkey was when it was 1/4 as old as it is now. How long is the banana?

*Yes, I know this has been around for ages, but every once in a while I rummage around in my old stuff to see what's still there. This was, and I had some fun with it. Maybe you will too. —DHA*

## Daily Bread

A garrison had bread for 11 days. If there had been 400 more men, each man's daily share would have been two ounces less; if there had been 600 less men, each man's daily share could have been increased by two ounces, and the bread would then have lasted 12 days. How many pounds of bread did the garrison have?

## Perfect Numbers

Have you ever wondered what a "perfect number" was? Did you ever *care* what a perfect number was? A perfect number is defined as one which equals the sum of its factors. Thus, six is a perfect number - the factors of six are one, two, and three, and 1+2+3=6 (This doesn't count 6 as being a factor of itself).

Since these numbers are few and far between, they are a natural for a computer to find. A good problem for a novice computer programmer is to write a routine to find perfect numbers. Or better yet, to find them efficiently. Here's one such program with which you can amaze your friends by telling them that your computer is working on an ancient math problem discovered by the Greeks.

Stephen P. Renwick
10 Pine Street
Portland, ME 04102

## Thinkers' Corner

**Layman E. Allen © 1977**

### SET THEORY PUZZLES

How many of the problems (a) through (f) below can be solved by forming an expression that will name the number of cards in the universe that is listed as the GOAL? (Suppose that each letter and other symbol in the problems below is imprinted on a disc.)

The expression must use:

(1) all of the discs in the REQUIRED column,
(2) as many of the discs in PERMITTED as you wish, and
(3) at most one of the discs in RESOURCES.

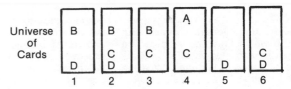

Examples:
The expression A names 1 card (4).
The expression A' names 5 cards (1,2,3,5,6).
The expression B ∩ C names 2 cards (2,3).
The expression B ∪ C names 5 cards (1,2,3,4,6).
The expression C-B names 2 cards (4,6).

| Problem | GOAL | REQUIRED | PERMITTED | RESOURCES |
|---------|------|----------|-----------|-----------|
| [a]     | 3    |          | A D --    | A C ∩ -   |
| [b]     | 2    | A        | B U       | A B C - ' |
| [c]     | 5    |          | B C -     | B C D ∩ ' |
| [d]     | 6    | U        | A B B D   | A B D U ∩ |
| [e]     | 0    | D -      | B C       | B C U ∩   |
| [f]     | 1    | B        | C D U ∩   | B C D U ∩ |

If you enjoy this kind of puzzle, you might like playing ON-SETS: The Game of Set Theory. Free information about this and other instructional games is available upon request from Learning Games Associates, 1490-S South Blvd., Ann Arbor, Michigan 48104.

a) C-A
b) (B∪A)'
c) (B-C)'
d) A∪B∪B
e) B-(C∪D)
f) B ∩ (D ∩ C)

A REMARKABLE MAGAZINE

# creative computing

*David Ahl, Founder and Publisher of Creative Computing*

*"The beat covered by Creative Computing is one of the most important, explosive and fast-changing."—Alvin Toffler*

You might think the term "creative computing" is a contradiction. How can something as precise and logical as electronic computing possibly be creative? We think it can be. Consider the way computers are being used to create special effects in movies—image generation, coloring and computer-driven cameras and props. Or an electronic "sketchpad" for your home computer that adds animation, coloring and shading at your direction. How about a computer simulation of an invasion of killer bees with you trying to find a way of keeping them under control?

**Beyond Our Dreams**

Computers are not creative per se. But the way in which they are used can be highly creative and imaginative. Five years ago when *Creative Computing* magazine first billed itself as "The number 1 magazine of computer applications and software," we had no idea how far that idea would take us. Today, these applications are becoming so broad, so all-encompassing that the computer field will soon include virtually everything!

In light of this generality, we take "application" to mean whatever can be done with computers, *ought* to be done with computers or *might* be done with computers. That is the meat of *Creative Computing*.

Alvin Toffler, author of *Future Shock* and *The Third Wave* says, "I read *Creative Computing* not only for information about how to make the most of my own equipment but to keep an eye on how the whole field is emerging.

*Creative Computing*, the company as well as the magazine, is uniquely light-hearted but also seriously interested in all aspects of computing. Ours is the magazine of software, graphics, games and simulations for beginners and relaxing professionals. We try to present the new and important ideas of the field in a way that a 14-year old or a Cobol programmer can understand them. Things like text editing, social simulations, control of household devices, animation and graphics, and communications networks.

**Understandable Yet Challenging**

As the premier magazine for beginners, it is our solemn responsibility to make what we publish comprehensible to the newcomer. That does not mean easy; our readers like to be challenged. It means providing the reader who has no preparation with every possible means to seize the subject matter and make it his own.

However, we don't want the experts in our audience to be bored. So we try to publish articles of interest to beginners and experts at the same time. Ideally, we would like every piece to have instructional or informative content—and some depth—even when communicated humorously or playfully. Thus, our favorite kind of piece is acessible to the beginner, theoretically non-trivial, interesting on more than one level, and perhaps even humorous.

David Gerrold of *Star Trek* fame says, "*Creative Computing* with its unpretentious, down-to-earth lucidity encourages the computer user to have fun. *Creative Computing* makes it possible for me to learn basic programming skills and use the computer better than any other source."

**Hard-hitting Evaluations**

At *Creative Computing* we obtain new computer systems, peripherals, and software as soon as they are announced. We put them through their paces in our Software Development Center and also in the environment for which they are intended—home, business, laboratory, or school.

Our evaluations are unbiased and accurate. We compared word processing printers and found two losers among highly promoted makes. Conversely, we found one computer had far more than its advertised capability. Of 16 educational packages, only seven offered solid learning value.

When we say unbiased reviews we mean it. More than once, our honesty has cost us an advertiser—temporarily. But we feel that our first obligation is to our readers and that editorial excellence and integrity are our highest goals.

Karl Zinn at the University of Michigan feels we are meeting these goals when he writes. "*Creative Computing* consistently provides value in articles, product reviews and systems comparisons... in a magazine that is fun to read."

**Order Today**

To order your subscription to *Creative Computing*, send $20 for one year (12 issues), $37 for two years (24 issues) or $53 for three years (36 issues). If you prefer, call our toll-free number, **800-631-8112** (in NJ 201-540-0445) to put your subscription on your MasterCard, Visa or American Express card. Canadian and other foreign surface subscriptions are $29 per year, and must be prepaid. We guarantee that you will be completely satisfied or we will refund the entire amount of your subscription.

Join over 80,000 subscribers like Ann Lewin, Director of the Capital Children's Museum who says, "I am very much impressed with *Creative Computing*. It is helping to demystify the computer. Its articles are helpful, humorous and humane. The world needs *Creative Computing*."

## creative computing

P.O. Box 789-M
Morristown, NJ 07960
Toll-free **800-631-8112**
(In NJ 201-540-0445)

## Reprints & Previews

**Sorting and Shuffling** is a 20-page booklet of reprints from *Creative Computing*. It includes in-depth discussions of five sorting techniques (bubble, heapsort, Shell-Metzner, delayed replacement and Woodrum). It also covers file structures and shuffling techniques. Most textbooks either ignore or gloss over these techniques. The booklet is a vital necessity for those doing any programming at all. 50 cents.

**Guide to Computer Music Systems** by Phil Tubb primarily dicusses the design philosophy behind the ALF computer music system. It also covers the principles of computer music reproduction and compares three popular systems for the Apple II. $2.00.

**Stocks and Listed Options** by Alfred Adler is a collection of five articles about using a small computer for analysis of a stock portfolio with an emphasis on listed options. The booklet serves as the instruction manual for a 5-program package for the TRS-80 marketing by Creative Computing Software. $1.00.

**Odell Woods**. This is a program listing in Basic and instruction booklet for a popular MECC program. In it, the user plays the role of a fox, mouse or wolf and attempts to survive in the northwoods. The listing is in Applesoft Basic but uses few special features so it could be converted easily to other systems. $1.00.

## Complete 6-year Index

# Find it Fast!

Our new 6-year cumulative index lists every article, program and review that has appeared in Creative Computing from its inception in November 1974 to December 1980. The index lists not only the issue in which an article appeared but a cross reference to The Best of Creative Computing, Volumes 1, 2 and 3. It also lists all the articles in ROM magazine.

Articles are classified by subject area and listed by title and author. Over 3500 separate items are included. Note: the index does not include a cross reference to author.

Looking for information on computers in education? You'll find 76 articles and 155 application programs. How about art and graphics? You'll find 44 entries. In the market for a computer? You find 82 hardware evaluations and 94 of software.

Price of this huge index is just $2.00. Even if you've been a reader for only a year or two you'll find the index of great value. Orders yours today.

**creative computing**

Morris Plains, NJ 07950

---

**Special editions for Apple, Atari and TRS-80 Computers.**

Hey kids, are the folks out of the room? Good, 'cause I've got a secret to tell you. You know that computer they fuss over? Well, kid, between you and me, this whole programming thing is a lot simpler than they realize.

What's that? Sure, you can learn. Just get a copy of **Computers For Kids**. It's a super book, and it tells you everything you need to know. Huh? You have an Apple? No problem. There's a version just for the Apple. One for the TRS-80 and one for the Atari too, with complete instructions for operating and programming.

The book will take you through everything programmers learn. Its easy to understand and the large type makes it easy to read. You'll find out how to put together a flowchart, and how to get your computer to do what you want it to do. There's a lot to learn, but **Computers For Kids** has 12 chapters full of information. You'll even learn how to write your own games and draw pictures that move.

Just so the folks and your teachers won't feel left out, there's a special section for them. It gives detailed lesson ideas and tells them how to fix a lot of the small problems that might pop up. Hey, this book is just right for you. But you don't have to take my word on that. Just listen to what these top educators have to say about it:

Donald T. Piele, Professor of Mathematics at the University of Wisconsin-Parkside says, "**Computers For Kids** is the best material available for introducing students to their new computer. It is a perfect tool for teachers who are learning about computers and programming *with* their students. Highly recommended."

Robert Taylor, Director of the Program in Computing and Education at Teachers College, Columbia University states, "it's a good idea to have a book *for chidren*."

Not bad, huh? Okay, you can let the adults back in the room. Don't forget to tell them **Computers For Kids** by Sally Greenwood Larsen cost only $3.95. And tell them you might share it with them, if they're good. Specify edition on your order: TRS-80 (12H); Apple (12G); Atari (12J).

Your local computer shop should carry **Computers For Kids**. If they don't ask them to get it or order by mail. Send $3.95 payment plus $1.00 shipping and handling to Creative Computing Press, P.O. Box 789-M, Morristown, NJ 07960.

**creative computing press**

# The most complex computer circuit can be explained with just nine cents

# Common Cents

The "penny switch." It sounds strange. But it's not.

Joe Weisbecker, the designer of the RCA 1802 microcomputer, was trying to explain to some children just how a computer works. He wasn't having much success.

**Computers Aren't Magic**

Joe's hobby is magic. He thought, "maybe I can use some kind of illusion to show how a computer works." But he didn't really want to use an illusion. He didn't want the children to think of a computer as magic.

So he hit upon the idea of a simple flip-flop switch (the most common circuit in a computer) represented by the head or tail of a penny. This flip-flop circuit uses just one penny. Every time it receives an impulse it changes from head to tail or tail to head. Simple.

But then Joe went on and put two of these simple flip flops together to make a circuit that adds two numbers together. And another that subtracts numbers. Kids loved these circuits and played with them like games.

**Games With Pennies**

Before long, Joe devised circuits to play more complicated games like Tic Tac Toe, Guess A Number and Create A Pattern. Pretty soon he had 30 circuits (or games) that explained everything about computers from a basic adder to complex error correction. The most complex circuit uses just nine pennies (or dimes for the big spender).

These circuits, each one with a full size playing diagram, have been collected together in a book called *Computer Coin Games*. With this book children or adults can easily understand the workings of even the most complex computer circuits.

*Games Magazine* said, "whether or not you have any experience with computer technology, you'll be both amazed and delighted with the simplicity of the format and the complexity of the play. All you need is some common cents."

*Dr. Dobbs Journal* agreed, saying, "*Computer Coin Games* is a simple approach to a complicated concept. The book is liberally sprinkled with clever illustrations and diagrams, and provides a relatively painless route to understanding how computer circuits function."

**Money back Guarantee**

We're convinced that you'll understand the inner workings of a computer after playing these 30 games. If you don't, send the book back and we'll refund the complete price plus your postage to send it back.

To order your copy of *Computer Coin Games*, just send $3.95 plus $1.00 shipping and handling to Creative Computing Press, Morris Plains, NJ 07950. Visa, MasterCard and American Express orders may be called toll free to 800-631-8112 (in NJ, 201-540-0445).

With its wonderful illustrations by Sunstone Graphics, *Computer Coin Games* makes an ideal gift. *The Association for Educational Data Systems* calls the book "an ideal introduction to the concepts of computer circuitry."

Order your copy today.

## creative computing

Morris Plains, NJ 07950
Toll-free **800-631-8112**
(In NJ 201-540-0445)

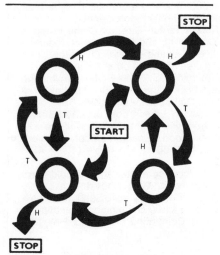

"Heads Up Game." Starting with tails in all positions, how many times through to get all four pennies heads up?

# An Important Reference

# Computers, Ethics and Society

Where is the computer leading us? Is it a menace or a messiah? What are its benefits? What are the risks? What is needed to manage the computer for society's greatest good? Will we become masters or slaves of the evolving computer technology?

This bibliography was created to help answer questions like these. The works cited can provide the range of facts and opinion necessary to your understanding of the role of the computer.

This is a bibliography of works dealing with the ways in which computers are being used in our society, the beneficial changes that are taking place in our lives as a result of computer technology, the social and ethical problems intensified by the improper use of computers, the dangers of computerized society, the safeguards and defenses against those dangers, the attempts to indicate what computerized direction the future will take, and the responsibilities of computer professionals. It contains 1920 alphabetical entries of books, magazine articles, news items, scholarly papers and other works dealing with the impact of computers on society and ethics. Covers 1948 through 1979.

Compiled by Gary M. Abshire. Hardbound, 128 page. $17.95. (12E)

# Problem

## The world is full of intriguing problems that never got into a textbook.

**Problems for Computer Solution**
by Stephen Rogowski.

Ninety intriguing and fascinating problems, each thoroughly discussed and referenced, make an excellent source of exercises in research and preliminary investigation. Eleven types of problems are provided in the following areas: arithmetic, algebra, geometry, trigonometry, number theory, probability, statistics, calculus and science. Author Stephen Rogowski of SUNY-Albany has included several problems which have never been solved. He feels that some research and an attempt to solve these will sharpen students' insight and awareness.

Some of the problems are not new like the one asking how much the $24 the Indians were paid for Manhattan would be worth today had it been deposited in a bank. However, this problem was revised to have a variable interest rate so it would be a challenge to program. Of course, many of the problems are new and have never been in print before.

The student edition has 106 pages and includes all 90 problems (with variations), 7 appendices and a complete bibliography. Cost is $4.95.

The 182-page teacher edition contains solutions to the problems, each with a complete listing in Basic, sample runs, and in-depth analyses explaining the algorithms and theory involved. Cost is $9.95.

To get one or both books send payment plus $2.00 shipping and handling per order to *Creative Computing*. Credit card orders may be called in toll-free to the number below.

Order yours today. If you are not completely satisfied, return it for a full refund plus your return postage.

## creative computing

Morris Plains, NJ 07950
Toll-free **800-631-8112**
(In NJ 201-540-0445)

# Sourcebook of Ideas

## Many mathematics ideas can be better illustrated with a computer than with a text book.

Consider Baseball cards. If there are 50 cards in a set, how many packs of bubble gum must be purchased to obtain a complete set of players? Many students will guess over 1 million packs yet on average it's only 329.

The formula to solve this problem is not easy. The computer simulation is. Yet you as a teacher probably don't have time to devise programs to illustrate concepts like this.

Between grades 1 and 12 there are 142 mathematical concepts in which the computer can play an important role. Things like arithmetic practice, X-Y coordinates, proving geometric theorems, probability, compounding and computation of pi by inscribed polygons.

### Endorsed by NCTM

The National Council of Teachers of Mathematics has strongly endorsed the use of computers in the classroom. Unfortunately most textbooks have not yet responded to this endorsement and do not include programs or computer teaching techniques. You probably don't have the time to develop all these ideas either. What to do?

For the past six years, *Creative Computing* magazine has been running two or three articles per issue written by math teachers. These are classroom proven, tested ideas complete with flowcharts, programs and sample runs.

Teachers have been ordering back issues with those applications for years. However, many of these issues are now sold out or in very short supply.

So we took the most popular 134 articles and applications and reprinted them in a giant 224-page book called *Computers in Mathematics: A Sourcebook of Ideas*.

### Ready-to-use-material

This book contains pragmatic, ready to use, classroom tested ideas on everything from simply binary counting to advanced techniques like multiple regression analysis and differential equations.

The book includes many activities that don't require a computer. And if you're considering expanding your computer facilities, you'll find a section on how to select a computer complete with an invaluable microcomputer comparison chart.

Another section presents over 250 problems, puzzles, and programming ideas, more than are found in most "problem collection" books.

*Computers in Mathematics: A Sourcebook of Ideas* is edited by David Ahl, one of the pioneers in computer education and the founder of *Creative Computing*.

The book is not cheap. It costs $15.95. However if you were to order just half of the back issues from which articles were drawn, they would cost you over $30.

### Satisfaction Guaranteed

If you are teaching mathematics in any grade between 1 and 12, we're convinced you'll find this book of tremendous value. If, after receiving it and using it for 30 days you do not agree, you may return it for a full refund plus your return postage.

To order, send your check for $15.95 plus $1.00 postage and handling to Creative Computing Press, Morris Plains, NJ 07950. Visa, MasterCard, and American Express orders may be called in toll-free to 800-631-8112 (in NJ 201-540-0445). School purchase orders should add an additional $1.00 billing fee for a total of $17.95.

Don't put it off. Order this valuable sourcebook today.

## creative computing

Morris Plains, NJ 07950
Toll-free **800-631-8112**
(In NJ 201-540-0445)

# Computer Lawnmower

Can a computer mow your lawn? Not yet. But a flowchart can show you how to make money cutting five lawns a day. The flowchart is easy. Mowing the lawns is still hard work.

Dr. Sylvia Charp and Marion Ball wanted a way to introduce basic computer concepts to children in grades 5 to 9 of the Philadelphia City Schools. So they identified some tasks that kids understood like mowing lawns, issuing paychecks and controlling traffic lights. They showed how computers are used in these tasks.

### Flowcharts - A basic concept

They devised flowcharts. They located scores of photos. And they found an artistic high school student to illustrate these concepts with lively full-color drawings.

They then wrote a light-hearted but informative text to tie it all together. It talked about kinds of computers, what goes on inside the machine, the language of the computer and how computers work for us.

They took the problem of averaging class grades and showed how a simple program could be written to do this job.

### Well-qualified authors

Marion Ball has written other books on computer literacy. Sylvia Charp is the director of educational compuuting for Philadelphia City Schools. They pooled their talents to produce this book, *Be A Computer Literate*.

This easy-to-read book explains how computers are used in medicine, law enforcement, art, business, transportation and education. It's interesting and understandable.

### Too much demand

The Bell System distributed 50,000 copies to schools throughout the U.S. but they couldn't meet the continuing demand. So Creative Computing Press now distributes the book. It's just $3.95 plus $1.00 shipping and handling. Send name and address plus payment or credit card number and expiration date to Creative Computing Press, Morris Plains, NJ 07950. Visa, MasterCard and American Express orders may also be called in toll-free to 800-631-8112 (in NJ 201-540-0445).

Order yours today. If, after reading it, you do not feel that you are "computer literate," return it for a full refund plus your postage to send it back.

## creative computing

Morris Plains, NJ 07950
Toll-free **800-631-8112**
(In NJ 201-540-0445)

---

# new friends for your child...
## Katie and the Computer

Fred D'Ignazio and Stan Gilliam have created a delightful picture book adventure that explains how a computer works to a child. Katie "falls" into the imaginary land of Cybernia inside her Daddy's home computer. Her journey parallels the path of a simple command through the stages of processing in a computer, thus explaining the fundamentals of computer operation to 4 to 10 year olds. Supplemental explanatory information on computers, bytes, hardware and software is contained in the front and back end papers.

Thrill with your chidren as they join the Flower Bytes on a bobsled race to the CPU. Share Katie's excitement as she encounters the multi-legged and mean Bug who lassoes her plane and spins her into a terrifying loop. Laugh at the madcap race she takes with the Flower Painters by bus to the CRT.

"Towards a higher goal, the book teaches the rewards of absorbing the carefully-written word and anticipating the next page with enthusiasm..."
**The Leader**

"Children might not suspect at first there's a method to all this madness—a lesson about how computers work. It does its job well."
**The Charlotte Observer**

"...the book is both entertaining and educational."
**Infosystems**

The book has received wide acclaim and rave reviews. A few comments are:

"Lively cartoon characters guide readers through the inner chamber of the computer."
**School Library Journal**

"...an imaginative and beautifully conceived children's story that introduces two characters—the Colonel and the Bug—who already seem to have been classic children's story book characters for generations."
**The Chapel Hill Newspaper**

Written by Fred D'Ignazio and illustrated in full color by Stan Gilliam. 42 pages, casebound, $6.95. (12A)

A t-shirt with the Program Bug is available in a deep purple design on a beige shirt. Adult size S, M, L, XL. Children's size S, M, L. $6.00.